Institutions as Conscious Food Consumers

Institutions as Conscious Food Consumers

Leveraging Purchasing Power to Drive Systems Change

Edited by

Sapna E. Thottathil

Annelies M. Goger

Academic Press is an imprint of Elsevier
125 London Wall, London EC2Y 5AS, United Kingdom
525 B Street, Suite 1650, San Diego, CA 92101, United States
50 Hampshire Street, 5th Floor, Cambridge, MA 02139, United States
The Boulevard, Langford Lane, Kidlington, Oxford OX5 1GB, United Kingdom

© 2019 Elsevier Inc. All rights reserved.

No part of this publication may be reproduced or transmitted in any form or by any means, electronic or mechanical, including photocopying, recording, or any information storage and retrieval system, without permission in writing from the publisher. Details on how to seek permission, further information about the Publisher's permissions policies and our arrangements with organizations such as the Copyright Clearance Center and the Copyright Licensing Agency, can be found at our website: www.elsevier.com/permissions.

This book and the individual contributions contained in it are protected under copyright by the Publisher (other than as may be noted herein).

Notices
Knowledge and best practice in this field are constantly changing. As new research and experience broaden our understanding, changes in research methods, professional practices, or medical treatment may become necessary.

Practitioners and researchers must always rely on their own experience and knowledge in evaluating and using any information, methods, compounds, or experiments described herein. In using such information or methods they should be mindful of their own safety and the safety of others, including parties for whom they have a professional responsibility.

To the fullest extent of the law, neither the Publisher nor the authors, contributors, or editors, assume any liability for any injury and/or damage to persons or property as a matter of products liability, negligence or otherwise, or from any use or operation of any methods, products, instructions, or ideas contained in the material herein.

Library of Congress Cataloging-in-Publication Data
A catalog record for this book is available from the Library of Congress

British Library Cataloguing-in-Publication Data
A catalogue record for this book is available from the British Library

ISBN 978-0-12-813617-1

For information on all Academic Press publications
visit our website at https://www.elsevier.com/books-and-journals

Publisher: Andre Gerhard Wolff
Acquisition Editor: Nancy Maragioglio
Editorial Project Manager: Barbara Makinster
Production Project Manager: Omer Mukthar
Cover Designer: Matthew Limbert

Typeset by SPi Global, India

Contents

Contributors	xiii
Foreword	xv
–Anim Steel	

Part I
The Role of Institutions in Food Systems Change

1. **Introduction: Institutions as Conscious Food Consumers**
   Sapna E. Thottathil

What Is Institutional Foodservice?	4
Targeting Consumption and Procurement to Change the Food System	6
The Unexamined Role of Foodservice in Reforming Industrialized Agriculture	9
How the Book Is Structured: An Overview of Chapters	11
Challenges Faced by Institutional Foodservice	13
Opportunities: Institutional Foodservice's Future	14
Additional Reading and Resources	16
References	17

2. **Trends in the Global Food System and Implications for Institutional Foodservice**
   Kristal Jones, Kimberly Pfeifer, Gina Castillo

Introduction	21
Food Systems Trends Past and Present	23
Environmental Sustainability: Trends, Drivers, and Challenges	25
Livelihoods: Trends, Drivers, and Challenges	26
Nutrition: Trends, Drivers, and Challenges	28
Nexus of Sustainability, Livelihoods, and Health Challenges	29
Emerging Approaches to Addressing Food Systems Challenges	30
Third-Party Certifications and Increasing Demand for Information	31
Shortening the Food Chain: Going Local and Fresh	32
Social Movements Addressing Equity and Sustainability	33

Institutions as Key Drivers of Food System Transformation	34
Improving Health and Sustainability through Menu Decisions	35
Improving Equity and Sustainability Through Engagement With Certifications and Standards	36
Improving Health, Equity, and Sustainability by Reducing Food Waste	37
Conclusion	39
References	39

3. Situating Institutional Foodservice in Agro-Food Value Chains: Overcoming Market Power and Structure With Values-Based Procurement
Annelies M. Goger

Introduction	47
Policy Context: How the Global Food System Became So "Market-Driven"	50
The Global Value Chain Approach: Rendering Obscure Supply Chains More Visible	52
What Is GVC Analysis?	53
Governance	54
Upgrading	55
The Structure and Governance of Agro-Food Value Chains	57
Key Features of Institutional Foodservice Value Chains	59
Shifting Institutional Foodservice to Values-Based Procurement	62
Growth of Agro-Food Intermediaries That Enhance Market Access, Information Sharing, and Capacity	64
Limitations of the Values-Based Procurement Movement	65
Conclusion	67
References	69

Part II
Setting Purchasing Standards Through Policy

4. From Foodservice Management Contracts to U.S. Federal Legislation: Progress and Barriers in Values-Based Food Procurement Policies
Raychel E. Santo, Claire M. Fitch

Introduction	78
Using Institutional Foodservice Policies to Address Food System Challenges in the United States	79
Market Structure and Rural Economy Considerations	81
Environmental Considerations	85
Health Considerations	86
Labor and Equity Considerations	87
Animal Welfare	88

	Foodservice Management Contracts as a Barrier to Increasing Regional and Sustainable Food Procurement	88
	Foodservice Contracts and the Rebate Pricing System	89
	Reflections on Uncertainties in the Larger Political Context	93
	Conclusion	95
	References	96

5. The Good Food Purchasing Program: A Policy Tool for Promoting Supply Chain Transparency and Food System Change

Lindsey Day Farnsworth, Alexa Delwiche, Colleen McKinney

Introduction	104
Using Policy to Leverage Public Spending for Healthier, Fairer Food	105
Defining "Good Food"	106
Implementation of the Good Food Purchasing Policy	107
Nuts and Bolts: How the Good Food Purchasing Program Works	108
Implementation and Data Collection	110
Case Studies: Institutions Adopting the Good Food Purchasing Program for Food System Change	111
Evaluating and Institutionalizing Values-Based, Data-Driven Purchasing Practices: Oakland, CA	112
Supply Chain Innovations and Growing Pains in Good Food Purchasing in Los Angeles	114
Creating Opportunities for Nontraditional Suppliers Through Requests for Proposals (RFPs) and Contracts	117
Challenges and Lessons: The Crucial Roles of Transparency and Multilevel Advocacy in Institutionalizing the Good Food Purchasing Program	121
Misguided Perceptions of Present Sourcing Practices	121
Data Quality and Availability	121
Difficulties With Contracts and Bidding Processes	122
Future Directions	**122**
Conclusion	**123**
References	**124**
Further Reading	**125**

6. From Navigating the Regulatory Environment to Designing a Good Food Supply for Institutions: Cases From Philadelphia

Jonathan Deutsch, Alexandra Zeitz, Benjamin Fulton, Brandy-Joe Milliron, Catherine Bartoli

Introduction	127
Background: Philadelphia's Food Systems and Levers for Institutional Change	129
Literature Review: Procurement Policy Implementation	130
Project 1: Education About Compliance and Cost	131

Project 2: Support for Compliance	135
Project 3: Changing the Upstream Value Chain to Increase Compliance	138
Conclusion	142
References	143
Further Reading	145

Part III
Creating Shared Identity Using Place-Based Connections

7. Making Local Sourcing Standard Practice: Lessons From Michigan
Kathryn Colasanti, Colleen Matts, Kaitlin K. Wojciak

Introduction	149
Growth in Farm-to-Institution Activity	151
History of Farm-to-Institution Organizations in Michigan	151
State Government Policy	158
Non-Profit and MSU Extension Facilitation	160
Supply Chain Partnerships	163
Case Studies	164
Case Study 1: Baxter Child Development Center	165
Case Study 2: Calumet Center	166
Case Study 3: Bronson Methodist Hospital	168
Conclusion	170
References	171
Further Reading	173

8. Farm to Institution New England: Mobilizing the Power of a Region's Institutions to Transform a Region's Food System
Nessa J. Richman, Peter H. Allison, Hannah R. Leighton

Introduction	175
History of Farm to Institution New England	177
The FINE Network Today	178
Core Functions: Convening, Communication, and Metrics	180
Farm-to-School	182
Regional FTS Efforts	183
Case Study: Northeast Farm to School Institute	184
Farm-to-Hospital	185
Case Study: Sea-to-Campus at Boston Medical Center	185
Farm-to-College	186
Case Study: The "Maine Food for UMaine System" Project	188

FINE's Research into the Supply Chain		188
Distributors		189
Producers		189
FINE's Impact as a Regional Network		191
Strategic Objectives of FINE Moving Forward		191
References		**192**
Further Reading		**194**

9. Montana's Beef-to-School Project: Making Connections to Enhance Local Agriculture

Carmen Byker Shanks, Thomas M. Bass, Joel B. Schumacher

Introduction	195
Background	197
Cattle Production Basics	197
General Description of Processing and Yield	198
Local Beef in Montana's School Food Programs	199
Governance of Beef-to-School Activities	199
Federal and State Nutrition Standards	199
Federal Food Safety Standards	201
Case Studies	201
Beef-to-School Decisions: School and District Levels	201
Hinsdale, Montana: Vertically Integrated Model	205
Livingston, Montana: Producer Contract Model	206
Flathead Valley Region: Processor Contract Model	207
Somers, Montana: Processor Contract Model	207
Whitefish, Montana: Processor Contract Model	208
Kalispell, Montana: Processor Contract Model	209
Dillon, Montana: Donation and Community Member Involvement Models	210
Case Study Analysis: Overarching Themes	211
Motivations for Beef-to-School	211
Challenges in Implementing Beef-to-School Models	211
Conclusion	215
References	**216**
Further Reading	**217**

10. Institutional Markets Supporting Midsized Farms: A Case Study of Iowa

Kranti Mulik

Introduction	219
Midsized Farming Operations in U.S. Food Systems	222
A Case Study From Iowa: Local Food Consumption by Institutions as a Niche for Midsized Farmers	224
Estimating the Impacts if Midsized Local Farms Supplied Iowa's Food-Buying Institutions	225
Conclusions and Policy Recommendations	229
References	**230**
Further Reading	**234**

Part IV
New Directions for Institutional Foodservice

11. **Sustainable Food Purchasing in the Health Care Sector: From Ideals to Institutionalization**
    Kendra Klein, Jenna Newbrey, Emma Sirois

Introduction	239
Redefining Healthy Food: An Environmental Nutrition Approach	241
History and Uptake of Environmental Nutrition	243
History of Health Care Without Harm's Healthy Food in Health Care Movement	244
Early Adopters	246
Putting Environmental Nutrition Into Practice in Hospital Foodservice	247
Defining Local and Sustainable Food	247
Purchasing Practices in Hospitals	249
Modeling Food Choices	249
Supply Chain Challenges and Opportunities for Farm-to-Hospital	250
Distributors	251
Group Purchasing Organizations	252
Supply Chain Innovation	253
Case Study: University of Washington Medical Center Purchasing Local Food	253
Purchasing Pathway 1: Northwest Agricultural Business Center	254
Purchasing Pathway 2: The University of Washington Student Farm	255
Challenges and Lessons Learned	255
Conclusion	256
References	257

12. **Bringing School Foodservice Staff Back in: Accounting for Changes in Workloads and Mindsets in K-12 Values-Based Procurement**
    Amy Rosenthal, Christine C. Caruso

Introduction	261
Background	263
Purchasing Changes in the National School Lunch Program	263
Context of Institutional Foodservice Work	266
Findings	267
The Cafeteria Worker Experience	268
Procurement Changes	271
Conclusion	277
References	279

Contents xi

13. Food Banks as Local Food Champions: How Hunger Relief Agencies Invest in Local and Regional Food Systems
Megan Bucknum, Deborah Bentzel

Introduction	286
Food Banking 101	286
The Organization of the Food Bank Network	287
Aggregation and Distribution by Food Banks	288
Food Bank Value Chains: Donations and Procurement	289
Donations	290
USDA Foods	291
Food Bank-Led Purchasing	292
Making the Case for Farm-to-Food-Bank: Existing Food Bank Assets	293
Procurement Expertise	293
Strategic Partnerships Across Sectors	294
Infrastructure	294
Stimulating Demand for Fresh Food	295
Policy Support for Farm-to-Food-Bank Efforts	295
Farm-to-Food Bank Examples	296
Methodology	296
Greater Cleveland Food Bank	298
Community Food Bank of Southern Arizona	298
Food Bank of South Jersey	298
Houston Food Bank and Feeding Texas	299
Northern Illinois Food Bank	299
Second Harvest Food Bank of Greater New Orleans and Acadiana	299
Emerging Themes	299
Current Challenges to the Growth of Farm-to-Food-Bank	301
Coordination Around Perishables	301
Price	302
Finding Suppliers	302
Conclusion	302
References	303
Further Reading	304

14. Plant Proteins Move to Center-Plate at Colleges and Universities
Kristie Middleton, Elise Littler

Introduction	307
Understanding the Plant-Based Food Trend	309
Animal Welfare	309
Health	309
The Environment	310
Cost Savings	311

Case Studies: College and University Foodservice	312
Self-Operating Foodservice Facilities	312
Foodservice Management Companies	314
Common Themes	317
The Role of Non-Profit Intermediaries and Peer Networks	317
Emergence of New Meatless Product Lines Among Private Manufacturers	319
Conclusion	**321**
References	**321**
Further Reading	**324**
Index	327

Contributors

Numbers in Parentheses indicate the pages on which the authors' contributions begin.

Peter H. Allison (175), Farm to Institution New England, Hartland, VT, United States

Catherine Bartoli (127), Get Healthy Philly, Philadelphia Department of Public Health, Philadelphia, PA, United States

Thomas M. Bass (195), Animal & Range Sciences Extension Service, Montana State University Extension, Bozeman, MT, United States

Deborah Bentzel (285), The Food Trust, Philadelphia, PA, United States

Megan Bucknum (285), Rowan University, Glassboro, NJ, United States

Carmen Byker Shanks (195), Food and Health Lab, Department of Health and Human Development, Montana State University, Bozeman, MT, United States

Christine C. Caruso (261), University of Saint Joseph, West Hartford, CT, United States

Gina Castillo (21), Oxfam America, Boston, MA, United States

Kathryn Colasanti (149), Center for Regional Food Systems, Michigan State University, East Lansing, MI, United States

Lindsey Day Farnsworth (103), Center for Integrated Agricultural Systems, University of Wisconsin-Madison, Madison, WI, United States

Alexa Delwiche (103), Center for Good Food Purchasing, Berkeley, CA, United States

Jonathan Deutsch (127), Center for Food and Hospitality Management; Department of Nutrition Sciences, Drexel University, Philadelphia, PA, United States

Claire M. Fitch (77), Farm Forward, Portland, OR, United States

Benjamin Fulton (127), Center for Food and Hospitality Management, Drexel University, Philadelphia, PA, United States

Annelies M. Goger (47), Social Policy Research Associates, Oakland, CA, United States

Kristal Jones (21), National Socio-Environmental Synthesis Center, University of Maryland, Annapolis, MD, United States

Kendra Klein (239), Friends of the Earth, Berkeley, CA, United States

Hannah R. Leighton (175), Farm to Institution New England, Hartland, VT, United States

Elise Littler (307), Farm Animal Protection, The Humane Society of the United States, Washington, DC, United States

Colleen Matts (149), Center for Regional Food Systems, Michigan State University, East Lansing, MI, United States

Colleen McKinney (103), Center for Good Food Purchasing, Berkeley, CA, United States

Kristie Middleton (307), Farm Animal Protection, The Humane Society of the United States, Washington, DC, United States

Brandy-Joe Milliron (127), Department of Nutrition Sciences, Drexel University, Philadelphia, PA, United States

Kranti Mulik (219), Fairfax, Virginia, United States

Jenna Newbrey (239), Health Care Without Harm, Reston, VA, United States

Kimberly Pfeifer (21), Oxfam America, Boston, MA, United States

Nessa J. Richman (175), Rhode Island Food Policy Council, Kingston, Rhode Island, United States

Amy Rosenthal (261), Rutgers University, New Brunswick, NJ, United States

Raychel E. Santo (77), Johns Hopkins Center for a Livable Future, Department of Environmental Health and Engineering, Bloomberg School of Public Health, Baltimore, MD, United States

Joel B. Schumacher (195), Department of Agricultural Economics & Economics, Montana State University Extension, Bozeman, MT, United States

Emma Sirois (239), Health Care Without Harm, Reston, VA, United States

Sapna E. Thottathil (3), University of California, Office of the President, Oakland, CA, United States

Kaitlin K. Wojciak (149), Michigan State University Extension, East Lansing, MI, United States

Alexandra Zeitz (127), Center for Food and Hospitality Management, Drexel University, Philadelphia, PA, United States

Foreword

My first experience of politics was in middle school. My parents, a black Ghanaian woman and a white American man who had relocated the family from West Africa to Washington D.C., took my sister and me to a protest in front of the South African embassy. Circling the sidewalk that cold night, with the candles melting onto our fingers, we became part of a worldwide movement to support the freedom struggle in South Africa. Later that year, the University of California Board of Regents was compelled by student protests to withdraw billions of dollars of assets from South Africa, joining over a hundred other universities and municipalities in the divestment campaign.

This was the 1980s. Apartheid then seemed as intractable as the Cold War and as durable as Ronald Reagan, whose administration was no friend to the anti-apartheid movement. And so, just five years later, my family was riveted by the television images of Nelson Mandela waving to the crowds from the balcony of Cape Town City Hall after nearly three decades in prison.

Six months after his release, Nelson Mandela capped off a tour of the United States with a stop in Oakland, California, where he pointedly thanked the divestment activists in the San Francisco Bay Area and beyond for the pivotal pressure they placed on the apartheid regime. "You are our comrades in the struggle," he told the crowd.

That marked my first awareness of the power of institutions in the service of social change – the power of active individuals amplified by the economic and cultural force of the schools, establishments, and associations that join us in common purpose.

The lesson was a touchstone for me and the young leaders who went on to create a national student-driven campaign for food justice called the Real Food Challenge. Already attuned to the multiple food-driven health, environmental, racial, and economic crises present in our communities and our bodies, we were in search of a lever. The anti-apartheid divestment movement was an inspiration. If we could "divest" university cafeteria budgets – about five billion dollars of annual food spending – from industrial agribusiness, we could help to catalyze a shift towards community-based, fair, ecologically-sound, and/or humane foods – what we called "real food." We didn't think it would be easy; the barriers are both logistical and political, with "Big Ag's" big presence on many university boards, departments, and programs. But we knew it was possible, and it'd be worth it. We felt that we could add a powerful new dimension to a similar wave that was just beginning in hospitals, cities, and K-12 schools.

Ten years after our 2008 launch, we've made a lot progress. To begin with, we've radically increased transparency about the food chain, empowering over 1,000 students at hundreds of universities to measure their schools' dining purchases with our Real Food Calculator. What's more, Real Food Challenge student leaders have won "real food procurement policies" at 83 institutions in every region of the country, stretching from Johns Hopkins to Western Washington University; from the University of Oklahoma to Northwestern. And we're seeing the benefits, as $80 million of re-directed spending has served to create a market for holistic ranch management, anchored new distribution channels for small farmer cooperatives, and allowed neighborhood food businesses to expand employment and hours, among other impacts. We are witnessing a new economy – fairer, healthier, and greener – rising out of the old.

Indeed, the whole movement is accelerating in the way the anti-apartheid divestment movement did in the early 1980s, and we haven't come anywhere close to exhausting our potential. How timely it is, then, to step back and consider, with this edited volume, where we have come from, where we are, and where we need to go in this multi-faceted effort around institutional food purchasing. This volume brings together some of the leading organizations and individuals working on institutional food purchasing in the U.S., and represents initiatives from a variety of perspectives to inspire this reflection. The following chapters deal with the key issues of the day, ones that the Real Food Challenge and almost all of our peers are grappling with, including: the particular challenges and benefits of local and ethical meat sourcing; the strategic advantages of shaping contracts with distributors and foodservice management companies; the critical importance of including cafeteria workers in the planning and piloting of farm-to-school initiatives; and the question of who decides what is good and culturally-appropriate food.

These chapters can help a wide range of people – from veteran foodservice professionals to new activists – who are looking both for broad context *and* cutting-edge ideas. I especially hope it will influence many government and institutional officials who are in prime positions to accelerate this movement. Because behind the technical questions is a moral one: Will our food system in general continue to enrich the corporate few while exploiting or depleting food workers, rural communities, and the soil and water we depend on? Or will it be a source of well-being and dignity for all? Even if Washington is as passive on this question as it was to the anti-apartheid cause – *especially* if Washington is passive – we understand this as a generation-defining effort that our institutions cannot sit out.

Anim Steel

ABOUT THE AUTHOR

Anim Steel is the Executive Director and co-founder of the Real Food Challenge, a national campaign to re-direct one billion dollars of college food purchases towards local, fair, and sustainable sources. Prior to Real Food Challenge, Anim led national initiatives at The Food Project in Boston, consulted with Economic Development Assistance Consortium, and developed employment training programs at the Bowery Residents Committee. Anim holds a BA in Astrophysics and History from Williams College and a Master's Degree in Public Policy from Harvard's Kennedy School of Government. He is the recipient of a Prime Mover Fellowship for movement-building and an Echoing Green award for social entrepreneurship.

Part I

The Role of Institutions in Food Systems Change

Chapter 1

Introduction: Institutions as Conscious Food Consumers

Sapna E. Thottathil
University of California, Office of the President, Oakland, CA, United States

Chapter Outline

What is Institutional Foodservice?	4	How the Book Is Structured: An Overview of Chapters	11
Targeting Consumption and Procurement to Change the Food System	6	Challenges Faced by Institutional Foodservice	13
The Unexamined Role of Foodservice in Reforming Industrialized Agriculture	9	Opportunities: Institutional Foodservice's Future	14
		Additional Reading and Resources	16
		References	17

Hospital food, dorm food, and school food have one major theme in common: They do not typically conjure up images of fresh and seasonal fare, crafted with minimally processed, artisanal, or local ingredients. Entire websites are devoted to documenting school food slop and YouTube channels abound with individuals grimacing when tasting or even looking at hospital food. Institutions, from hospitals to public schools at the kindergarten through twelfth-grade levels (K-12), do not have a good reputation in popular culture when it comes to culinary offerings.

The disdain and disgust is not without merit. Foodservice operations—places outside the home where prepared meals can be purchased and eaten—have been responsible for almost twice as many food-borne illnesses compared with eating at home (Center for Science and Public Interest, 2014). The Center for Science and Public Interest (2007) unearthed that several school cafeterias are lacking in food safety standards and protocols. Hospitals have been at the center of controversies for allowing fast food chains, notorious for unhealthy food and deep fryers, into their buildings, when hospitals are expected by the public to promote healthful practices (Ravella, 2016). Colleges and universities have had long-standing reputations for offering unappealing meals as well, and students have been embroiled in battles over meal offerings with fast food chains and to protect the rights and fair wages of farm workers (PBS, n.d.).

These stories reflect that institutional foodservice is rife with unhealthy offerings, and source food from long, complicated value chains, which are geographically dispersed networks of firms through which goods and services flow from inputs to the final product (at the point of consumption).[1] Institutional value chains are typically dominated by large corporate food companies and agribusiness intermediaries that have come under scrutiny for their environmental, business, and labor practices (Fuchs et al., 2009). Institutional food, therefore, embodies traits reflecting what our food system has become: industrialized and lacking in transparency, built to supply uniform ingredients at a large scale at the expense of environmental and social values, as well as culinary taste and nutrition.

Institutional food is changing, however. Several prominent organizations are now focusing their programming efforts on interventions in institutional foodservice. For example, the United States Department of Agriculture's (USDA) Farm to School program encourages school districts to support local farmers through their food purchases, and Bon Appetit Company's "Farm to Fork" local food purchasing plans have goals ranging from serving healthier food to customers to enhancing the livelihoods of farmers.

This edited volume aims to bring together, for the first time, in-depth insights into several such examples and promising practices of sustainable, healthy, and regional food purchasing by institutions.[2] It documents growing interest among nonprofit organizations and activists in institutional food interventions and highlights emerging evidence about the impacts of institutional food purchases on the supply chain, while also being attentive to the challenges and barriers to successful food systems change. Food manufacturers, government officials, policymakers, scholars, and funders can do more to support foodservice initiatives that are playing an increasingly large role in shaping the long-term health, equity, and local economic power of our food systems. Ultimately, the goal of this volume is to show that institutional foodservice represents an important market sector and political force. Institutions have become conscious food consumers, worthy of attention.

WHAT IS INSTITUTIONAL FOODSERVICE?

According to the USDA's Economic Research Service (2016), the "food marketing system"—food that is marketed and sold directly to consumers—supplied about $1.46 trillion worth of food to U.S. consumers in 2014. Just

---

1. The terms "supply chain" and "value chain" will be used interchangeably at times throughout this volume to help ensure accessibility to wide audiences. As editors, Annelies Goger and I acknowledge that these terms and "global production networks" come from separate academic traditions that place different emphasis on firms and other institutions involved in food systems. For a thorough discussion of these academic debates, see Bair (2008) and Hess (2008). Chapter 3 also has more details on using a value chain approach to analyze the food system.
2. I'd like to thank my coeditor, Annelies Goger, for her help in framing sections of this chapter.

under half of these food products were sold by the retail segment of the market, while the other half came from foodservice (more on this as follows).[3] Retailers include "traditional" food stores such as supermarkets, convenience stores, and specialty shops such as seafood and meat markets, and "non-traditional" food stores such as farmers' markets. What retailers have in common is that food is purchased from them by consumers to be prepared and eaten off-site. Since the 1990s, the retail sector has become increasingly consolidated, through acquisitions and mergers, and 20 companies (including Walmart) control more than 66 percent of the retail market, valued at $515.3 billion in sales in 2016 (USDA Economic Research Service, 2017a).[4] Given their market share, these few companies have enormous power in influencing raw material producers and food processors around new product development, supply chains, and pricing (Howard, 2016). Food value chains are highly "buyer-driven," as these retail firms exert extensive control over suppliers (Freidberg, 2004; Gibbon and Ponte, 2005).

Foodservice, on the other hand, refers to a collection of establishments that offer prepared foods for consumers to eat on-site (away from home). It is worth $731 billion, over half the value of the entire food marketing system. Foodservice is made up of *commercial* and *institutional* foodservice (the latter is also referred to as *non-commercial* foodservice). Commercial foodservice refers to fast food and full-service restaurants, places where selling food is the main business. Institutional foodservice includes (but is not limited to) private and public hospitals, university dining halls, correctional facilities, nursing homes, government agencies, corporate cafeterias, and school meal programs at K-12 schools (USDA, 2016). The institutional foodservice market accounted for more than $200 billion in sales in 2012. Its food and beverage sales are expected to keep growing as consumers continue to eat more meals and snacks away from their homes (Foodservice Director, 2013; Pullman and Wu, 2012).

The budgets for individual institutions vary. The New York City Department of Education, for example, the largest school district in the country, spent an estimated $469,267,990 in food and beverages for almost one million people in 2015. Carolinas Medical Center spent more than $25 million on food expenditures, and the University of Colorado, Boulder spent more than $31 million in 2015, representing the largest spending in their respective sectors of hospitals and colleges and universities (Foodservice Director, 2017). With this purchasing power, institutions have the ability to impact animal welfare, workers' rights, and sustainable growing practices in the food system and supply chain at a scale that is significantly larger than one consumer buying organic or Fair Trade foods at retail outlets such as grocery stores.

---

3. Based on USDA (2016) data, foodservice expenditures overtook retail expenditures for the first time in 2014.
4. As of publication, the USDA did not yet have available data about how the 2017 acquisition of the grocery store Whole Foods Market by internet commerce store Amazon would affect the makeup of food retail sales (USDA, 2017).

Of course, each institutional sector is different in how it purchases food. K-12 public schools in the United States, for example, must abide by federal nutrition standards if they choose to participate in the National School Lunch program, a federally assisted meal program where schools receive cash subsidies and food from the USDA per meal served (USDA, 2013). Hospitals are required to purchase up to 80–90 percent of their food through their Group Purchasing Organizations (GPOs), supply chain entities that coordinate the purchasing of a variety of hospital goods that develop two- to three-year contracts with specific food distributors (Klein, 2014). Foodservice at colleges, universities, and prisons tends to be run by foodservice management companies, which are private businesses that institutions have contracted with to purchase, prepare, and serve food. Other institutions may be "self-operating," and prepare food with their own staff.

Yet, there are many similarities among all of these institutions. With their large budgets and populations served, institutions purchase food at bigger volumes (compared with individual consumers) and contract with food distributors for transporting and storing food (Pullman and Wu, 2012). Institutional foodservice can also provide a unique market opportunity for food suppliers throughout the supply chain, from raw material producers to processors and distributors. For example, as a different market segment, institutions can be a part of diversifying a food supplier's sales strategy. Food suppliers often share that they enjoy working with the institutional market, because they can have more direct interactions with large-volume food buyers, and new product ideas can be turned around and feedback can be obtained more quickly from these customers (Kowitt, 2016).

TARGETING CONSUMPTION AND PROCUREMENT TO CHANGE THE FOOD SYSTEM

Our mainstream food system is plagued with environmental, social, and economic problems. It is a complicated system filled with contradictions, inefficiencies, and many activities that lay out of sight of the average consumer. Despite our human need for food to nourish ourselves, agriculture is rife with cases of harmful human impacts, from pesticide poisonings to instances of drinking water becoming contaminated from animal waste or pesticide runoff (Rosin et al., 2012).

Close to 800 million people living around the world today are hungry or malnourished, yet half of the food that is produced globally is wasted (Food and Agriculture Organization (FAO) of the United Nations, 2015; Institution of Mechanical Engineers, 2013). There are more mouths to feed today than ever before, as our world population grows, and yet farm debt is also increasing, and more and more farmers are leaving the profession. The USDA's most recent Agriculture Census documented a 4.3 percent drop in principal farm operators, those responsible for farm operations, between 2007 and 2012

(USDA, 2016). Farm income continues to decline—down to 42 percent in the United States since 2001 (Wilson and Durisin, 2016). Meanwhile, labor conditions in many agricultural and food processing sectors are shockingly horrific. The International Labour Organization (2014) estimated that 3.5 million people are working under forced labor conditions globally, and Oxfam America (2015) documented that workers in several poultry processing facilities are routinely denied breaks, leaving them to resort to wearing diapers during their shifts.

Another illogical example from our food system includes international trade in animal products. For example, the United States imported more than three billion pounds of beef and veal in 2016 while simultaneously exporting 2.5 billion pounds of beef and veal that very year (USDA Economic Research Service, 2017a,b). While industry analysts would argue that the imported beef products are different from those that are exported (domestic beef is usually grain fed and finished, while imports are typically grass fed), the United States frequently imports food items that can be grown and processed within the country, from tomatoes to apple juice.[5] At the same time, cases of fraud within the international trade of food items, from the high-profile horse meat scandal in Europe (found in products labeled as beef) to mislabeled seafood in the United States, are growing (Pimentel, 2014).

One solution to these problems, championed by organizations and activists in the food movement, is that individuals should shop for different and better food at the grocery store—food that is produced and processed more sustainably, humanely, and equitably, is healthier, and comes from regional sources. Educational and consumer campaigns have sprung up to encourage individuals to purchase food that is grown without chemical inputs or that is locally produced. The Environmental Working Group (EWG), for example, publishes annually its "Dirty Dozen List" for consumers, a guide that lists the top twelve fruits and vegetables with high levels of pesticide residues. EWG argues that these are the items that consumers should prioritize buying as organic, to minimize their exposure to harmful chemicals. Other campaigns include the American Society for the Prevention of Cruelty to Animals' (ASPCA) "Shop With Your Heart," which recommends that shoppers purchase meat, poultry, dairy, and egg products that carry a label vetted by the organization, such as the Animal Welfare Approved label, which guarantees a high standard of animal welfare in food production.

These market-based solutions, and the philosophy that consumers, as private citizens, can and should "vote with their dollars" by buying organic, humane, local, or Fair Trade food to reform the food system, unfortunately have a few shortcomings. One set of critiques developed by scholars who study the food movement focuses on the neoliberal idea of "the individual" and one person's economic action as the stimulus for social change. Specifically, these scholars

---

5. Several chapters in this volume, including Chapter 2, discuss the footprint of food and farming in more detail.

argue that the food movement has increasingly viewed consumers as critically important for sustaining and growing certain markets, such as the organic food market, or significantly responsible for impacting their own health outcomes through their food choices. However, this emphasis on individual private citizens and the assumption that consumers can easily choose healthy and sustainable food tends to ignore surrounding social and environmental factors that impact and influence consumer decisions and their health. For example, fast food marketing targets children at a young age to purchase food high in sodium and sugar (Best, 2017). Likewise, pesticide drift lingers in the air and water, imperiling nearby communities regardless of who is buying organic food from within them or elsewhere. A consumer-oriented approach to social change also overlooks the fact that many people cannot afford to purchase more sustainable and healthy food options or have limited access to them close to where they live (Alkon and Agyeman, 2011; Minkoff-Zern, 2014).

And still others have pointed out that many food campaigns are developed implicitly with a narrow framing of the ideal consumer, typically white and affluent, leaving out a majority of the populace and ignoring critical elements of race, class, and culture that influence access to food and preference for certain flavors and methods of cooking. Many food campaigns that champion "local food" and "healthy food" even contain elements of xenophobia (Alkon and Agyeman, 2011; Guthman, 2010; Slocum and Saldanha, 2016; Winter, 2003).[6] Moreover, the past few years have seen a surge in various label claims and standards for food production, confusing consumers about what to purchase when faced with a variety of choices over organic, local, or other sustainable and healthy food items (Thottathil, 2014). It is not always easy to purchase better food; nor is there necessarily one right choice to make when buying food.

Given the extent of change that is needed in the food system globally, many scholars also argue that focusing on the individual consumer level, and targeting private citizens, is not the most efficient and effective scale at which the food movement should invest resources, especially as parts of the food movement

---

6. Michael Pollan's (2006) first rule for "eating wisely" is to not "eat anything your great, great grandmother would not recognize as food." This is a rule that has been repeated to me in my own food advocacy and work in the United States, and yet I know that my grandmothers, both born and raised in India, and accustomed to eating rice-based breads, tapioca, and Indian cucurbits, would have been hard-pressed to recognize vegetables such as broccoli or lean proteins such as poached chicken breasts, items frequently upheld as healthy by the food movement in the United States. Additionally and undoubtedly, the food movement's fixation on purchasing "local" food often contains xenophobic elements (Winter, 2003), which can alienate many consumers of color—it certainly has made me, a first-generation Indian American, uncomfortable, when in situations where food imports from Asia are criticized by food activists without a clear reason. However, I recognize that broader political economic factors complicate these racial and cultural conversations and also play a role in influencing what people consider "healthy food." For example, researchers have well documented the sugar industry's lobbying efforts in downplaying sugar's contributions to heart disease compared with fat, which led to a demonization of fat by nutritionists in American diets without the similar villainizing of sugar (Kearns et al., 2016).

itself come to mirror industrial agriculture. For example, advocates of organic farming lament how many certified organic crops and animals are coming from agribusiness sources, which produce food on large, capital-intensive scales, in monocrops, and often employ poorly compensated or undocumented workers, replicating an agricultural system that exploits the environment, animals, and humans. Thus, private citizens each individually purchasing organic food from these sources perpetuate a form of industrial agriculture. Further, the organic farming movement to date has not done away with conventional agriculture, either. Government oversight, on the other hand, could better address the negative environmental and social impacts of industrial farming, by requiring that farms meet certain labor conditions (for example) at the state or federal levels, providing a larger public good that benefits all citizens. Targeting individual consumers, however, can leave out these and other solutions for redressing the harms of our industrialized food system (Guthman, 2006).

The Unexamined Role of Foodservice in Reforming Industrialized Agriculture

Institutional foodservice purchasing initiatives are similar, in some respects, to the campaigns that target individual consumers. Foremost, they rely on a similar logic around choice and consumption (Guthman, 2006). Many advocacy organizations that lead consumer campaigns are also increasingly foraying into institutional foodservice campaigns, and are replicating existing narratives about ideal consumers, purchasing, and even food.

However, there are several notable differences between individuals as consumers and institutions as consumers. The first is scale. The buying power of one college, hospital, or government agency is significantly larger than one person's purchasing power; many of these places alone spend tens of thousands more on food and beverages on an annual basis. Each institution also serves hundreds to thousands of people each year and is shaping the food environments within their walls, as opposed to targeting one person's behavior and choice alone.[7]

Further, national and local policymakers have greater opportunities to shape institutional food purchasing than they have at the individual level. As shown in several chapters in this volume, policymakers can use a wide variety of policy tools that affect institutions (especially government-supported institutions) at multiple scales, such as food purchasing agreements endorsed by school boards, city-level policies for government-funded institutions, and funding to support local agro-food business development and extension services that help connect small-sized and midsized farmers and their associations with institutional markets.

---

7. See Chapter 3 for an in-depth analysis of how foodservice value chains differ from those of retail.

This change is not just top-down in nature; it is dialectical. Institutional food purchasing is also influencing laws, policy, and regulations. The USDA's Farm to School program, for example, was developed with input from activists, nonprofit organizations, and schools already working on local food procurement from throughout the country and with the clear intent to protect the public good by tackling obesity, diabetes, and hunger (USDA, 2011). In 2015, the USDA also worked with K-12 public school districts to create a new, third-party certification around antibiotic use in poultry production (called Certified Responsible Antibiotic Use, or CRAU) that schools could rely on when buying chicken and turkey products (USDA, n.d.).[8] Lawmakers are also keenly interested in learning about how constituent institutions are conspicuously purchasing healthy, sustainable, and regional food; these initiatives and emerging research on the effects of food policy on health are providing politicians with the political support and evidence base to support legislation that will address the market failures and price distortions currently built into our food and agricultural system.[9]

When organized, consumers have impacted food systems for the better. Boycotts are one such example. The United Farm Workers' successful consumer grape boycott of the 1960s and 1970s in California eventually led to union contracts for workers. Imagine the power of institutions, if these efforts are coordinated and collectively organized, for changing social relations and how food is produced and processed—and indeed, several non-profit advocacy groups, from School Food Focus to Farm to Institution New England (FINE), which are featured in Chapters 8 and 11 of this volume, are bringing together several institutions to pool their purchasing power around food. There is a countermovement of institutions underway, which has the potential to change food and agriculture, and the case studies within this edited volume purport to unearth the impacts of values-based institutional food purchasing.

---

8. Eighty percent of antibiotics sold in the United States are for use in animal agriculture and 70 percent of these are important for human medicine; this antibiotic use has been linked to antibiotic-resistant infections that threaten human health (Martin et al., 2015). With the help of organizations featured in this edited volume, including the advocacy group School Food Focus, institutions are increasingly purchasing food animal products that have been produced without the use of antibiotics to protest the inordinate amount of antibiotic use in agriculture.

9. While I was a Senior Program Associate in Health Care Without Harm's food program in 2014, a group of representatives from constituent hospitals and I would regularly participate in legislative visits with key lawmakers to discuss concerns over antibiotic misuse in food animal agriculture and how hospitals were actively changing their food procurement choices to purchase meat raised without antibiotics. Many of these elected officials went on to support legislation that addressed antibiotic use in animal agriculture, or they asked key questions for the record at Congressional hearings to examine how antibiotics are used in food production. Approximately 80 percent of antibiotics sold in the United States are sold for use in animal agriculture, and evidence demonstrates that there is a link between antibiotic use in the food animal sector and antibiotic-resistant infections in human populations (Martin et al., 2015).

HOW THE BOOK IS STRUCTURED: AN OVERVIEW OF CHAPTERS

This volume targets a wide audience of foodservice professionals, policymakers, scholars, consumers, and non-profit practitioners. The chapters are written by a mix of researchers and high-profile food systems leaders working in the foodservice sector, and many of them have written about their first-hand experiences with institutional food purchasing. As editors, our intention in bringing together this broad mix of contributors is to not only document the emerging trends that are coalescing in the space of institutions as they become more conscious consumers, but to also inspire and support others who are just starting to embark upon values-based institutional foodservice initiatives by promoting the exchange of ideas, strategies, and lessons learned.

The proceeding chapters tackle the following questions: How are institutional foodservice activists reforming industrial agriculture? How have these initiatives begun to make food systems more sustainable, healthy, and equitable? To address these questions, the book is divided into four parts.

Part I ("The Role of Institutions in Food Systems Change") contextualizes how and why institutions can change the food system. Chapter 2 provides a broad overview of our current food system by discussing the trends around the "triple threat" of environmental degradation, dangers to livelihoods, and malnutrition that our global agro-food system must now confront. The authors then pull from case studies from around the world to demonstrate how institutional foodservice can focus on third-party certifications and standards, menu decisions, and food waste to address this triple-threat head on. Chapter 3 provides a conceptual foundation to situate institutional foodservice initiatives within the complex structure and dynamics of agro-food supply chains. Using a global value chain approach, this chapter highlights what aspects of the value chain and purchasing process that values-based procurement initiatives have and have not leveraged for food systems change.

Part II ("Setting Purchasing Standards Through Policy") examines how policy—from within institutions to the laws and regulations created by governments—interacts with institutional food purchasing. Chapter 4 provides an overview of the policy environment from the institutional to the federal levels, and discusses how it can help and hinder values-based procurement. It also critically examines the role of contracts with foodservice management companies and GPOs, which can act as barriers to implementing many policies around sustainable and local food purchasing. Chapter 5 tells the origin story of the Good Food Purchasing Policy, and how it has been used by the city of Los Angeles, Los Angeles Unified School District, and Oakland Unified School District to change procurement practices. The values-based purchasing by these institutions, which Los Angeles has codified into its new procurement contract, is also stimulating change in the supply chain, from the creation of new and healthier food items to better labor conditions for transportation workers. Chapter 6 takes readers to the other side of the country, as it describes the implementation of the City of Philadelphia's nutrition standards. The authors of this chapter, including those at the Drexel Food Lab at Drexel University, worked with the city and food manufacturers to reformulate existing food items to comply with Philadelphia's new standards, making it easier for institutions themselves to comply with the city's nutrition regulations.

Part III ("Creating Shared Identity Using Place-Based Connections") takes a close look at how several farm-to-institution initiatives have leveraged a sense of regional identity (rooted in a specific place) to catalyze a network of institutions to re-embed the food system in the "local" economy and instill "local" values. Chapter 7 describes the history of farm-to-institution work in the state of Michigan, and analyzes the factors that have made the state one of the national leaders in farm-to-institution initiatives, including policy leadership at the state level. Chapter 8 details the efforts by institutions across sectors in six states in New England that have come together to support the region's local food (and local seafood) movements through collective policy work, data collection, and knowledge sharing. Chapter 9 looks at how public schools in Montana are increasingly supporting local producers and processors by buying beef that is raised, finished, and slaughtered in the state—this is truly innovative because Montana's beef supply chain is structured primarily around shipping young calves (born in the state) to be raised for meat out of state. Chapter 10 then provides a case study of Iowa and uses survey data to explore the potential for local food purchasing by institutions to increase the number of jobs and midsized farms within the state.

Part IV ("New Directions for Foodservice") explores innovative directions in foodservice research and activism that are deserving of more attention and resources. Chapter 11 begins this part of the book by bringing to light an underexamined issue in institutional foodservice: The role that food banks can play in supporting regional farmers, redistributing excess food to underserved communities, and promoting healthier eating. The authors of this chapter list the assets that food banks have around supply chain distribution infrastructure, which could be leveraged further to support farm-to-institution and other values-based procurement programs. Next, Chapter 12 showcases how hospitals are increasingly using a broader definition of health, one that encompasses values about external environmental factors, to justify purchasing more local and sustainable food; the uptake of this "environmental nutrition" definition of health has been facilitated by the non-profit, Health Care Without Harm. The adoption of more plant-based foods in colleges and universities has also been supported by non-profits; Chapter 13 showcases this work that the Humane Society and other, similar organizations are doing in institutions of higher education. This chapter notes that the interest of students in being served more plant-based foods is growing, and college and university foodservice, in addition to food manufacturers, are responding. Finally, Chapter 14 closes the book with the recognition of the additional work and skills that values-based procurement initiatives can place on cafeteria workers in K-12 schools. The authors report that frontline cafeteria workers are experiencing increased workloads and more stress to accommodate the implementation of new recipes and less processed food that may require more scratch-cooking skills, new equipment to learn, new food safety procedures, and more preparation time. Institutional activities around values-based procurement must better address the needs of foodservice workers, the authors argue, from appropriate job training, to professional development opportunities, management education, and higher wages.

CHALLENGES FACED BY INSTITUTIONAL FOODSERVICE

Despite progress to date on values-based food purchasing, institutions face several challenges as conscious consumers, many of which are discussed in the proceeding chapters. One major challenge is cost. The price of food is continuing to increase globally (FAO, 2017). Because purely profit-oriented markets do not account for the negative externalities associated with exploitative or unsustainable food production, buying more healthy, local, and sustainable foods (such as organic produce) is often more expensive than their conventional counterparts, dissuading foodservice operators from purchasing them, especially if budgets are tight (York, 2017).

As such, institutions are still buying and serving terrible food. Many school districts are continuing to sell fast food and sweetened beverages to middle and high school children, for example. And the influence that corporate food and beverage companies have over young people with their advertising continues to be a contested topic (Nestle, 2013).

As food prices keep rising (FAO, 2017), there are also implications for equity around which institutions will be able to afford to buy more nutritious, sustainable, equitable, and local food. The spread of resources and awareness around institutional food purchasing is already uneven—concentrated in urban and suburban areas, private institutions, and educational facilities (from school districts to colleges and universities). Yet, many other institutions, such as correctional facilities, nursing homes, and day cares, could also benefit from values-based procurement initiatives. One recent example that stands out includes accusations that the corporation Aramark—which provides contracted foodservices to several hundred correctional facilities across the country—has been serving prisoners substandard food with maggots and rocks (Kelkar, 2017). (Aramark, by the way, has yearly sales in the billions of dollars and also provides amenities at several National Parks, colleges and universities, hospitals, and K-12 school districts.) Government entities and public institutions in particular have been privatizing and outsourcing several activities, including foodservice, with the rationale of trimming budgets (Chassy and Amey, 2011; Gonzalez and Kemp, 2016).

The limited food budgets that institutions possess remain one of many barriers to market entry faced by new, historically marginalized, and sustainable suppliers who would like to create financially viable businesses. Supply chain intermediaries, such as aggregators, possess more power in value chains, and retailers such as Walmart continue to wield enormous market power in influencing value chains, driving down prices and encouraging consolidation; these are difficult realities with which new, independent, innovative, and small suppliers have to compete (Howard, 2016). And, as Chapters 3–5 also demonstrate, foodservice supply chains are structured through existing contracts, which makes values-based procurement by institutions more difficult than purchasing conventional food (through their broad line distributors or foodservice management companies, for example). Further complicating these supply chains is the lack of transparency within them and the lack of definitions around terms such as "local," "regional," and "sustainable," which can lead to a race to the bottom around standards.

As for standards around working conditions, there are only a few organizations that are dedicating themselves to thinking about the relationship between institutional foodservice and labor. The Food Chain Workers Alliance, in coordination with Solidarity Research, recently released a report called *No Piece of the Pie*, which documents the poor conditions that workers in the food industry face across the supply chain, from food processing and distribution to foodservice. The report found that while employment in the food chain has been growing at a rate more than double than that of all other industries over the past fourteen years, the sector overall pays the lowest hourly median wage to workers compared with workers in other industries. (The annual median wage for food chain workers is $16,000, and the hourly median wage is $10, below the median wages across all industries.) Further, significant racial and gender differences exist within these numbers, many of these jobs provide little opportunities for advancement, and food supply chain workplaces offer little or no professional development (Food Chain Workers Alliance and Solidarity Research Cooperative, 2016).[10] Aside from the research and advocacy that the Food Chain Workers Alliance is doing, as well as what the Good Food Purchasing Program is attempting to codify (detailed in Chapter 5), institutions have a long way to go to address issues around equity and labor with their purchasing dollars (which Chapter 14, about cafeteria workers in schools, examines very well in the foodservice context).

OPPORTUNITIES: INSTITUTIONAL FOODSERVICE'S FUTURE

Clearly, institutional foodservice is an arena in need of more research, resources, and attention. There are several opportunities common among the proceeding chapters that could potentially be replicated to spread the nascent success of

---

10. I want to thank Jose Oliva and Zachary Herman with the Food Chain Workers Alliance for productive conversations and writing on this topic for this volume.

values-based purchasing by institutions. These can be grouped into two broad and related categories centered on local public policy and the market structures that govern relationships and interactions between supply chain entities.

The first opportunity is policy based, particularly at the local level, around standards and messaging that are supportive of values-based procurement by institutions. Specific examples include local government policy, discussed in the chapters about the Good Food Purchasing Policy in Los Angeles (Chapter 5), the nutrition standards adopted by the City of Philadelphia (Chapter 6), and the state of Michigan (Chapter 7), which can spur innovation in supply chains and provide the environment and resources within which institutions acting as conscious consumers can flourish more easily. Non-profit campaigns, funded largely through private foundations and individual donations, have been key drivers in supporting these public policies. They have also supported foodservice personnel in incorporating values into food purchasing decisions and institutional-level policies, as Chapters 12 and 13 about the healthcare sector and plant-based meals in higher educational facilities demonstrate.

The second opportunity builds on the first. It focuses on the market and addresses structural challenges within supply chains. The work of non-profit organizations such as the Center for Good Food Purchasing (Chapter 5) and FINE (Chapter 8) has been instrumental in overcoming barriers to values-based procurement that may be created by the provisions within contracts with distributors and foodservice management companies.[11] Additionally, agriculture extension agencies and land grant universities, as highlighted in the chapters about Michigan and Montana (Chapters 7 and 9), have the potential to play powerful roles in connecting institutions to suppliers of local, healthy, and sustainable food within their states, and in building local food processing and distribution capacity. As Chapter 3 also points out, other intermediaries such as food hubs and producers' cooperatives can be key players in institutional foodservice value chains that address the market failures of our industrialized agro-food system, particularly those borne by smaller and sustainable producers and processors.

Of course, I recognize that when we first began soliciting interest in our edited volume in 2016, the Obama administration was still in place at the federal level in the United States. Under the Trump administration, several progressive food and farming initiatives that were instituted from even prior to the Obama-era have been rolled back, halted, or delayed, such as the animal welfare standards for certified organic egg-laying hens, the long-expected ban on the chemical pesticide chlorpyrifos, and nutrition standards for school lunches (Lipton, 2017; USDA Agricultural Marketing Service, 2017; USDA, 2017). These developments lead to questions about the permanency of the policy initiatives taken on by government institutions, such as the USDA's Farm to School

---

11. Chapter 4 discusses some of these barriers in foodservice management contracts in detail.

programs and CRAU standards, mentioned earlier in this chapter.[12] These developments also suggest that U.S. lawmakers are unlikely to pass federal policy that addresses the contradictions and inefficiencies of the agro-food system in the near future.

As such, in this politicized environment, this market-based work that institutional foodservice is taking on, and the local policy work that has emerged to support values-based procurement, is more important now than ever before. In showcasing examples in the following chapters of K-12 public schools, hospitals, colleges, universities, and other institutions incorporating values into food procurement, we hope that we have successfully made the case that institutions as conscious consumers can indeed be key drivers of environmental, social, and economic change in food value chains.

ADDITIONAL READING AND RESOURCES

The following chapters contain examples of institutions and supporting organizations that are pursuing values-based procurement. If you are a foodservice professional or practitioner, I recommend that you check out the websites of the places highlighted in each chapter. Additionally, I recommend a few other resources that practitioners might find useful when embarking upon institutional foodservice initiatives.

- The Los Angeles Unified School Board successfully approved the Good Food Purchasing Policy and values around the environment, labor, animal welfare, local economies, and nutrition, which the Los Angeles Unified School District then incorporated into a recent food procurement Request for Proposals (RFP). Examples of School Board policies that promote values-based procurement are available from the Center for Good Food Purchasing: https://goodfoodpurchasing.org/resources/#example-policies
- Farm to Institution New England (FINE) has several examples of contract language that institutions can use to communicate interest in local food with their distributors: https://www.farmtoinstitution.org/food-service-toolkit (see sections six, seven, and eight)

---

12. I would be remiss to not point out other, notable institutional food procurement work that took place at the federal level during the Obama administration. For example, due to long-term relationship-building and educational efforts, the Humane Society successfully introduced "Meatless Monday" (a program highlighted in relation to universities in Chapter 13) into several U.S. military facilities, which are offering more vegetarian options at mealtimes and even cutting meat consumption (Huffman, 2016). Other branches of the U.S. government are also changing their procurement practices. In a 2015 Presidential Memorandum, federal cafeterias that the General Services Administration (GSA) manages were required to "develop and implement a strategy that creates a preference for awarding contracts to vendors that offer, as an option, meat and poultry produced according to responsible antibiotic-use policies for sale in domestic Federal cafeterias to civilian Federal employees and visitors, to the extent such an option is available and cost effective" (The White House Office of the Press Secretary, 2015).

- FINE also hosts several job descriptions for foodservice personnel that incorporate experience around local food purchasing: https://www.farmtoinstitution.org/food-service-toolkit (see Appendix E)
- The Humane Society offers free, plant-based recipes that are scaled up for foodservice: http://www.humanesociety.org/assets/pdfs/farm/meatless-monday-food-service-recipes-2014.pdf
- Health Care Without Harm also offers free, plant-based recipes: https://noharm-uscanada.org/issues/us-canada/balanced-menus-recipe-toolkit
- The USDA's Farm to School website contains helpful information about local food purchasing for school districts: https://www.fns.usda.gov/farmtoschool/farm-school

REFERENCES

Alkon, A.H., Agyeman, J. (Eds.), 2011. Introduction: the food movement as polyculture. In: Cultivating Food Justice: Race, Class, and Sustainability. The MIT Press, Cambridge, MA.

Bair, J., 2008. Analysing economic organization: embedded networks and global chains compared. Econ. Soc. 37 (3), 339–364.

Best, A.L., 2017. Fast Food Kids: French Fries, Lunch Lines, and Social Ties. New York University Press, New York.

Center for Science and Public Interest, 2007. What danger lurks in the school cafeteria? Available from: https://cspinet.org/new/200701301.html. Accessed January 29, 2018.

Center for Science and Public Interest, 2014. Restaurants pose twice the risk of foodborne outbreaks as homes, data show. Available from: https://cspinet.org/new/201404071.html. Accessed January 29, 2018.

Chassy, P., Amey, J., 2011. Bad business: billions of taxpayer dollars wasted on hiring contractors. Project on Government Oversight. Available from: http://pogoarchives.org/m/co/igf/bad-business-report-only-2011.pdf. Accessed January 29, 2018.

FAO, 2017. World food situation. Available from: http://www.fao.org/worldfoodsituation/foodpricesindex/en/. Accessed January 29, 2018.

Food and Agriculture Organization (FAO) of the United Nations, 2015. The state of food insecurity in the world 2015. Available from: http://www.fao.org/hunger/en/. Accessed January 29, 2018.

Food Chain Workers Alliance and Solidarity Research Cooperative, 2016. No piece of the pie: U.S. food workers in 2016. Available from: http://foodchainworkers.org/wp-content/uploads/2011/05/FCWA_NoPieceOfThePie_P.pdf. Accessed January 29, 2018.

Foodservice Director, 2013. Technomic sees growth in non-commercial foodservice. Available from: http://www.foodservicedirector.com/archive/archived-content/articles/technomic-sees-growth-non-commercial-foodservice. Accessed January 29, 2018.

Foodservice Director, 2017. Top 100 noncommercial operators. Available from: http://www.foodservicedirector.com/research/top-100-noncommercial-operators. Accessed January 29, 2018.

Freidberg, S., 2004. French Beans and Food Scares: Culture and Commerce in an Anxious Age. Oxford University Press, New York.

Fuchs, D., Kalfagianni, A., Arentsen, M., 2009. Retail power, private standards, and sustainability in the global food system. In: Clapp, J., Fuchs, D.A. (Eds.), Corporate Power in Global Agrifood Governance. MIT Press, Cambridge, MA.

Gibbon, P., Ponte, S., 2005. Trading Down: Africa, Value Chains, and the Global Economy. Temple University Press, Philadelphia.

Gonzalez, J.J., Kemp, R.L. (Eds.), 2016. Privatization in Practice: Reports on Trends, Cases and Debates in Public Service by Business and Nonprofits. second revised edition. McFarland, Jefferson, NC.

Guthman, J., 2006. From "old school" to "farm-to-school": neoliberalization from the ground up. Agric. Hum. Values 23 (4), 401–405.

Guthman, J., 2010. Fast food/organic food: reflexive tastes and the making of 'yuppie chow'. Soc. Cult. Geogr. 4 (1), 45–58.

Hess, M., 2008. Governance, value chains and networks: an afterword. Econ. Soc. 37 (3), 452–459.

Howard, P.H., 2016. Concentration and Power in the Food System: Who Controls What We Eat? Bloomsbury Academic, London.

Huffman, J., 2016. Humane society looks to spread meatless mondays. Available from: https://www.politico.com/tipsheets/morning-agriculture/2016/05/humane-society-looks-to-spread-meatless-mondays-this-week-ag-panels-stroll-through-barnyard-organic-takes-the-hill-tpps-slow-roll-in-japan-214437. Accessed January 29, 2018.

Institution of Mechanical Engineers, 2013. Global food. Waste not, want not. Available from: http://www.imeche.org/knowledge/themes/environment/global-food. Accessed January 29, 2018.

International Labour Organization, 2014. Profits and poverty: the economics of forced labour. Available from: http://www.ilo.org/global/publications/ilo-bookstore/order-online/books/WCMS_243391/lang--en/index.htm. Accessed January 29, 2018.

Kearns, C.E., Schmidt, L.A., Glantz, S.A., 2016. Sugar industry and coronary heart disease research: a historical analysis of internal industry documents. JAMA Intern. Med. 176 (11), 1680–1685.

Kelkar, K., 2017. Prison strike organizers to protest food giant Aramark. PBS. Available from: https://www.pbs.org/newshour/nation/prison-strike-protest-aramark. Accessed January 29, 2018.

Klein, K., 2014. Values-based food procurement in hospitals: the role of health care group purchasing organizations. Agric. Hum. Values. https://doi.org/10.1007/s10460-015-9586-y.

Kowitt, B., 2016. Startups are finding a powerful partner in this hidden corner of the food economy. Fortune. Available from: http://fortune.com/2016/10/05/back-to-the-roots-food-service/. Accessed January 29, 2018.

Lipton, E., 2017. E.P.A. chief, rejecting agency's science, chooses not to ban insecticide. The New York Times. Available from: https://www.nytimes.com/2017/03/29/us/politics/epa-insecticide-chlorpyrifos.html?_r=0. Accessed January 29, 2018.

Martin, M.J., Thottathil, S.E., Newman, T.B., 2015. Antibiotics overuse in animal agriculture: a call to action for health care providers. Am. J. Public Health 105 (12), 2409–2410.

Minkoff-Zern, L., 2014. Hunger amidst plenty: farmworker food insecurity and coping strategies in California. Local Environ. 19 (2), 204–219.

Nestle, M., 2013. Eat Drink Vote: An Illustrated Guide to Food Politics. Rodale Books.

Oxfam America, 2015. Lives on the line. Available from: https://www.oxfamamerica.org/livesontheline/. Accessed January 29, 2018.

PBS, The coalition of Immokalee workers vs. Taco Bell. Available from: http://www.pbs.org/now/society/ciw.html. Accessed January 29, 2018.

Pimentel, P., 2014. Trends and solutions in combating global food fraud. Food Saf. Mag. Available from: https://www.foodsafetymagazine.com/magazine-archive1/februarymarch-2014/trends-and-solutions-in-combating-global-food-fraud/. Accessed January 29, 2018.

Pollan, M., 2006. Six rules for eating wisely. TIME Mag. Available from: http://michaelpollan.com/articles-archive/six-rules-for-eating-wisely/.

Pullman, M., Wu, Z., 2012. Food Supply Chain Management: Economic, Social and Environmental Perspectives. Routledge, New York/London.

Ravella, S., 2016. When the hospital serves McDonald's. The Atlantic. Available from: https://www.theatlantic.com/health/archive/2016/02/unhealthy-hospital-food/461898/. Accessed January 29, 2018.

Rosin, C., Stock, P., Campbell, H. (Eds.), 2012. Introduction: shocking the global food system. In: Food Systems Failure: The Global Food Crisis and the Future of Agriculture. Earthscan. Milton Park, Abingdon, Oxon, New York.

Slocum, R., Saldanha, S. (Eds.), 2016. Geographies of race and food: an introduction. In: Geographies of Race and Food: Fields, Bodies, Markets. Routledge, New York/London.

The White House Office of the Press Secretary, 2015. Presidential Memorandum—creating a preference for meat and poultry produced according to responsible antibiotic-use policies. Available from: https://obamawhitehouse.archives.gov/the-press-office/2015/06/02/presidential-memorandum-creating-preference-meat-and-poultry-produced-ac. Accessed January 29, 2018.

Thottathil, S., 2014. India's Organic Farming Revolution: What it Means for Our Global Food System. University of Iowa Press, Iowa City.

United States Department of Agriculture (USDA), Certified responsible antibiotic use. Available from: https://www.ams.usda.gov/services/auditing/crau. Accessed January 29, 2018.

USDA, 2011. USDA farm to school team, 2010 summary report. Available from: https://www.fns.usda.gov/sites/default/files/2010_summary-report.pdf. Accessed January 29, 2018.

USDA, 2013. National school lunch program. Available from: https://www.fns.usda.gov/sites/default/files/NSLPFactSheet.pdf. Accessed January 29, 2018.

USDA, 2016. 2012 census highlights. Available from: https://www.agcensus.usda.gov/Publications/2012/Online_Resources/Highlights/Farm_Demographics/. Accessed January 29, 2018.

USDA, 2017. USDA publishes school meals rule, expands options, eases challenges. Available from: https://www.fns.usda.gov/pressrelease/2017/015717. Accessed January 29, 2018.

USDA Agricultural Marketing Service, 2017. National organic program (NOP); organic livestock and poultry practices—withdrawal. Fed. Regist. Available from: https://s3.amazonaws.com/public-inspection.federalregister.gov/2017-27316.pdf?utm_campaign=pi%20subscription%20mailing%20list&utm_source=federalregister.gov&utm_medium=email. Accessed January 29, 2018.

USDA Economic Research Service, 2016. Market segments. Available from: https://www.ers.usda.gov/topics/food-markets-prices/food-service-industry/market-segments.aspx. Accessed January 29, 2018.

USDA Economic Research Service, 2017a. Retail trends. Available from: https://www.ers.usda.gov/topics/food-markets-prices/retailing-wholesaling/retail-trends/. Accessed January 29, 2018.

USDA Economic Research Service, 2017b. Livestock and meat international trade data. Available from: https://www.ers.usda.gov/data-products/livestock-and-meat-international-trade-data/. Accessed January 29, 2018.

Wilson, J., Durisin, M., 2016. Betting the farm and losing: banks seek collateral for debts. Bloomberg. Available from: https://www.bloomberg.com/news/articles/2016-11-13/betting-the-farm-and-losing-banks-seek-collateral-as-debts-rise. Accessed January 29, 2018.

Winter, M., 2003. Embeddedness, the new food economy and defensive localism. J. Rural. Stud. 19 (1), 23–32.

York, T., 2017. Organics fail to justify extra cost in foodservice. The Packer. Available from: https://www.thepacker.com/article/organics-fail-justify-extra-cost-foodservice. Accessed January 29, 2018.

Sapna E. Thottathil is the co-editor of *Institutions as Conscious Food Consumers*. She has worked on institutional food purchasing and supply chains with hospitals and K-12 public schools while employed at the organizations Health Care Without Harm and School Food Focus, and now works in higher education with the University of California as an Associate Director of Sustainability for the University of California's Office of the President. She is continuing her food advocacy work as a board member for Pesticide Action Network (North America), as a council member for the Oakland Food Policy Council, and as an advisor to the Plant Based Foods Association. She is also a faculty member in the University of the Pacific's Food Studies Program, and is the author of *India's Organic Farming Revolution: What it Means for Our Global Food System*. She holds a PhD in geography from the University of California-Berkeley, an MSc from Oxford University, and a BA from the University of Chicago.

Chapter 2

Trends in the Global Food System and Implications for Institutional Foodservice

Kristal Jones[*], Kimberly Pfeifer[†], Gina Castillo[†]
[*]National Socio-Environmental Synthesis Center, University of Maryland, Annapolis, MD, United States, [†]Oxfam America, Boston, MA, United States

Chapter Outline

Introduction	21
Food Systems Trends Past and Present	23
Environmental Sustainability: Trends, Drivers, and Challenges	25
Livelihoods: Trends, Drivers, and Challenges	26
Nutrition: Trends, Drivers, and Challenges	28
Nexus of Sustainability, Livelihoods, and Health Challenges	29
Emerging Approaches to Addressing Food Systems Challenges	30
Third-Party Certifications and Increasing Demand for Information	31
Shortening the Food Chain: Going Local and Fresh	32
Social Movements Addressing Equity and Sustainability	33
Institutions as Key Drivers of Food System Transformation	34
Improving Health and Sustainability Through Menu Decisions	35
Improving Equity and Sustainability Through Engagement With Certifications and Standards	36
Improving Health, Equity, and Sustainability by Reducing Food Waste	37
Conclusion	39
References	39

INTRODUCTION

Actors in contemporary food systems have the potential to contribute to or hinder a broad range of outcomes related to human and environmental health and well-being, as well as to increase equity and sustainability along food value chains. Globally, agriculture produces sufficient calories to feed the world, and is doing so more efficiently than ever. However, greenhouse gas emissions along the food chain, as well as the depletion of the natural resource base that sustains agriculture threaten future food production (Hunter et al., 2017). For producers and workers in food value chains, livelihoods can be undermined by these changes in the environment, as well as by structural dynamics within food systems that prioritize standardization over diversity, and efficiency over equity (Abdulsamad and Gereffi, 2018; Friedmann, 2005; Lobao and Meyer, 2002; Pechlaner and Otero, 2010). In addition, the current structure of the global food system has generated a paradoxical situation in which close to one billion people are hungry (lacking access to sufficient amounts of macronutrients), another billion suffer from nutrition insecurity (lacking crucial micronutrients), and more than half a billion people are substantially overconsuming. This is spawning a new public health epidemic involving chronic conditions such as Type 2 diabetes and cardiovascular disease (FAO, 2015; von Grebmer et al., 2014; WHO, 2016).

Health professionals, food and agriculture businesses, environmental and poverty activists, and policymakers have begun to acknowledge a core set of interrelated food systems challenges, namely: How to produce enough food sustainably, how to support those who produce our food to earn a viable livelihood, and how to access and maintain a healthy diet. Transformation of the global food system is clearly needed if we wish to embed equity, sustainability, and health as priorities in food provision and consumption. Some of these transformations will be facilitated through new technologies, while others will require public policy shifts, changes in the private agro-food industry, actions by civil society, and behavioral changes by individuals.

This chapter presents an overview of alternative food initiatives led by nonprofit organizations, public and private institutions, and consumers, all of which are proactively working to generate positive changes and impacts along food value chains. These efforts are transforming not only what people eat, but also how they think of food in relation to issues such as climate change, environmental degradation, and social justice (Hinrichs and Allen, 2008; Lyson, 2007; West et al., 2014). In this dynamic context, institutional foodservice actors and the consumers they serve sit at an important nexus within the food system, and have the potential to make decisions that cut across the challenges and opportunities to improve food system outcomes for people and the environment.

In this chapter, we use a "triple threat" framework to characterize the major challenges confronting contemporary food systems (Oxfam America, 2018; FAO, 2008):

- The environmental sustainability challenge: We must produce enough nutritious food for nine billion people by 2050, while reducing vulnerability and contributions of food systems to climate change and environmental degradation.
- The livelihoods equity challenge: We must value food in such a way that supports local and regional economies, including stable and equitable

livelihoods for small-scale producers, farm labor, processors, and distributors, while maintaining access to healthy choices for consumers.
- The nutrition challenge: We must address inequities along with policies and incentive structures in the food system that limit production of, access to, and consumption of adequate, healthy, and appropriate foods to meet people's nutritional needs.

"Food system change" refers to the movements and actions to overcome the challenges described herein. We define change in terms of equity and sustainability. An equitable food system is one that provides access to nutritious foods and adequate calories for consumers, and to economically stable and viable rural livelihoods. A sustainable food system is one that produces in ways that minimize degradation to the natural resource base required to produce food and sustains viable livelihoods.

The chapter begins by reviewing the history of the modern global food system,[1] and we highlight key actors and actions that have contributed to the trends and challenges seen today. We then examine stakeholder behavior, policies, and incentive structures that are responding to these challenges. The approaches, successes, and limitations of these approaches offer insight into possible actions and activities that food system actors looking to contribute to improving human and environmental health and well-being can take. The final section explores how changes in foodservice purchasing could contribute to addressing the underlying drivers and negative outcomes of contemporary food systems. We highlight procurement and management decisions that foodservice institutions can make to help address the complex sustainability, equity, and nutrition challenges presented by the global food system.

FOOD SYSTEMS TRENDS PAST AND PRESENT

From the times of Thomas Malthus to today, efforts to explain and address drivers of trends in the global food system reflect different and evolving paradigms about what the key problems and potential solutions are. For example, an emphasis in the sustainability literature on population growth and the idea of the limits to carrying capacity of the Earth for food production points to the imbalance between supply of and demand for food (Cohen, 1995; Rockström et al., 2017). Other scholars, such as Boserup (1965), have highlighted the complex relationship between population growth, technological change, and shifting

---

1. In this chapter, the term "global food system" refers to all the activities involved in the production, processing, transporting, and consumption of food. Generally, global food system refers to an export-driven, industrial model driven by the logic of efficiency and profit. A major characteristic of the contemporary global food system and more localized food systems nested within it is the high degree of complexity and interdependence over space and among different actors along the supply chain (Antle et al., 2014; Abdulsamad and Gereffi, 2018).

potential supply in the food system, and they focus on both intensification of food production and access to the means to produce and distribute food efficiently. Arguments about the relationship between poverty and hunger made by Sen (1981) wholly shift the emphasis toward unequal access to and distribution of food, and put much of the focus on national policy and international relationships that reinforce power dynamics within and around food systems. These paradigms have all contributed to the more recent systems perspectives on the causes and consequences of changes in food systems.[2] Over the past several decades, practitioners, policymakers, and academics have framed the practices in food systems as related to an increasingly globally integrated food system that reflects a history of colonialism and uneven development, as well as a capitalist emphasis on efficiency over equity, sustainability, and nutrition (Buttel, 2001; Gomez et al., 2013; Holt-Giménez and Peabody, 2008).

In the post-World War II era, many newly independent and developing countries faced challenges because plantation and peasant farming systems competed with pressures to modernize agricultural production and produce cash crops for global markets (rather than subsistence crops). By the 1970s, there was a policy focus on achieving yield increases, as what came to be called the Green Revolution emphasized providing farmers with seeds and technology packages and focused on a relatively small number of global commodities such as wheat and rice (Das, 2002; Evenson and Gollin, 2003). Structural adjustment policies in the 1980s and 1990s provided further incentives to transition households out of subsistence agricultural production and into larger-scale agro-food production, subsidizing the use of technology over labor and promoting the goal of increased efficiency (Birner and Resnick, 2010). In some places, especially in South and Southeast Asia, the Green Revolution and subsequent policy shifts that emphasized the use of synthetic inputs to produce food commodities for global markets did contribute to increasing production and lessening hunger for both rural and urban populations (Evenson and Gollin, 2003). However, these and other discrete approaches to dealing with the shortcomings of the industrial food system did not see the global food system as a complex system comprised of many social and environmental variables, with interconnected activities and actors that move food from the field to the table. The prospect of higher yields and increased incomes has played out in some places, but it has also led to resource degradation and economic insecurity in volatile global markets (Abdulsamad and Gereffi, 2018).

Following is a summary of some of the key trends and drivers of equitable and sustainable outcomes (or the lack thereof) from contemporary food systems at the global level. We highlight both persistent problems and recent changes

---

2. See, for example, Friedmann's (2005) updated article on food regimes, which identifies food systems as one manifestation of socio-political systems, as well as analysis from the food sovereignty movement and organizations such as La Vía Campesina of the exploitation and disenfranchisement inherent in neoliberal food systems (Desmarais, 2002).

that reflect the three interrelated challenges of supporting sustainability, livelihoods, and nutrition through food system activities.

Environmental Sustainability: Trends, Drivers, and Challenges

There are multiple environmental sustainability challenges facing the food system. Changes to the environment, such as toxic contamination and climate change, pose serious risks to all food systems, and large-scale food systems themselves threaten and degrade the environment. Most climate models predict that rising temperatures will have mixed effects on agricultural production, with the potential for yield increases in some temperate regions and severe yield declines in dryland and rain-fed systems (Gornall et al., 2010). Climate change may result in unpredictable growing periods, including more intense rainfall events between prolonged dry periods, as well as increase the frequency and severity of extreme weather events such as heat waves, droughts, and floods, which can destroy harvests instantly and lead to food shortages (Lobell and Field, 2007). In addition, recent work by Smith et al. (2017) suggests that climatic change could negatively affect the nutritional composition of many staple grains. These and other changes are already impacting food systems in parts of the world, such as the Middle East and North Africa, where growing populations and increasing water scarcity are generating pressure to further intensify agricultural production at the same time that climate variability is decreasing yields for rain-fed grains and legumes (Sowers et al., 2011).

On the other hand, agriculture also depletes and pollutes the environment. For example, agriculture is the largest contributor of non-CO_2 greenhouse gas emissions, with an outsized proportion (per calorie emissions) coming from the production of animal-based protein (FAO, 2017b). Agriculture also accounts for 70 percent of global fresh water use, and is both a driver and increasingly a victim of water scarcity (AQUASTAT, 2017). In particular, the growing demand for livestock products has raised concerns about the land and water requirements for meat production. Furthermore, agricultural activities are the largest contributor to land degradation—it is estimated that agriculture is the proximate driver for approximately 80 percent of deforestation worldwide (FAO, 2017b). Although patterns of land use change and land degradation are highly variable globally (Beddow et al., 2010), intensive land management practices (epitomized in the Green Revolution), such as the practice of monoculture, drain the soil of nutritional value, which is then replenished with additional synthetic fertilizers (Khan and Hanjra, 2007). In the Midwest United States, for example, the historical Dust Bowl devastation and more recent trends toward intensification of corn-soybean systems have engendered highly input-intensive agricultural systems that affect soil health as well as water quality (Robertson and Swinton, 2005).

In addition to the impacts of agricultural production, life-cycle analyses of carbon and water footprints along the food value chain suggest that energy use occurs mostly in storage facilities and in transport (FAO, 2014; Khan and Hanjra, 2007). By contrast, water use in processing, distribution, and retail is minimal compared with water used in agricultural production (FAO, 2014). Multilateral organizations, national government agencies, and researchers focused on sustainability in food and agriculture now emphasize decreasing food loss and food waste as food travels from farm-to-table as a way to improve both environmental sustainability and food availability (FAO, 2014; Lipinski et al., 2013; USDA, 2013). The FAO (2014) estimates that one third of human food production is lost or wasted globally, around 1.3 billion tons per year.[3] The limited data available suggest that in developing countries, losses are much higher at the immediate postharvest stages of production, resulting from a lack of access to appropriate storage, infrastructure, and markets and leading to wasted natural resources. In medium and high-income countries, environmental impacts tend to be highest at the point of consumption (in the form of food waste and greenhouse gas emissions), which is related to consumer behavior, and a set of policies and regulations that favor other priorities in the food system.

Livelihoods: Trends, Drivers, and Challenges

The social and environmental contexts of agricultural production vary widely around the world. However, over the past three decades, national policies and international investment have led to the consolidation of farm ownership and the firms that supply the inputs, as well as consolidation in the retail outlets for sales of agricultural products (ETC, 2015; Reardon and Timmer, 2012; Worldwatch, 2008). There has been a general and consistent trend across developed and developing countries of the consolidation of agriculture into fewer, larger farms, making it increasingly difficult for small-scale producers to compete on the basis of cost. In the United States, for example, fewer than 8 percent of farm owners account for 70 percent of farmland (NASS, 2014).

Commercial farmers in many countries and those working in commodity crop production have also experienced a consolidation of input providers, including those for seed and fertilizer (ETC, 2015). For producers, this vertical consolidation often leads to fewer input choices, lower farm-gate prices, and a lack of agency in decision-making throughout their production and marketing process (Busch and Bain, 2004). More recently, consolidation has come to mean full "financialization," with large financial actors investing in activities all across the food chain (Murphy et al., 2012). Consolidation can lead to both homogenization of products, and increased market share and influence for a small number of firms, which have negative impacts on market access, value share, and bargaining power for smaller producers in the food value chain.

---

3. To put this in context, if food loss were its own country, it would be the third-largest greenhouse gas emitter after China and the United States (Hanson et al., 2016).

Beyond farming, consolidation in pre- and postfarm activities has increased, driven by the rise of large supermarkets in the distribution and retail segments; dominant food and beverage manufacturers in the midstream segments; and vertically integrated global companies in processing and trade (Euromonitor, cited by ERS, 2012; Reardon, 2015). Retail food markets have consolidated within the "modern supermarket" format, with a few major players spreading across North American and West European countries during the 1980s and 1990s, and more recently in different phases and often through cross-border mergers and acquisitions in various regions of the world (Abdulsamad and Gereffi, 2018; Herger et al., 2008; Reardon et al., 2003).[4] In contrast with supermarkets, there is a general consensus that the top five to ten global food companies account for less than 10 percent of the value of revenue generated within the processing and distribution segments of the global food value chain, while the top 50 account for less than 20 percent of total value (ERS, 2012; IMAP, 2010). Consolidation is more pronounced for specific types of processed or specialty foods, including alcohol and candy, as well as for commodity grains.[5,6] National policies that allow for the vertical integration of agro-food firms across the value chain have further shifted power away from producers and toward other value chain actors (Magdoff et al., 2000). At an aggregated global level, farmers accrued less than 14 percent of the total value-added share generated in agro-food global value chains in 2011, in contrast with the retail segment's value share of more than 30 percent (Abdulsamad and Gereffi, 2018).

Agricultural production in both developed and developing countries has also been increasingly consolidated in terms of ownership of land, and requires less human labor as a result of the increase in automation and use of technology. There has been a steady and significant decrease both in agriculture's economic contribution relative to other sectors and in overall employment in agriculture.[7] The decrease in agriculture's overall contribution to global GDP, from 8 percent in 1995 to less than 4 percent in 2014, is reflective of both increases in high-profit technology and service sectors, diversification away from agriculture, and

---

4. In 2016, approximately 70–90 percent of all grocery retail sales were channeled through modern retail markets (Euromonitor, 2015; Abdulsamad and Gereffi, 2018), and the top fifteen firms generate over 30 percent of global retail revenue (ERS, 2012).
5. For a summary of top firms' market share by type of product, see Coriolis (2013); see also Murphy et al. (2012).
6. For example, chocolate manufacturing, mergers and acquisitions have left five firms—Mondelez International, Mars Inc., Nestle, Ferrero, and Hershey Co.—to control 56 percent of the global chocolate markets in 2013 (Abdulsamad and Gereffi, 2018). These firms have divested and outsourced processing activities, and as a consequence, three firms (Cargill, Barry Callebaut, and Olam International) controlled two-thirds of the global grinding and processing market in 2016 (Abdulsamad and Gereffi, 2018).
7. These figures are relative to other parts of the global economy, and reflect both decreases in absolute numbers of individuals engaged in agriculture in some regions, and shifts out of agriculture and into other economic sectors, decreasing agriculture's share in terms of employment.

policies such as trade liberalization that create unfavorable environments for domestic production in many countries (World Bank, 2017).

Increased concerns for food safety and demand for higher quality agro-food products have also increased the skills and knowledge required for value-added postfarming activities (Reardon and Timmer, 2012). Changes in consumption patterns and subsequent changing demand within food systems contribute to this trend toward expertise in the middle stages of the food supply chain, and have induced a shift in employment and livelihoods within the food system. As a result, fewer people work in agriculture and more work in transport, wholesaling, retailing, food processing, and vending (Cohen and Garrett, 2009).

Trade policies have shaped the livelihoods of small producers globally as well. Over the past 30 years, for producers of all sizes, national policies and private investments have been oriented toward free trade of all goods, including food products (Friedmann, 2005). Free trade policies have obliged national governments to enact policy that exposes agricultural producers to global economic pressures, undermining the comparative advantage of national agricultural systems in favor of promoting individual competitiveness. At the same time, countries belonging to the Organisation of Economic Cooperation and Development (OECD) receive high levels of farm subsidies, which depress world prices for several key commodities (especially sugar, cotton, milk, and beef) and undermine growth of domestic agricultural production in low-income countries (Shelton, 2014). National agricultural policies can also emphasize the production of global commodity crops by providing subsidies for inputs and crop insurance for only certain kinds of production, which can in turn distort global market prices for those crops (Birner and Resnick, 2010; Busch, 2010; McMichael, 2009). For example, subsidy supports used to protect agricultural businesses in many high-income countries and some emerging economies set the rules of the game for trade and pricing of many global commodities such as rice, corn, and beef, on which producers in developing countries rely to generate income (Bureau et al., 2013).

Nutrition: Trends, Drivers, and Challenges

The nutritional challenge for global food systems today stems less from insufficient production—globally we produce nearly 3000 cal per person per day (FAO, 2017a)—than from production of the wrong kinds of food and inadequate access to diversified diets. For example, in 2016, there was an overall increase in the global food supply but this was driven largely by maize and wheat production (IFPRI, 2017). Over the past 20 years (from 1995 to 2015), availability (per capita production) of total calories and of protein has increased as agricultural technologies such as seeds and pesticides have led to increased yields, and the share of the global population that is undernourished has decreased (FAO, 2017a). However, the absolute number of undernourished people has remained largely the same, and micronutrient deficiency ("hidden hunger") has remained stable as well, due in part to an increased emphasis on improving yields for grain crops that are not nutritionally diverse (von Grebmer et al., 2014).

Concurrently, global rates of obesity and overweight status have increased, and mortality rates from Type 2 diabetes have doubled over the past fifteen years (WHO, 2016). Obesity and its health consequences are generated by both overnourishment; by consuming more calories than needed, and malnourishment; by consuming "empty" calories that lack micronutrients and undermine dietary diversity. The emphasis in the industrialized global food system has been on increasing yields for a small number of crops such as wheat, maize, and rice, which are not nutrient dense, but which are amenable to a wide range of processing approaches and end products. The outcomes of this food system are not oriented toward delivering nutritionally optimal and environmentally sustainable diets, but rather toward maximizing profits (Wallinga, 2009).

In short, the contemporary global food system has facilitated a strange situation in which the poor are more likely to both go hungry and suffer from obesity (Popkin et al., 2012). In the United States, for example, people who live in the most poverty-dense counties are those most prone to obesity (Segal et al., 2016). Research on "food deserts" in the urban United States has highlighted an inverse relationship between access to food outlets and obesity, with increased obesity associated with further distances from grocery stores and other sources of fresh food (Walker et al., 2010). The relationship between food access and obesity in high-income countries points to the notion of "food swamps," areas in places like urban New Orleans, where the food to which individuals have access is largely calorically dense snack food and fast food, and where there is a lack of fresh and unprocessed foods available (Rose et al., 2009). In addition, in many middle- and low-income countries, food inadequacy now co-exists with obesity, as what is known as the "nutrition transition" occurs (Drewnowski and Popkin, 1997). In many of these countries, some consumers are adding significant amounts of fats, oils, and sugars to their diets through processed foods, even while others (especially those in rural areas) continue to lack access to adequate calories.

Nexus of Sustainability, Livelihoods, and Health Challenges

The challenge of developing more sustainable food systems is entangled in complex ways with the need to also promote nutrition and livelihoods (West et al., 2014). Climate variability, water availability, and associated declining increases in agricultural productivity will likely impact food availability and accessibility, as production volumes become insufficient to meet demand, and prices therefore rise (FAO, 2017b). Smale and Alpert (2009) show that when the combined variables of population pressure, land degradation, and environmental variability are accounted for, many farmers in the developing world rely on marginal lands for their livelihoods, linking livelihood challenges to those of environmental sustainability. National subsidies for certain crops that are a part of many processed foods, such as corn for high-fructose corn syrup, are also facilitating expanded production of just a few commodity crops at the expense of fruits, vegetables, and other diversified products. Even so, as a recent review by Samberg et al. (2016) suggests, the majority of calories, as well as diversification in the global food system, continues to come from smallholder producers, for whom agriculture is a primary livelihood strategy. Collectively, these trends have had doubly negative consequences for some rural communities, where there is less opportunity to work in agriculture and small-scale food systems, and less stability in agricultural contributions to livelihoods and economic well-being. Increasing instability in agriculture as a livelihood strategy has hit women especially hard, as in many parts of the world they are less mobile and therefore less able to transition out of agriculture and into alternative livelihood strategies (Sachs, 1996).

For consumers, if these triple threat trends persist, the impact is likely to be higher food prices and decreased access to nutritious foods. Most experts agree that demand for food will grow over the coming decades due to population growth, urbanization, and per capita increases in income (Hunter et al., 2017). Globally, more people now live in cities than in rural areas, and urbanization has accelerated the burdens of malnutrition and overnutrition, as it is associated with the shift in consumption to animal-based protein, sugar, fats, salt, and processed foods (IFPRI, 2017). Processed foods are often cheaper than fresh produce and plant-based protein, because the economics of scale associated with food processing, as well as national subsidies, often lead to lower prices. Processed foods are also more convenient to prepare. This affects gender equity, because women have historically held disproportionate responsibility for cooking and other forms of household labor, and processed foods help many women—especially working women—save time on food preparation (Scott-Villier et al., 2016). These advantages have contributed to a rapid spread of "Western" diets, comprised of more animal-based protein and high-calorie processed foods (Fields, 2004). However, the comparatively low price and easy preparation of processed foods creates a doubly negative nutritional outcome for low-income consumers (Jaffe and Gertler, 2006). These trends also have negative consequences for the environment and livelihoods along the food system. For example, large-scale commodity food production has increased soil, water, and air contamination, and it has driven deforestation and land use change (Foley et al., 2005; Magdoff et al., 2000).

EMERGING APPROACHES TO ADDRESSING FOOD SYSTEMS CHALLENGES

As many consumers and food system advocacy organizations have begun to push for changes in the drivers and outcomes of sustainability and equity in local and national food systems, solutions and alternatives to some of the trends and challenges overviewed in this chapter have begun to emerge. In this section, we highlight a few of these solutions and note how they can be integrated into institutional foodservice decision-making.

Third-Party Certifications and Increasing Demand for Information

On the supply side, metrics and monitoring are increasingly focused on the environmental impacts of food production, with firms increasing transparency and consumers demanding information about where and how their food is produced. Fair Trade and similar certification programs focus on commitments to equitable livelihoods for producers and farm labor, and communicate this to consumers through labeling (Hinrichs and Allen, 2008). The United States Department of Agriculture's (USDA) "Know Your Farmer, Know Your Food" program is one example of a push via public policy to improve transparency and shift the drivers of both sustainability and equity for producers (USDA, 2012).

On the demand side, consumers are increasingly pressuring private businesses and institutions to purchase ethically and sustainably produced products, as well as healthy food (Busch and Bain, 2004). In order to assess whether food produced and processed in specific places is in fact sustainable and equitable, it is important to keep in mind the complexity demonstrated in the results presented herein, and to make decisions that reflect the food system challenges of particular locations.

In North America and Europe, consumers are demanding more information about their food products, and this had led food system actors to develop and apply a variety of tools and metrics that can assist in ensuring that purchasing decisions support both sustainability and livelihoods. For example, nongovernmental organizations are leading several efforts at scorecards that rank food companies or food products in terms of the environmental and social impacts of production.[8] Diverging models of retail transparency highlight a shift toward a multiplicity of personified, value-laden, and niche-based retail attempts to connect with food purchasers and consumers around their values and fears. Smaller farmers, millers, and food makers, and companies such as Lay's potato chips, aim to personalize food relationships through online tools, attempting to call on ideals of personal trust. Others, such as Whole Foods Market, are developing or employing certification labels to communicate the values behind their foods by proxy (Avery et al., 2010). Foodservice procurement has options to buy produce from producers and suppliers who uphold certain environmental (e.g., organic, biological, etc.) and social practices (e.g., Fair Trade, fair wages, no slavery, etc.).

---

8. See for example Ceres' water use report (Roberts and Barton, 2015) and Oxfam's (2016) "Behind the Brands" campaign.

Third-party certifications and related standards require either strong government management or private efforts, and the lack of regulatory capacity in many countries for setting and effectively delivering quality control services related to social, environmental, and food safety standards have equity and sustainability implications (Abdulsamad et al., 2015). As a result, smallholder farmers often bear the costs associated with meeting standards through private certification agencies, or those associated with the deterioration of productivity in land, water, and labor inputs (Potts et al., 2014; Abdulsamad and Gereffi, 2018). As an example, weak environmental management regulations have led to overexploitation of underground water resources in Peru's asparagus industry (Abdulsamad and Gereffi, 2018). Given that the United States constitutes the major market for Peruvian green asparagus, we can begin to see how institutional foodservice procurement decisions in the United States can have sustainability and social equity implications another part of the world.

Shortening the Food Chain: Going Local and Fresh

In addition to supporting producer livelihoods, movements that stress the local origins of food increase the likelihood that production and transport will be sustainable. On a practical level, supply chains now are also leaner and more brittle, as the retail trend is to offer more ingredients that are perishable and fewer frozen or shelf-stable processed products. In the United States, "fresh" has become one of the most valued quality distinction markers, along with descriptors such as "locally grown." As a result, retailers and distributors are increasingly exposed to the same risks that farmers face: weather, water, and pests (Born and Purcell, 2006). Prices for fresh foods can change as often as the weather, and change as well minute-by-minute based on speculation about how the harvest will turn out. Such price volatility is only increasing as climate change and water limits have a greater impact on farmers. This price volatility has implications for institutional foodservice's ability to make decisions about not only what they source, in the form of more fresh products, but also about the relative cost of prioritizing food that comes from different types of actors and locations. Again, footprint calculators and assessment of food production and processing impacts can help institutions determine which producers and companies are making place-appropriate decisions about, for example, water use (Roberts and Barton, 2015; Oxfam, 2016).

A number of companies have recently launched or expanded efforts to make it easier to source locally. Sourcery, Dine Market, and Blue Cart, for example, all offer online purchasing platforms that enable restaurants and food distributors to order from multiple local suppliers through a central dashboard. By streamlining the sourcing, invoicing, and payment process, these services make it easier to manage purchases from local producers who might not otherwise be set up to handle large commercial orders. Institutional foodservice can work directly with these services or with distributors that use them to identify and support locally sourced food in a manner that remains economically efficient.[9] At larger scales, institutional foodservice can continue to make sourcing decisions that support producer livelihoods by working with food hubs, agricultural cooperatives, and similar organizations that prioritize equitable profits for producers (Anderson et al., 2014; Blay-Palmer et al., 2013).

Social Movements Addressing Equity and Sustainability

In the United States, there is a diversity of organizations and movements working on reforming the food system in terms of sustainability, equity, and nutritional quality. Some organizations and activists have been leading the calls for quality, environmental sustainability, and safety of food (e.g., fresh, organic, local, etc.) as well as for the reaffirmation of environmental values and community relationships. It finds its expression through support for farmers' markets, Community Supported Agriculture (CSAs), and high-end organic, "locavore" restaurants. Whether it is locavore restaurants, CSAs, or farmers' markets, these efforts support the local economy in many ways and respond to consumer demands. For example, farmers' markets are able to cater to niche and specialty markets. They also generate the "multiplier effect" as markets are located next to local businesses and there are often spillovers to the local businesses from market patrons. Beyond economic capital gains, what is also important is the "social capital"—the networks created from the interactions between producers and consumers at farmers' markets and CSAs.[10]

Equity is not a foregone conclusion in local food systems, however. It is often noted that the majority of people who participate in these local food systems movements are primarily white, educated, affluent citizens (Goodman and DuPuis, 2002; Hesterman, 2011; Hinrichs, 2003). Conversely, those most affected by access to food are typically low income and racial minorities, who often do not have access to nutritional food within their communities (Hesterman, 2011). Convenience, accessibility, and affordability are all factors that make sustainable consumption more difficult (Coit, 2008).

In addition, government and institutional policies tend to limit alternative network participants by favoring conventional agricultural networks that reduce small producer and consumer's power. For example, the United States subsidizes commodities such as corn, which skews competition and makes prices for those goods artificially low (and cheaper than fresh vegetables). Governmental

---

9. As Hinrichs and Allen (2008) note, local sourcing does not guarantee equitable distribution of profits to producers, and any food system actors with a commitment to supporting livelihoods must assess which value chains, in fact, prioritize these values. In addition, local food is not necessarily more sustainable than food procured from further away, and efforts to address and balance various food system challenges must account for the trade-offs that can exist between, for example, support for local farmers and sustainable water use when considering from where to access fresh produce (Born and Purcell, 2006).

10. For a summary of the local food movement, see Coit (2008).

institutions also tend to adopt policies that lack consideration of diversity within the food system and, due to political pressure to minimize expenditures, prioritize the lowest cost options (which are more likely to come from subsidized commodity crops and large-scale industrial producers). In this way, American food culture prioritizes price and convenience rather than the issues of justice, sustainability, and even nutrition. Policy-based initiatives such as the option to redeem Supplemental Nutrition Assistance Program (SNAP) coupons (often referred to as "food stamps") at local farmers' markets are an important step to address the economic divide that limits access to healthier and more sustainable food.

Although these food movements offer a path to transform the way that we approach agricultural production and food consumption, local alternative food networks represent a small share of the U.S. food system. There is a strong interest among certain parts of the consumer base in developed countries in local food, organic production and Fair Trade; however, the economic contribution of these types of foods remains tiny in comparison with the overall food system supply and demand (for analysis in the United States, see ERS, 2015).

INSTITUTIONS AS KEY DRIVERS OF FOOD SYSTEM TRANSFORMATION

In many parts of the world, both in high-income countries and in increasingly urbanized middle-income countries, people consume much of their food outside the home, so the foodservice industry broadly has a critical role to play in aligning health, sustainability, and value. Institutional foodservice is in some ways even better positioned to influence consumer knowledge by promoting just and sustainable foods, because in many institutional settings, consumers are a bit of a "captive audience" in terms of alternative food choices. While institutions should not ignore consumer demand, these types of settings—in schools, hospitals, and public offices—do create the opportunity to educate consumers about the implications of where food comes from and the impacts of different types of food decisions (Vogt and Kaiser, 2008).

Because public and private institutions also purchase such large amounts of food, for a variety of purposes and outlets, they offer important and dynamic sites of food system transformation. Equitable and sustainable procurement policies have the potential to increase the overall demand for healthier, fairly priced, and sustainable products, drive the reformulation of foods by producers and food manufacturers, and increase the availability of healthier foods to the general public. They can also encourage environmental and livelihood sustainability in production with policies that require sourcing food from local growers at fair prices, or food produced in ways that are more ecologically sound. Therefore, in purchasing large volumes of food for consumption, there is a key opportunity for institutional consumers to raise awareness, shape individual preferences in new ways, and ultimately generate new demand for healthier and sustainable food choices.

Institutional foodservice can directly address and contribute to improving sustainability, livelihood, and nutrition outcomes of food systems through actions at several points along the food supply chain. To be this lever of change, actors within institutional foodservice must first be aware of the kinds of changes needed and opportunities for effecting such change. Foodservice actors today can use a growing number of initiatives and tools to obtain information and inspiration to make procurement and management decisions that can express and generate greater demand for healthy food produced in ways that are sustainable and equitable. In this section, we highlight three categories of decision-making that institutional foodservice can take to address current challenges in the food system: changing menu decisions, engagement with certifications and standards, and reducing food waste.[11]

Improving Health and Sustainability through Menu Decisions

Institutional foodservice can engage with many of the concerns highlighted by social movements, including those that emphasize local food (including many of the farm-to-institution efforts such as those highlighted in this volume), organics, and plant-based foods. As one specific variable in focusing on an overall balanced diet, foodservice settings are slowly shifting the amount of animal versus plant-based protein that they offer. The average American over age 20 consumes between 48 percent and 76 percent more protein than is recommended, for women and men respectively, and animal sources account for approximately two-thirds of this dietary protein (NCHStats, 2010). Yet, plants such as nuts, seeds, beans, peas, legumes, grains, and cereals are also important sources of protein. The amount and types of protein consumed can have significant effects on the environment and the risk of chronic diseases and premature death (Richter et al., 2015). The Meatless Monday movement, for example, has been adopted by many types of institutions, including hospitals, schools, restaurants, and governments as a way to promote consuming less meat and increasing awareness for consumers about the benefits of plant-based protein (Meatless Monday, 2017). Plant-based proteins can also be cheaper than animal-based ones, making them more economically accessible to institutions as well as for consumers and society as a whole.[12]

Along with changes in their menu offerings, many institutions, such as those highlighted in this book, are adding complementary educational materials to help consumers understand the impacts of the changes and to provide guidance on how to make similar decisions outside of the institutional setting.

---

11. For a similar assessment, see Begun (2015).
12. Recent projections by Springmann et al. (2016) suggest that the cost savings associated with decreasing animal protein production and consumption could be as high $1.5 trillion in healthcare and climate change-related costs by 2050.

For example, there has long been an effort to incorporate nutrition education into schools, but often without the financial and information support necessary to help school foodservice purchase healthy food (Briggs et al., 2003; Glanz and Mullis, 1988). More recently, institutions, including schools, have been compelled by policy and by citizens' demand to examine how they can more fully incorporate nutritional considerations into their procurement decisions. In 2008, for example, England introduced a national regulation that requires all primary schools to use a healthy food procurement standard for foods throughout the school day. These regulations improved the purchases of fruits, vegetables, and salads by 15 percent, and reduced processed foods high in sodium, fats, and sugars by 12 percent (e.g., french fries, pizza, and cookies) (Niebylski et al., 2014). Programs that bring together foodservice decision-making and public education are most visible in school-based programs that complement farm-to-school efforts, and there is an increasing emphasis on making fresh and nonprocessed foods more accepted by students, in an effort to make unfamiliar foods familiar and requested by students at home as well (Chen, 2016). Hospitals and government cafeterias are also increasingly highlighting consumer education as a part of improving the sustainability of their food systems (GSA, 2016; NHS, 2009).

Improving Equity and Sustainability Through Engagement With Certifications and Standards

A second major focal area of engagement for foodservice institutions is with third-party certification programs and industry-led standards that increase transparency and assure certain standards of production. Certifications can address the treatment of farmers in such global food values chains that produce fruits and vegetables that continue to gain importance in institutional foodservice offerings. An example of third-party certifications in the United States include USDA Organic (Guthman, 2004). Globally, Fair Trade, Rainforest Alliance, Trustea, and many other certifications provide evidence to consumers as well as to other actors along the value chain that the products address the concerns of those actors.[13] As consumer demand has grown for food that meets a variety (and often a combination) of values-based standards, certifications and labelling efforts have provided both producers and retailers with a value-added product (Bloom, 2015; Seyfang, 2008). Similar efforts to develop standards have been led by groups of private firms for specific products and values chains. Examples of these include the Roundtable on Sustainable Palm Oil, GLOBALG.A.P., and Starbuck's CAFÉ Practices (Argenti, 2004; GLOBALG.A.P., 2017). At the same time, incorporating diverse foods from distant locales into global supply

---

13. To provide a sense of how supply has grown in offering products with certification around equity and sustainability, the Fair Trade certification has more than 1000 partners selling foods, apparel, homes goods, and body care products.

chains, even when a premium is paid for them, can have the perverse effect of making these healthy foods less available to consumers in the locations in which they are produced. Quinoa is one example of this challenge; recent moves toward expanding sales of fonio, a West African grain, are raising similar concerns (Cáceres et al., 2007; Roberts, 2014). Decision makers in institutional foodservice must be aware of these kinds of issues and factor this into procurement.

Although the push to develop these types of standards has come largely from consumer demand, they also present an opportunity for institutions and organizations that provide food to highlight their commitments to sustainability and equity in the food system. However, although institutions, such as schools and hospitals, are increasingly making food-purchasing decisions that respond to the consumer demand generated by food movements such as those discussed herein, they do not necessarily highlight these decisions. Organic food is often the exception (DuPuis, 2000; Seyfang, 2008), and the certification systems for organic production are well established and promoted all along the value chain. Institutions interested in supporting and articulating visions for alternative food systems that are just and equitable can work to increase consumer demand for specific types of products by highlighting the information provided by these third-party certifications (Feenstra et al., 2011). Certifications are especially important as a way for buyers to address the complex issues facing production in specific locations. Seyfang (2006) highlights the example of Banana Link, a program that supports sustainable procurement of bananas from Central America, as an example of the need for certifications when trying to meet consumer demand for a varied diet while accounting for place-based food system challenges such as the need to support rural livelihoods.

Improving Health, Equity, and Sustainability by Reducing Food Waste

Last, the increased emphasis on food waste reflects the potential impacts that reducing food waste can have on all three food systems challenges highlighted in this chapter. The inclusion of food waste in the recently updated Sustainable Development Goals from the United Nations, for example, makes explicit the connections among dimensions of the food system that reductions in food waste can provide (Lipinski et al., 2013). Most immediately, reducing food waste can improve sustainability in the food system by ensuring that the water and energy inputs to food production, processing, and distribution are utilized efficiently and are not wasted. Improving efficiency and reducing waste along food supply chains also has the potential to reduce costs for all actors, and potentially to increase the share of the food dollar received by farmers if other food system actors do not need to pad their bottom lines to account for food waste losses. Finally, by not wasting fresh produce and animal products (the bulk of food waste due to high perishability), more healthy food can be made available and accessible to consumers and thereby improve nutritional outcome, such as through food recovery programs. This is especially relevant for educational institutions; for example, there is a growing realization of the presence of hunger among students on university campuses.[14]

Recent campaigns to reduce food loss and waste highlight the ways that institutional foodservice and food advocacy organizations can proactively engage in transformation of the global food system (FAO, 2014; USDA, 2013). On the procurement side, engaging in short food supply chains as have been promoted in Japan, North America, and Europe, can help reduce food waste within the distribution process (Canfora, 2016; Mundler and Laughrea, 2016). Short food supply chains are primarily oriented toward restructuring the economics of food systems, by including as few actors as possible, and so maintaining transparency and profitability within the supply chain, and can have a large impact on reducing food waste by shortening transportation and storage time. However, many institutions have limited storage and processing facilities themselves, and often lack the human capacity to process the fresh food that is often sourced through short food supply chains. Although fresh food is more nutritious, processing and serving it requires changes to institutions' foodservice infrastructure and training of employees (Vogt and Kaiser, 2008). To contribute to reducing food waste in ways that maximize benefits across the food system, institutions will need to invest in kitchen and storage facilities that are appropriate for preparing fresh food, and will need to train their employees in appropriate preparation methods to maximize the nutritional benefits of that food.

There is also the issue of what to do with excess food to prevent it from ending up in a landfill. Despite the fact that in all countries around the world, some portion of the population experiences hunger and undernutrition, it is rarely straightforward to transfer "wasted" food from institutions or restaurants to these consumers. Specific barriers that limit food recovery and donation include a legal inability to donate mislabeled or mis-formulated food as well as officially expired food, a lack of refrigerated storage for donations, limited employee awareness of donation programs, and a lack of organizations or resources to accept the donated food (for further discussion, see Food Recovery Project, 2013). However, there is a fast-growing trend to bridge potential fresh and perishable food waste with charitable organizations in need of nutritious foods to give to people struggling with food insecurity in communities (e.g., see the Food Waste Reduction Alliance, 2017). With the use of web-based platforms, food distribution companies and retailers can make known what foods will go to waste and offer pick-up times. Institutions interested in reducing their own food waste footprint and contributing to food recovery can and should connect to this infrastructure, which can provide the legal and distributive expertise needed to

---

14. A group of organizations raising awareness on hunger on university campuses found that 48 percent of students surveyed reported food insecurity and 22 percent with levels of food security that qualify them as hungry (Dubick et al., 2016).

divert food waste to address nutritional needs in local communities, as well as to minimize the environmental impacts of wasted food ending up in landfills.

CONCLUSION

Institutional foodservice has the potential to address the "triple threat" confronting contemporary food systems around environmental sustainability, livelihoods equity, and nutrition, by incorporating an awareness of and values around the triple threat into procurement and management decisions. For example, institutions can proactively reduce land-based animal-protein portions (with a special focus on reducing offerings of red meat) and devote more of their creativity to vegetables and plant-based proteins (e.g., legumes and nuts, and products made from these) to reduce their climate footprint.

Institutional foodservice actors can also use their market influence to understand and raise awareness around existing opportunities to increase equitable and sustainable food procurement decisions. The standards and certification systems overviewed in this chapter have been primarily led by private-sector actors pushing for their own standards and accountability mechanisms under the auspices of sound business practices that prioritize a "triple-bottom line" approach to sustainability. Institutions have the opportunity to extend these efforts by making procurement decisions based on these certifications and by increasing their visibility and the public education around their importance.

Finally, because institutional foodservice sits at the nexus between upstream food system actors and end consumers, there is a unique opportunity to leverage foodservice decision-making and investments to address the negative impacts of food waste across many stages of the food system. By engaging with short supply chains and making investments in their own processing capacity, institutional foodservice actors can improve the sustainability and equity of their procurement decisions. And by working to divert wasted but still edible food to other types of food system actors, institutional foodservice can contribute to improving food security and health outcomes for vulnerable populations.

Institutions that provide foodservices as a relatively small part of their overall operations might seem an unlikely actor in advocating for food system change. However, the scope and breadth of institutions, as well as the fact that many are partially or wholly publicly funded, creates an opportunity for institutional foodservice to identify and articulate an incentive structure that will enable them to be levers of change for a more equitable and sustainable food system.

REFERENCES

Abdulsamad, A., Gereffi, G., 2018. Measurement in a World of Globalized Production. Duke University Center on Globalization, Governance & Competitiveness, Durham, NC.
Abdulsamad, A., Frederick, S., Guinn, A., Gereffi, G., 2015. Pro-Poor Development and Power Asymmetries in Global Value Chains. Duke University Center on Globalization, Governance & Competitiveness, Durham, NC.

Anderson, C.R., Brushett, L., Gray, T.W., Renting, H., 2014. Working together to build cooperative food systems. J. Agric. Food Syst. Community Dev. 4 (3), 3–9.

Antle, J.M., Stoorvogel, J.J., Valdivia, R.O., 2014. New parsimonious simulation methods and tools to assess future food and environmental security of farm populations. Philos. Trans. R. Soc. Britain 369 (1639). art. 20120280.

AQUASTAT, 2017. AQUASTAT database 2017. United Nations Food and Agriculture Organization, Rome. Available from: http://www.fao.org/nr/water/aquastat/data/query/index.html?lang=en. Accessed November 28, 2017.

Argenti, P.A., 2004. Collaborating with activists: how Starbucks works with NGOs. Calif. Manag. Rev. 47 (1), 91–116.

Avery, M.L., Kreit, B., Falcon, R., 2010. Food Web 2020: Forces Shaping the Future of Food. Institute for the Future, Palo Alto, CA. Available from: http://www.iftf.org/uploads/media/SR1255B_FoodWeb2020report_1_.pdf. Accessed November 28, 2017.

Beddow, J.M., Parde, P.G., Koo, J., Wood, S., 2010. The changing landscape of global agriculture. In: Alston, J.M., Babcock, B.A., Pardey, P.G. (Eds.), The Shifting Patterns of Agricultural Production and Productivity Worldwide. Iowa State University Press, Ames, pp. 7–38.

Begun, R., 2015. 5 foodservice trends for 2016. 30 December 2015. Food Management. Available from: http://www.food-management.com. Accessed November 28, 2017.

Birner, R., Resnick, D., 2010. The political economy of policies for smallholder agriculture. World Dev. 38 (10), 1442–1452.

Blay-Palmer, A., Landman, K., Knezevic, I., Hayhurst, R., 2013. Constructing resilience, transformative communities through sustainable "food hubs". Int. J. Justice Sustain. 18 (5), 521–528.

Bloom, J.D., 2015. Standards for development: food safety and sustainability in Wal-Mart's Honduran supply chains. Rural Sociol. 80 (2), 198–227.

Born, B., Purcell, M., 2006. Avoiding the local trap: scale and food systems in planning research. J. Plan. Educ. Res. 26 (2), 195–207.

Boserup, E., 1965. The Conditions of Agricultural Growth. George Allen & Unwin, London.

Briggs, M., Safaii, S., Beall, D.L., 2003. Position of the American Dietetic Association, Society for Nutrition Education, and American School Food Service Association—Nutrition services: an essential component of comprehensive school health programs. J. Am. Diet. Assoc. 103 (4), 505–514.

Bureau, J.C., Laborde, D.D., Orden, D., 2013. US and EU farm policies: the subsidy habit. In: 2012 Global Food Policy Report. International Food Policy Research Institute, Washington, DC, pp. 58–67.

Busch, L., 2010. Can fairy tales come true? The surprising story of neoliberalism and world agriculture. Sociol. Rural. 50 (4), 331–351.

Busch, L., Bain, C., 2004. New! Improved? The transformation of the global agrifood system. Rural. Sociol. 69 (3), 321–346.

Buttel, F., 2001. Some reflections on late 20th century agrarian political economy. Sociol. Rural. 41 (2), 165–181.

Cáceres, Z., Carimentrand, A., Wilkinson, J., 2007. Fair Trade and quinoa from the southern Bolivian Altiplano. In: Raynolds, L.T., Murray, D.L., Wilkinson, J. (Eds.), Fair Trade: The Challenges of Transforming Globalization. Routledge, New York, pp. 180–199.

Canfora, I., 2016. Is the short food supply chain an efficient solution for sustainability in food market? Agric. Agric. Sci. Procedia 8, 402–407.

Chen, G., 2016. Eliminating processed foods in public school cafeterias: the new trend. 21 June 2016. Public School Review. Available from: https://www.publicschoolreview.com/blog/eliminating-processed-foods-in-public-school-cafeterias-the-new-trend. Accessed November 28, 2017.

Cohen, J.E., 1995. Population growth and Earth's human carrying capacity. Science 269 (5222), 341–346.

Cohen, M.J., Garrett, J.M., 2009. The food price crisis and urban food (in)security. Human Settlements Working Paper. International Institute for Environment and Development & United Nations Population Fund. Available from: http://www.fao.org/fileadmin/user_upload/fsn/docs/The_food_price_crisis_and_urban_food__in_security.pdf. Accessed November 28, 2017.

Coit, M., 2008. Jumping on the next bandwagon: an overview of the policy and legal aspects of the local food movement. J. Food Law Policy 4 (1), 45–70.

Coriolis, 2013. Top 200 global food & beverage firms. Coriolis Research. Available from: http://www.coriolisresearch.com/pdfs/coriolis_global_top_200_food_&_beverage_firms_1.00.pdf. Accessed November 28, 2017.

Das, R., 2002. The Green Revolution and poverty: a theoretical and empirical examination of the relation between technology and society. Geoforum 33 (1), 55–72.

Desmarais, A.A., 2002. The Vía Campesina: consolidating an international peasant and farm movement. J. Peasant Stud. 29 (2), 91–124.

Drewnowski, A., Popkin, B.M., 1997. The nutrition transition: new trends in the global diet. Nutr. Rev. 55 (2), 31–43.

Dubick, J., Matthews, B., Cady, C., 2016. Hunger on campus: the challenge of food insecurity for college students. National Student Campaign Against Hunger & Homelessness. Available from: https://studentsagainsthunger.org/wp-content/uploads/2016/10/Hunger_On_Campus.pdf. Accessed November 28, 2017.

DuPuis, E.M., 2000. Not in my body: rBGH and the rise of organic milk. Agric. Hum. Values 17 (3), 297–309.

ERS, 2012. Global food industry. Economic Research Service, United States Department of Agriculture. Available from: http://www.ers.usda.gov/topics/international-markets-trade/global-food-markets/global-food-industry.aspx.

ERS, 2015. Trends in U.S. local and regional food systems. Economic Research Service, United States Department of Agriculture. Available from: https://www.ers.usda.gov/webdocs/publications/ap068/51173_ap068.pdf?v=42083. Accessed November 28, 2017.

ETC, 2015. Breaking bad: big Ag mega-mergers in play. ETC Group. Available from: http://www.etcgroup.org/sites/www.etcgroup.org/files/files/etc_breakbad_23dec15.pdf. Accessed November 28, 2017.

Euromonitor, 2015. Does modern grocery expansion lead to industry consolidation. Euromonitor. Available from: https://blog.euromonitor.com/2015/03/does-modern-grocery-expansion-lead-to-industry-consolidation.html. Accessed 19 March 2018.

Evenson, R.E., Gollin, D., 2003. Assessing the impact of the Green Revolution. Science 300 (5620), 758–762.

FAO, 2008. Climate change and food security: a framework document. United Nations Food and Agriculture Organization. Available from: http://www.fao.org/forestry/15538-079b31d45081f-e9c3dbc6ff34de4807e4.pdf. Accessed November 28, 2017.

FAO, 2014. Food wastage footprint full-cost accounting: final report. United Nations Food and Agriculture Organization. Available from: http://www.fao.org/3/a-i3991e.pdf. Accessed November 28, 2017.

FAO, 2015. State of food insecurity in the world. Meeting the 2015 international hunger targets: taking stock of uneven progress. United Nations Food and Agriculture Organization. Available from: http://www.fao.org/3/a-i4646e.pdf. Accessed November 28, 2017.

FAO, 2017a. FAOSTAT. United Nations Food and Agriculture Organization, Rome. Available from: http://www.fao.org/faostat/en/#data. Accessed November 28, 2017.

FAO, 2017b. The future of food and agriculture: trends and challenges. United Nations Food and Agriculture Organization. Available from: http://www.fao.org/3/a-i6583e.pdf. Accessed November 28, 2017.

Feenstra, G., Allen, P., Hardesty, S., Ohmart, J., Perez, J., 2011. Using a supply chain analysis to assess the sustainability of farm-to-institution programs. J. Agric. Food Syst. Community Dev. 1 (4), 69–85.

Fields, S., 2004. The fat of the land: do agricultural subsidies foster poor health? Environ. Health Perspect. 112 (14), A820–A823.

Foley, J.A., DeFries, R., Asner, G.P., Barford, C., Bonan, G., Carpenter, S.R., Chapin, F.S., Coe, M.T., Daily, G.C., Gibbs, H.K., Helkowski, J.H., Holloway, T., Howard, E.A., Kucharik, C.J., Monfreda, C., Patz, J.A., Prentice, I.C., Ramankutty, N., Snyder, P.K., 2005. Global consequences of land use. Science 309 (5734), 570–574.

Food Recovery Project, 2013. Food recovery: a legal guide. In: Food Recovery Project. University of Arkansas School of Law. Available from: https://law.uark.edu/service-outreach/food-recovery-project/. Accessed November 28, 2017.

Food Waste Reduction Alliance, 2017. Food Waste Reduction Alliance. Available from: http://www.foodwastealliance.org. Accessed November 28, 2017.

Friedmann, H., 2005. From colonialism to green capitalism: social movements and the emergence of food regimes. Res. Rural Sociol. Dev. 11, 227–264.

Glanz, K., Mullis, R.M., 1988. Environmental interventions to promote healthy eating: a review of models, programs, and evidence. Health Educ. Behav. 15 (4), 395–415.

GLOBALG.A.P., 2017. GlobalG.A.P. Available from: http://www.globalgap.org/uk_en/. Accessed November 28, 2017.

Gomez, M., Barrett, C., Raney, T., Pinstrup-Anderson, P., Meerman, J., Croppenstedt, A., Carisma, B., Thompson, B., 2013. Post-revolution food systems and the triple burden of malnutrition. Food Policy 42, 129–138.

Goodman, D., DuPuis, E.M., 2002. Knowing food and growing food: beyond the production-consumption debate in the sociology of agriculture. Sociol. Rural. 42 (1), 5–22.

Gornall, J., Betts, R., Burke, E., Clark, R., Camp, J., Willett, K., Wiltshire, A., 2010. Implications of climate change for agricultural productivity in the early twenty-first century. Philos. Trans. R. Soc. B 365, 2973–2989.

GSA, 2016. Concessions and cafeterias: healthy food in the federal workplace. United States General Services Administration. Available from: https://www.gsa.gov/portal/content/104429. Accessed November 28, 2017.

Guthman, J., 2004. Agrarian Dreams: The Paradox of Organic Farming in California. University of California Press, Berkeley.

Hanson, C., Lipinski, B., et al., 2016. Food loss and waste accounting and reporting standard. World Resources Institute. Available from: http://www.wri.org/publication/flwstandard. Accessed November 28, 2017.

Herger, N., Kotsogiannis, C., McCorriston, S., 2008. Cross-border acquisitions in the global food sector. Eur. Rev. Agric. Econ. 35 (4), 563–587.

Hesterman, O.B., 2011. Fair Food: Growing a Healthy, Sustainable Food System for All. Perseus Books Group, New York.

Hinrichs, C., 2003. The politics and practice of food system localization. J. Rural. Stud. 19 (1), 33–45.

Hinrichs, C.C., Allen, P., 2008. Selective patronage and social justice: local food consumer campaigns in historical context. J. Agric. Environ. Ethics 21 (4), 329–352.

Holt-Giménez, E., Peabody, L., 2008. From food rebellions to food sovereignty: urgent call to fix a broken food system. In: Food First Backgrounder. Institute for Food and Development Policy, Oakland, CA.

Hunter, M.C., Smith, R.G., Schipanski, M.E., Atwood, L.W., Mortensen, D.A., 2017. Agriculture in 2050: recalibrating targets for sustainable intensification. Bioscience. https://doi.org/10.1093/biosci/bix010.

IFPRI, 2017. 2017 Global Food Policy Report. International Food Policy Research Institute. Available from: http://www.ifpri.org/publication/2017-global-food-policy-report. Accessed November 28, 2017.

IMAP, 2010. Food and beverage industry global report—2010, IMAP Consumer Staples Report. Available from: http://www.proman.fi/sites/default/files/Food%20%26%20beverage%20global%20report%202010_0.pdf.

Jaffe, J., Gertler, M., 2006. Victual vicissitudes: consumer deskilling and the (gendered) transformation of food systems. Agric. Hum. Values 23 (2), 143–162.

Khan, S., Hanjra, M.A., 2007. Footprints of water and energy inputs in food production—global perspectives. Food Policy 34 (2), 130–140.

Lipinski, B., Hanson, C., Waite, R., Searchinger, T., Lomax, J., Kitinoja, L., 2013. Reducing food loss and waste: Creating a sustainable food future, installment two. World Resources Institute. Available from: http://www.wri.org/publication/reducing-food-loss-and-waste. Accessed November 28, 2017.

Lobao, L.M., Meyer, K., 2002. The great agricultural transitions: crisis, change and social consequences of 20th century U.S. farming. Annu. Rev. Sociol. 27, 103–124.

Lobell, D.B., Field, C.B., 2007. Global scale climate-crop yield relationships and the impacts of recent warming. Environ. Res. Lett. 2 (1). art. 014002.

Lyson, T., 2007. Civic agriculture and the North American food system. In: Lyson, T., Hinrichs, C.C. (Eds.), Remaking the North American Food System: Strategies for Sustainability. University of Nebraska Press, Lincoln, pp. 19–32.

Magdoff, F., Foster, J.B., Buttel, F. (Eds.), 2000. Hungry for Profit: The Agribusiness Threat to Farmers, Food, and Environment. Monthly Review Press, New York.

McMichael, P., 2009. A food regime analysis of the world food crisis. Agric. Hum. Values 26 (4), 281–293.

Meatless Monday, 2017. Meatless Monday Global. Available from: http://www.meatlessmonday.com/. Accessed November 28, 2017.

Mundler, P., Laughrea, S., 2016. The contributions of short food supply chains to territorial development: a study of three Quebec territories. J. Rural. Stud. 45, 218–229.

Murphy, S., Burch, D., Clapp, J., 2012. Cereal secrets: the world's largest grain traders and global agriculture, Research report. Oxfam. Available from: https://www.oxfam.org/sites/www.oxfam.org/files/rr-cereal-secrets-grain-traders-agriculture-30082012-en.pdf. Accessed November 28, 2017.

NASS, 2014. 2012 Census of Agriculture highlights: Farms and Farmland. National Agricultural Statistical Service, United States Department of Agriculture. Available from: https://www.agcensus.usda.gov/Publications/2012/Online_Resources/Highlights/Farms_and_Farmland/Highlights_Farms_and_Farmland.pdf. Accessed November 28, 2017.

NCHStats, 2010. Adults' daily protein intake much more than recommended. 3 March 2010. National Center for Health Statistics Blog. Available from: https://nchstats.com/2010/03/03/adults%E2%80%99-daily-protein-intake-much-more-than-recommended/. Accessed November 28, 2017.

NHS, 2009. Sustainable food: a guide for hospitals. National Health Service, United Kingdom Department of Health. Available from: https://www.sustainweb.org/pdf2/295087_sustainable-foodguide_acc.pdf. Accessed November 28, 2017.

Niebylski, M.L., Lu, T., Campbell, N.R.C., Arcand, J., Schermel, A., Hua, D., Yeates, K.E., Tobe, S.W., Twohig, P.A., L'Abbé, M.R., Liu, P.P., 2014. Health food procurement policies and their impact. Int. J. Environ. Res. Public Health 11 (3), 2608–2627.

Oxfam, 2016. Behind the Brands. Available from: https://www.behindthebrands.org/. Accessed November 28, 2017.

Oxfam America, 2018. Food security, agriculture and livelihoods. Available from: https://policy-practice.oxfamamerica.org/work/food-agriculture-livelihoods/. Accessed November 28, 2017.

Pechlaner, G., Otero, G., 2010. The neoliberal food regime: neoregulation and the new division of labor in North America. Rural. Sociol. 75 (2), 179–208.

Popkin, B.M., Adair, L.S., Ng, S.W., 2012. Now and then: the global nutrition transition: the pandemic of obesity and developing countries. Nutr. Rev. 70 (1), 3–21.

Potts, J., Lynch, M., Wilkings, A., Huppé, G., Cunningham, M., Voora, V., 2014. The state of sustainability initiatives review 2014: Standards and the green economy. In: International Institute for Sustainable Development (IISD) and the International Institute for Environment and Development (IIED). 332 pp.

Reardon, T., 2015. The hidden middle: the quiet revolution in the midstream of agrifood value chains in developing countries. Oxf. Rev. Econ. Policy 31 (1), 45–63.

Reardon, T., Timmer, C.P., 2012. The economics of the food system revolution. Ann. Rev. Resour. Econ. 4 (1), 225–264.

Reardon, T., Timmer, C.P., Barrett, C.B., Berdegué, J., 2003. The rise of supermarkets in Africa, Asia and Latin America. Am. J. Agric. Econ. 85 (5), 1140–1146.

Richter, C.K., Skulas-Ray, A.C., Champagne, C.M., Kris-Etherton, P.M., 2015. Plant protein and animal proteins: do they differentially affect cardiovascular disease risk? Adv. Nutr. 6 (6), 712–728.

Roberts, N., 2014. Fonio: the grain that would defeat quinoa as king among foodies. The Guardian. 9 February. Available from: https://www.theguardian.com/money/2014/feb/09/fonio-quinoa-senegal-africa-harlem-restaurant. Accessed November 28, 2017.

Roberts, E., Barton, B., 2015. Feeding ourselves thirsty: how the food sector is managing global water risks. Ceres. Available from: http://www.ceres.org/issues/water/agriculture/water-risks-food-sector. Accessed November 28, 2017.

Robertson, G.P., Swinton, S.M., 2005. Reconciling agricultural producvitiy and environmental integrity: a grand challenge for agriculture. Front. Ecol. Environ. 3 (1), 38–46.

Rockström, J., Williams, J., Daily, G., Noble, A., Matthews, N., Gordeon, L., Wetterstrand, H., DeClerk, F., Shah, M., Steduto, P., de Fraiture, C., Hatibu, N., Unver, O., Bird, J., Sibanda, L., Smith, J., 2017. Sustainable intensification of agriculture for human prosperity and global sustainability. Ambio 46 (1), 4–17.

Rose, D., Bodor, J.N., Swalm, C.M., Rice, J.C., Farley, T.A., Hutchinson, P.L., 2009. Deserts in New Orleans? Illustrations of urban food access and implications for policy. Paper prepared for University of Michigan National Poverty Center/USDA Economics Research Service. Available from: https://pdfs.semanticscholar.org/abc8/b418aa0783c8f3b0a0c4fca8f137ad806e0a.pdf. Accessed November 28, 2017.

Sachs, C., 1996. Gendered Fields: Rural Women, Agriculture, Environment. Westview Press, Boulder, CO.

Samberg, L.H., Gerber, J.S., Ramankutty, N., Herroro, M., West, P.C., 2016. Subnational distribution of average farm size and smallholder contributions to global food production. Environ. Res. Lett. 11 (12). art. 124010.

Scott-Villier, P., Chisholm, N., Wanjiku Kelbert, A., Hossain, N., 2016. Precarious lives: work, food and care after the global food crisis. Institute of Development Studies and Oxfam. Available from: https://opendocs.ids.ac.uk/opendocs/bitstream/handle/123456789/12190/PrecariousLives_Online.pdf?sequence=6. Accessed November 28, 2017.

Segal, L.M., Rayburn, J., Martín, A., 2016. The state of obesity 2016: better policies for a healthier America. Trust for America's Health and Robert Wood Johnson Foundation. Available from: http://stateofobesity.org/files/stateofobesity2016.pdf. Accessed November 28, 2017.

Sen, A., 1981. Poverty and Famines. Oxford University Press, New York, NY.

Seyfang, G., 2006. Ecological citizenship and sustainable consumption: examining local organic food networks. J. Rural. Stud. 22 (4), 383–395.

Seyfang, G., 2008. Avoiding Asda? Exploring consumer motivations in local organic food networks. Int. J. Justice Sustain. 13 (3), 187–201.

Shelton, P., 2014. Can the US Farm Bill and EU Common Agricultural Policy address 21st century global food security? 23 July 2014. International Food Policy Research Institute Blog. Available from: http://www.ifpri.org/blog/can-us-farm-bill-and-eu-common-agricultural-policy-address-21st-century-global-food-security. Accessed November 28, 2017.

Smale, M., Alpert, E., 2009. Making investments in poor farmers pay: a review of evidence and sample of options for marginal areas. Research Backgrounder, Oxfam America. Available from: https://www.oxfamamerica.org/static/oa3/files/making-investments-in-poor-farmers-pay.pdf. Accessed November 28, 2017.

Smith, M.R., Golden, C.D., Myers, S.S., 2017. Potential rise in iron deficiency due to future anthropogenic carbon dioxide emissions. GeoHealth 1 (6), 248–257.

Sowers, J., Vengosh, A., Weinthal, E., 2011. Climate change, water resources, and the politics of adaptation in the Middle East and North Africa. Clim. Chang. 104 (3–4), 599–627.

Springmann, M., Godfray, H.C.J., Rayner, M., Scarborough, P., 2016. Analysis and valuation of the health and climate change cobenefits of dietary change. Proc. Natl. Acad. Sci. 113 (15), 4146–4151.

USDA, 2012. Introducing…the Know Your Farmer, Know Your Food compass. 29 February 2012. United States Department of Agriculture Blog. Available from: https://www.usda.gov/media/blog/2012/02/29/introducing-know-your-farmer-know-your-food-compass. Accessed November 28, 2017.

USDA, 2013. U.S. food waste challenge. United States Department of Agriculture. Available from: https://www.usda.gov/oce/foodwaste/. Accessed November 28, 2017.

Vogt, R., Kaiser, L.L., 2008. Still a time to act: a review of institutional marketing of regional-grown food. Agric. Hum. Values 25 (2), 241–255.

von Grebmer, K., Saltzman, A., Birol, E., Wiesmann, D., Prasai, N., Yin, S., Yohannes, Y., Menon, P., Thompson, J., Sonntag, A., 2014. 2014 Global Hunger Index: The challenge of hidden hunger. Welthungerhilfe, International Food Policy Research Institute and Concern Worldwide. Available from: https://doi.org/10.2499/9780896299580. Accessed November 28, 2017.

Walker, R.E., Keane, C.R., Burke, J.G., 2010. Disparities and access to healthy food in the United States: a review of food deserts literature. Health Place 16 (5), 876–884.

Wallinga, D., 2009. Today's food system: how healthy is it? J. Hunger Environ. Nutr. 4 (3–4), 251–281.

West, P.C., Gerber, J.S., Engstrom, P.M., Mueller, N.D., Brauman, K.A., Carlson, K.M., Cassidy, E.S., Johnston, M., MacDonalad, G.K., Ray, D.K., Siebert, S., 2014. Leverage points for improving global food security and the environment. Science 345 (6194), 325–328.

WHO, 2016. Obesity and overweight fact sheet. World Health Organization. Available from: http://www.who.int/mediacentre/factsheets/fs311/en/. Accessed November 28, 2017.

World Bank, 2017. World development indicators. World Bank, Washington, DC. Available from: http://databank.worldbank.org/data/home.aspx. Accessed November 28, 2017.

Worldwatch, 2008. Agribusinesses consolidate power. Worldwatch Institute Vital Signs Blog. Available from: http://vitalsigns.worldwatch.org/vs-trend/agribusinesses-consolidate-power. Accessed November 28, 2017.

Kimberly Pfeifer is the Research Director at Oxfam America. She serves as the editor of Oxfam America's Research Backgrounder Series, and is a cofounder and chair the Oxfam Research Network. She has managed numerous research projects on issues of biotechnology, food security, agricultural innovation, trade, extractive industries, and economic inequality. Kimberly also worked for the AFL-CIO with the Center for Strategic Research; and for the Aga Khan Foundation in Zanzibar, Tanzania. She received her MA and PhD from the University of Florida in Political Science. Mostly recently, she was an Executive in Residence with the Atkinson Center for Sustainable Future at Cornell University.

Gina E. Castillo is Senior Manager at Oxfam America. She provides technical leadership on agriculture projects, manages staff, and oversees research and investments across Oxfam's agriculture portfolio. She also co-leads Oxfam's Resilience Knowledge Hub. Prior to joining Oxfam America, she worked for six years at Oxfam Novib as senior policy advisor on livelihood issues and at Both Ends. She has worked extensively on food security, resilience, climate change adaptation, and environment and has written a number of Oxfam international policy briefing papers. Originally from Ecuador, Gina received her PhD in social anthropology from York University in Toronto, Canada.

Kristal Jones is an Assistant Research Scientist at the National Socio-Environmental Synthesis Center at the University of Maryland, where she studies the social dimensions of natural resource management and the environmental dynamics of food systems, and leads several center-wide programs to build capacity for data-driven interdisciplinary research. She has a PhD in Rural Sociology and International Agriculture and Development from the Pennsylvania State University. Previous research has focused on the approaches and impacts of international agricultural research for development efforts, and the specific effects of seed system development projects in Sub-Saharan Africa.

Chapter 3

Situating Institutional Foodservice in Agro-Food Value Chains: Overcoming Market Power and Structure With Values-Based Procurement

Annelies M. Goger
Social Policy Research Associates, Oakland, CA, United States

Chapter Outline

Introduction	47	Key Features of Institutional	
Policy Context: How the Global		Foodservice Value Chains	59
Food System Became So		Shifting Institutional Foodservice	
"Market-Driven"	50	to Values-Based Procurement	62
The Global Value Chain		Growth of Agro-Food	
Approach: Rendering		Intermediaries That	
Obscure Supply Chains More		Enhance Market Access,	
Visible	52	Information Sharing, and	
What Is GVC Analysis?	53	Capacity	64
Governance	54	Limitations of the Values-Based	
Upgrading	55	Procurement Movement	65
The Structure and Governance of		Conclusion	67
Agro-Food Value Chains	57	References	69

INTRODUCTION

Trade liberalization and the global industrialization of agriculture has generated a highly complex and geographically dispersed food system that has had a wide variety of negative social, environmental, and economic impacts (see Chapter 2; Berti and Mulligan, 2016; Clapp, 2016; McMichael, 2005). Many also argue that the food system is not on a sustainable path in several respects, such as food security, climate change, or debt relations (FAO, 2013;

Clapp, 2016; Mcmichael, 2013). The cases presented throughout this volume represent a growing movement to harness the purchasing power of institutions to shape upstream practices in agro-food value chains and make them more sustainable, transparent, and ethically responsible—referred to hereafter as "values-based procurement" initiatives. This movement appears to be successful at gradually expanding infrastructure in regional food systems and piloting new models of aggregation, product innovation, distribution, purchasing, and supply chain coordination, with many lessons learned along the way.

However, a growing body of evidence suggests that this movement is constrained from scaling up and reaching its full potential because it is inhibited by structural barriers, such as limited processing and distribution infrastructure, price volatility, and consistent supply (see Chapter 4; Feenstra et al., 2011; Hardesty et al., 2010; Vogt and Kaiser, 2008). Others have documented capacity and ideological barriers, such as the cost, buy-in, and awareness necessary to adhere to additional food safety regulations (Thompson et al., 2017), and the capacity and motivation of foodservice staff to implement new food purchasing and preparation practices (see Chapter 14; Askelson et al., 2016). As a result, most institutions still tend to channel their purchasing power in ways that support conventional agro-foods value chains and large-scale industrialized agriculture (Pullman and Wu, 2011; Berti and Mulligan, 2016).

Recent trends in global agro-food value chains appear to be making it even more difficult to effect change in the food system, whether as individuals, activists, non-profits, or institutions. These trends include the financialization of agro-food value chains (Clapp, 2014; Isakson, 2014; Burch and Lawrence, 2013), the dominance of supermarket retailers (Burch et al., 2013; Dolan and Humphrey, 2000), and increasing concentration in the hands of a few large corporations (Howard, 2016; Clapp and Fuchs, 2009). For example, financialization and concentration in agro-food value chains has led to higher price volatility, lower producer prices over time, and higher costs of production (Clapp et al., 2017; Clapp and Fuchs, 2009; Harrison and Getz, 2015). Consequently, Clapp (2014) notes that the geographic expansion ("distancing") of food chains from farm-to-table and the growing opacity of food chains makes it more challenging to trace who is responsible for negative impacts and how to hold them accountable.

The emergence of place-based institutional foodservice initiatives focused on cultivating a local identity—exemplified in Part Three of this volume—can be read as a double movement to counteract these forms of distancing by re-embedding[1] food purchasing practices with local (societal) values

---

1. This argument draws on Polayni's (1957) classic text on political economy, in particular his concepts of "embeddedness" and "double movement." Polanyi argued that the economy cannot be understood outside of the social relations in which it is embedded; in other words, that markets do not function external to society, but rather, are always operating through historically situated institutions, rules, and social norms. The "double movement" refers to how historical periods of economic liberalism tend to be followed by a social response and a push to re-embed markets and economic activities in societal norms and rules.

(Polanyi, 1957; Feenstra et al., 2011). Although social movements such as the institutional foodservice movement have the potential to be emancipatory, there is also a danger that some forms of localism are informed by a sense of "community" that is exclusionary, authoritarian, and nationalist/nativist (Scoones et al., 2018). Given this context, continuing to deepen and strengthen the values-based procurement movement in institutional foodservice means being cautious about the assumption that "local" is always better (Harrison and Getz, 2015; Zepeda and Reznickova, 2017). It also highlights a need to ground the emerging field in a deeper analysis of agro-food value chains to ensure the movement is moving beyond romanticized notions of "local" and "community" that ignore the complex ways in which both the food system and movements to change it are shaped by long-standing histories and ongoing dynamics of class, race, gender, and imperialism (Scoones et al., 2018; DeLind, 2011; Alkon and Guthman, 2017).

This chapter offers a conceptual framework, the global value chains (GVC) approach, that can help institutional foodservice movements begin to situate their work more thoroughly in the broader historical social and political dynamics at play and identify new imaginaries and potential focal points of change. The chapter draws on the GVC literature and other secondary research to address the following questions:

- What are the governance dynamics that influence institutional foodservice agro-food value chains?
- What are the common features of institutional foodservice value chains that distinguish them from other agro-food value chains?
- How can producers, foodservice manufacturers, distributors, and other intermediaries upgrade to more effectively connect with institutional food markets interested in values-based procurement?

I argue that values-based procurement initiatives can use value chain analysis to begin to unpack agro-food value chains as a complicated and shifting assemblage of actors and power dynamics, helping to make these relations less abstract and the paths forward more clear. Specifically, GVC analysis can be useful for cultivating an appreciation of the broader structural dynamics at work, re-assessing alliances and strategies for changing the food system, and identifying new forms of collaboration and coordination. Although the emergence of new intermediaries and regional infrastructure for values-based procurement initiatives is promising, the movement needs to address some broader structural dynamics such as a largely market-led policy context, capacity limitations of intermediaries and smaller manufacturers/growers, and equity problems in order to make more sustainable and equitable changes to the food system.

The chapter begins with an overview of the market-led policy context that governs agro-food value chains. The next section provides background on what GVC analysis is and how it can be used as a tool to examine the complex relations of global production and consumption in the food system. The third section reviews how agro-food value chains are generally structured, and the fourth section examines the key features of institutional foodservice value chains. The fifth section analyzes value-based procurement initiatives as an attempt to change the food system, reviewing the forms of upgrading that these initiatives have focused on and their limitations. The chapter concludes by taking stock of what the policy context and broader dynamics of agro-food value chains mean for shifting to values-based procurement, reflecting on the strategies, alliances, and risks that the movement faces now and into the future.

POLICY CONTEXT: HOW THE GLOBAL FOOD SYSTEM BECAME SO "MARKET-DRIVEN"

The late 1970s and early 1980s marked a major shift in the dominant ideologies of development from state-led import-substitution industrialization policies that emphasized protectionism to market-led export-oriented development policies that emphasized economic liberalism (Dicken, 2007; McMichael, 2007; Peck and Tickell, 2002). The distinctive economic features of this era of neoliberal globalization included trade liberalization, the global integration of finance capital and banking systems, and the decentralization of production activities into functionally coordinated, multilayered transnational subcontracting arrangements—commonly referred to as supply chains or value chains (Castells, 2000; Harvey, 1989; McMichael, 2007; Piore and Sabel, 1984; Porter, 1990). International trade agreements, such as the North American Free Trade Agreement (NAFTA), the creation of the World Trade Organization (WTO) in 1995, and structural adjustment policies were the key mechanisms for implementing these policy shifts (Gibbon and Ponte, 2005; Goodman and Watts, 1997; Dicken, 2007). New technologies have also played a critical role in supporting the expansion of global supply chains. For example, in the fresh fruit and vegetable sector, cooling and other logistical technologies made it feasible to extend value chains over longer distances (Pullman and Wu, 2011).

These dominant ideologies and policies of market-led development have had a dramatic effect on agriculture and rural life in the United States and around the world (Araghi, 2000). Large-scale deregulation of the agro-food sector in the 1980s led to the dramatic restructuring of these industries and the rise of large agro-food transnational corporations, industrial farming operations, and retailers (Goodman and Watts, 1997; Clapp and Fuchs, 2009; Dolan and Humphrey, 2000). The world has been facing continued intensification of these market-oriented trends since the late 2000s, as markets at several stages of food chains are increasingly concentrated in the hands of a few large corporations, and several major agribusiness firms have expanded their role in global finance and commodities trading (Clapp, 2014, 2016; Isakson, 2014; Salerno, 2017; Burch and Lawrence, 2013; Howard, 2016).

Although market-led development policies use the imagery of "free trade" (and reducing "barriers" to trade) to conjure up the notion that it is possible for markets to function externally from government regulation, in practice the "rolling back" of one type of regulation was replaced by the "rolling forward" of several new forms of regulation (Peck and Tickell, 2002). The result is a highly uneven terrain of trade relations and governance. For example, the export-oriented trade regime sought to constrain the extent to which nation states could protect domestic agricultural industries and farmers; for example, by stipulating the removal of agricultural subsidies and import tariffs as a condition for entry into trade agreements or for receiving development loans. However, in the United States, domestic subsidies for crops such as corn, wheat, and soybeans have endured. This has had significant impacts on global commodity food prices, and many countries have accused the United States of "dumping," in which one country sells food products at a price lower than the cost of production such that producers in other countries cannot compete (Clapp, 2016). Consequently, many small and midsized producers were dispossessed globally, and the liberalization of agriculture accelerated rural migration and urbanization (Araghi, 2000). For domestic markets in the United States, crop subsidies and other policy incentives have encouraged the overproduction of grains and other commodity crops and the underproduction of healthier and less processed foods, such as fresh vegetables, that are, therefore, relatively more expensive.

In addition, the roll-back of trade regulations at the level of the nation state made other forms of governance at various scales (public and private) more important for governing and influencing practices in agro-food value chains. For example, third-party certification schemes arose as a leading mechanism for governing global value chain practices and ensuring certain standards are maintained despite the fact that activities may be occurring in multiple countries with varying levels of state regulation. Among the most common third-party certifications globally are the International Organization for Standardization (ISO) certification schemes, Fair Trade, and organic certifications. In addition, lead firms in some sectors of agriculture operate in an oligopolistic market, and their market power allows them greater influence on firms and producers up and down the chain (Howard, 2016). For example, the market for international grain traders is extremely concentrated, and a handful of companies have exerted their dominance through the value chain, which has given them influence over global prices and farming practices (Murphy et al., 2012; Ahmed et al., 2017; Morgan, 1979; Salerno, 2017).

Although the influence of trade regulations in the United States has declined at the national scale since the late 1970s, other policy instruments and investments in the United States have become more important in shaping the food system directly and indirectly. These include policies and investments at the national, state, and local level; and, for this reason, values-based purchasing in institutional foodservice could be understood as a hybrid (public-private) form of consumer-driven governance rather than purely privatized. At the national level (aside from the subsidies noted herein) these include health and safety regulations, labor and immigration laws, agricultural extension supports, labeling requirements, intellectual property laws (e.g., patenting seeds or biotechnology used in agriculture), government research and development funding (in technology, biotechnology, and basic science), and government food purchasing policies and practices (e.g., the U.S. Department of Defense, the justice system, school food policies, etc.) State and local government and institutional governing boards (e.g., school boards) also have been active in creating policies to encourage values-based procurement. These types of initiatives are the main focus of and audience for this volume, but they also need to be situated in the global policy context outlined herein that generated the food system problems to which these policies are responding.

THE GLOBAL VALUE CHAIN APPROACH: RENDERING OBSCURE SUPPLY CHAINS MORE VISIBLE

The GVC approach is one subfield within a broader set of commodity chains literatures that has emerged to better understand the structure and dynamics of globalization and trade liberalization since the late 1970s.[2] GVCs are the globally dispersed, but functionally coordinated, networks of firms that produce goods and services for final consumption (Gereffi et al., 1994).[3] For example, the GVC for wheat involves a complex assemblage of multiple inputs, farmers, intermediaries, corporations, regulatory structures, and stages of production, processing, and distribution spread out across multiple countries and end uses (Ahmed et al., 2017).

GVC analysis shifts the unit of analysis from the nation state to the interfirm, cross-border linkages that form to produce commodities. This allows development scholars and practitioners to analyze sector-specific, meso-level case studies within and between countries, rather than folding all industry sectors into a macro-level analysis at a national scale (Gereffi and Memedovic, 2003). GVC analysis can also be useful for identifying major shifts in the geographic scope of production and consumption and thinking through how that effects activities up and down the value chain. For example, there was concern that tobacco regulations in Western countries were negatively impacting small tobacco producers

---

2. Another prominent strand of literature within the commodity chain literatures is the global production networks approach, which was formulated to foreground not just inter-firm relations, but also the role of other institutions, social norms, and relationships (Henderson et al., 2002; Coe et al., 2004). For a thorough discussion, see (Bair, 2008). I chose "value chains" rather than "production networks" to focus the scope of the chapter on structural inter-firm dynamics, although the key role that other actors, social norms, and relationships play is worthy of further analysis.

3. The phrase was originally termed "global commodity chains" and drew from the French *filiere* (chain) studies in agriculture. The terminology was later changed to "value chains" to take attention away from products and toward value-added as a key driving mechanism of competitiveness (Gibbon and Ponte, 2005; Sturgeon 2009), a move that was also influenced by Michael Porter's (1990) work in business management. A "firm" refers to a business organization that sells services or products to generate profit.

and their livelihoods, but a GVC analysis of tobacco documented that the global market for tobacco had actually grown, because increases in demand in China and other Asian countries surpassed decreases in demand in Western countries (Goger et al., 2014). Many scholars who use GVC analysis have sought to understand the conditions under which insertion into global commodity trade circuits can contribute to development that benefits workers, firms, and the region as a whole rather than just growth that leads to greater inequality (Barrientos et al., 2011b; Gereffi, 1995; Kaplinsky, 1998).[4]

What Is GVC Analysis?

Value chains analysis generally focuses on describing the stages of production (from input to final product); the value-added activities that take place at each stage; the firms, workers, and other actors involved; and sector-level trends. There are six key dimensions of analysis (Gereffi and Fernandez-Stark, 2016):

(1) *The input-output structure*: the stages of production from inputs to final product and the value-added activities taking place at each stage;
(2) *Geographic scope*: an analysis of global supply and demand, based on trade flows;
(3) *Governance structure*: lead firms, industry organization, and the dynamics of control over and coordination of activities through the value chain;
(4) *Upgrading*: strategies that firms or other stakeholders use to improve their position in the chain and capture a higher share of the value-added;
(5) *Local institutional context*: how local and national policies and social norms shape activities through the value chain;
(6) *Industry stakeholders*: a map of the actors involved in the value chain and their role.

Value chain studies have proliferated in a wide array of industry sectors, including various agricultural sectors (dairy, wheat, coffee, etc.).[5] In addition to academic research, multilateral development institutions are increasingly adopting a GVC approach, such as the Inter-American Development Bank, the

---

4. It is also important to acknowledge critiques of the value chain framework, which have come from various angles. Some argue that it treats the firm like a "black box" (Coe et al., 2008; Hess, 2008). Others argue that it insufficiently incorporates the role of labor and the state (Carswell and De Neve, 2013; Smith et al., 2002; Werner, 2012; Herod, 2001; Selwyn, 2011). Finally, several scholars have noted the absence of a coherent theory of value in the framework (Bair, 2005; Gibbon et al., 2008; Guthman, 2009; Smith et al., 2002).
5. A large body of technical reports on various sectors for specific clients are available on the Duke University Global Value Chain Center website (formerly the Center on Globalization, Governance & Competitiveness), https://globalvaluechains.org/. The Capturing the Gains project, http://www.capturingthegains.org/, is another collection of literature from GVC and global production network scholars.

Organization for Economic Co-operation and Development (OECD),[6] and the United Nations Conference on Trade and Development (UNCTAD).[7] As the framework has evolved, the focus has explicitly turned to the question of how participation in GVC can enable regions, firms, and workers can "capturing the gains" from this participation (Gereffi and Fernandez-Stark, 2016; Barrientos et al., 2012). This is an acknowledgment that participating in GVCs is not necessarily a guarantee of successful growth or equitable development.

The foundational concepts of governance and upgrading in the value chain literature have been particularly influential in development studies and practice. The emphasis on these concepts is based on the proposition within the GVC literature that a "high road" development trajectory is possible, in contrast to the "low road" of downgrading—chasing the lowest costs and least regulation to be globally competitive. In the high road trajectory, governance from lead firms in GVCs stimulates organizational learning and global competitiveness among suppliers, which in turn generate "inclusive" regional economic growth that enhances the well-being of firms, workers, and society as a whole (Gereffi, 1995; Bair, 2005; Gereffi et al., 2005; Gibbon et al., 2008). By contrast, in the low road trajectory, firm efforts to achieve competitiveness by squeezing workers or cutting other corners may benefit individual firms or capitalists, but fail to trigger widespread economic development—a phenomenon Kaplinsky (1998) calls "immiserating growth."[8]

Governance

In the GVC literature, governance, in the most general sense, refers to private governance—nongovernmental institutions that enable and constrain economic activities in the global economy (Mayer and Gereffi, 2010). Governance in value chain analysis refers mainly to the power dynamics of private firms in the chain, such as which firm controls market access and coordinates the chain and how attributes such as product specifications and capital intensity correspond to specific types of governance structures (Gereffi, 1994; Gereffi et al., 2005). Hess (2008) characterizes this conception of governance as one in which power is understood as centralized and deployed in a hierarchical, structural relation of dominance and subordination between firms in the chain.

---

6. The OECD has an Initiative on Global Value Chains, Production Transformation and Development. Available from: http://www.oecd.org/sti/ind/global-value-chains.htm (Accessed 28 December 2017).
7. UNCTAD developed an input-output database, the Eora multi-region input-output table (MRIO) database, which has a component that reports on Trade in Value Added and Global Value Chains. Available from: http://www.worldmrio.com/ (Accessed 28 December 2017).
8. This high/low road distinction is inspired by new institutional economics, which posited that capitalism can bring about mutually beneficial outcomes for capital and labor given the "right" institutional framework (Selwyn, 2011).

Value chain scholars use the term "lead firm" to identify which firm in the chain has the most influence over the activities up or down the chain. Originally, Gereffi (1994) distinguished between "buyer-driven" chains (such as clothing) that are driven by retailers and "producer-driven" chains (such as the automobile industry) that are driven by original equipment manufacturers (e.g., General Motors). Gereffi et al. (2005) further refined this into a typology of governance by putting transaction costs at the center of the analysis. They asserted that the governance model operationalized in a value chain depends on the complexity of information needed to complete transactions, the codifiability of knowledge used in transactions, and the fit between the capability of suppliers and needs of the lead firms (Gereffi et al., 2005). Their typology specified five types of GVC governance—hierarchy, captive, relational, modular, and market—to capture variation in the extent to which the value chains are explicitly coordinated and power dynamics are asymmetrical.

In agro-food sectors, the type of governance varies in different types of chains, because some sectors are more complex and labor-intensive than others. For example, the value chains for fruits and vegetables are increasingly buyer-driven as consolidation in the retail segments of the chain continues to intensify (Fernandez-Stark et al., 2011; Dolan and Humphrey, 2000; Burch et al., 2013). By contrast, value chains for commodity foods, such as wheat and other grains, are largely driven by international grain traders—a market segment that is highly concentrated and capital-intensive (Murphy et al., 2012; Ahmed et al., 2017; Salerno, 2017). Therefore, a GVC analysis can be most useful as a tool for analyzing governance dynamics if it is used for specific agro-foods subsectors rather than for the industry as a whole.

Ethical trade counter-movements in major consumer markets can also be understood as an important form of GVC governance, the Polanyian double movement to this wave of economic liberalism. Freidberg (2004, p. 513) refers to these consumer-driven movements as an "ethical complex." It represents a desire—largely on the part of consumers, domestic producers and trade unions in wealthy nations, and activist groups—to pressure buyers to improve labor and environmental standards globally in an attempt to resolve the perceived absence of governance. In agriculture, this movement has manifested in several "voluntary" ethical consumer trends, such as the locavore movement, slow food, ugly food, organics, non-GMO food, farmers' markets, community supported agriculture, etc. (Alkon and Guthman, 2017). The institutional foodservice initiatives presented in this volume fall within this larger movement of food activism in that institutions are consumers; but, on the other hand, many institutions are also making policy or are wholly or partially government-funded, so they represent a public policy approach to regulation of the food system as well.

Upgrading

Upgrading in the value chains literature is an attempt to improve a firm or nation's position in international trade networks by, for example, making better products, adopting more efficient production processes, or adding higher skilled functions (Gereffi, 1999; Humphrey and Schmitz, 2002). Critical to the high road hypothesis of private governance in value chain analysis is the notion that upgrading can lead to global competitiveness at a national or sectoral-scale, and that competitiveness then can contribute to regional development. Humphrey and Schmitz (2002) identified four types of upgrading: product, process, functional, and chain upgrading.

- *Product upgrading*: Shifting production to higher unit-value, higher margin, or higher social/environmental value products. Example: in this volume, the development of new dining hall menu items with higher nutritional value (Chapter 13).
- *Process upgrading*: Changing the production process to be more productive, efficient, or environmentally friendly through the use of technology or the re-organization of production. Example: using new machinery or robots to till, sow, or harvest certain types of food.
- *Functional upgrading*: Adding additional functions of the production process to a firm's existing capabilities. Example: a small or midsized farmer increases their own capacity to process raw produce in-house (vertical integration) rather than selling raw produce to an intermediary processor.
- *Chain upgrading*: Purchasing products from higher-value supply chains, such as niche or boutique product chains. Example: Institutions changing their purchasing practices to buy more organic or Fair Trade food products.

The value chain literature posits that upgrading is a potential pathway to broad-based development because of its potential to stimulate organizational learning, enhance workforce skills, generate higher wages, create additional jobs, and enhance the competitiveness of a region or industry sector relative to other regions.[9]

The notion that upgrading tends to lead to inclusionary growth has also been critiqued (Gibbon and Ponte, 2005; Pickles, 2006; Werner, 2012; Tokatli, 2012; Goger, 2013a). For example, earlier debates about upgrading tended to focus on how firms will move into higher value-added activities to survive intensifying competition, but Tokatli (2012) argues that, in practice, firms often use multiple strategies of diversification at once—some high-value, some low-value, etc. Others have critiqued the concept of upgrading in the value chain's literature for having an "inclusionary bias," focusing too much on successful firms and omitting cases where firm-level upgrading led to industrial consolidation and displacement of workers from international circuits of production (Bair and Werner, 2011; Gibbon and Ponte, 2005; Goger, 2013a).

---

9. This conceptualization of upgrading revolves around the Schumpeterian notion of economic rent, in which a firm innovates and maintains a short-term monopoly over the innovation to capture profits. Soon after, competitive firms mimic the innovation and erode the profitability of the original innovator, and this spurs further innovation (Kaplinsky, 1998; Werner, 2012).

GVC and production networks scholars have revised the concept of upgrading by adding a distinction between social, environmental, and economic upgrading (Barrientos et al., 2011a,b; Bernhardt and Milberg, 2011; de Marchi et al., 2013). "Social upgrading" is defined as the process of improving the rights of workers and enhancing the quality of their employment, which includes measureable standards and enabling rights (Barrientos et al., 2011a,b). Likewise, "environmental upgrading" refers to pursuing industrial upgrading strategies that enhance competitiveness and reduce environmental impacts (de Marchi et al., 2013; Goger, 2013a,b). These revisions to the concept of upgrading take into account that industrial upgrading can, but does not always, benefit workers or the environment. Many of the values-based procurement initiatives in this volume can be read as a form of upgrading.

THE STRUCTURE AND GOVERNANCE OF AGRO-FOOD VALUE CHAINS

There is a high level of variation in the structure and dynamics of agro-food chains at the product level, because of the particularities and geographies of climate (where it is feasible for a given crop to grow), trade regulations/subsidies, and perishability levels for different types of food. For example, both food imports and exports have risen over the last two decades, but the feasible trading distance for many food products is constrained by perishability, declining freshness (quality), and food safety risks. The United States exported roughly 20 percent of domestic agro-food products between 2011 and 2013, but more than half of the cotton, tree nuts, rice, and wheat produced in the United States was exported during that same period (USDA ERS, 2017a). The share of U.S. food consumed that comes from imports has risen by 4 percent each year since 2000 (USDA ERS, 2017a), and has increased from 11 percent in 2009 to 14.8 percent in 2016 (USDA ERS, 2018). However, more than 95 percent of the coffee/cocoa/spices and fish/shellfish products consumed in the United States are imported, followed by about 50 percent of fruit and fruit juices (USDA ERS, 2017a). Because of this level of variation, this section focuses on some common features of agro-food value chains rather than a detailed GVC analysis.

Agro-foods can be broadly classified into three industry subsectors (Pullman and Wu, 2011):

(1) *Commodity crop chains*: wheat, rice, corn, soybeans, sugar, coffee, and other agro-food products that can be stored for long periods of time without spoiling and produced at large scales;
(2) *Fresh fruit and vegetable chains*: perishable plant-based food products that require short turnaround times and refrigeration for storage and transportation;
(3) *Animal protein chains*: livestock, fish, and meat products that are frequently heavily processed and preserved through refrigeration, smoking, roasting, or other methods.

Each subsector of the U.S. agro-food system is influenced by a complex and ever-changing landscape of economic, political, cultural, environmental, and market dynamics. Institutions, food activists, scholars, and policymakers can use GVC analysis to examine these interfirm relationships for specific types of food to identify leverage points for change.

The input-output structure varies significantly across the subsectors described herein, as do the set of actors involved and activities that take place in each stage along the value chain. Broadly speaking, however, agro-food value chains follow general stages of production, including:

- *Inputs*: seeds, fertilizers, pesticides, irrigation systems, financing, crop insurance, and tilling machinery and technologies that go into producing plant-based food products for human or animal consumption.
- *Growing/production*: farming and livestock-raising activities to produce raw food products.
- *Processing and packaging*: raw food products are sorted, packaged, and processed (e.g., cut, trimmed, prepared) according to the specifications of specific destination markets.
- *Distribution*: food products (raw and processed) are distributed to different end markets: retail (e.g., grocery stores), wholesale, institutional foodservice, and direct to consumer (e.g., farmers' markets, community-supported agriculture).
- *Consumption*: food products are consumed.
- *Waste*: food waste is produced throughout the chain, including agrochemical runoff (pollution), animal waste, factory waste from the processing and preparation stage (plant or animal components), packaging waste, transportation emissions, and consumer waste.[10]

There are also several intermediaries (supporting institutions or organizations), forms of infrastructure, and supportive services that are critical to agro-food production at multiple stages, many of which are industries in and of themselves, such as banking (finance), insurance, and agrochemicals. Regional food hubs, farmer and manufacturer associations, cooperatives, and logistics operations also function as intermediaries that help coordinate and share information within agro-foods value chains.

Agro-food products may come from the same set of producers, but they are distributed through different channels, including retail (aka "food stores") and foodservice (USDA ERS, 2016):

(1) *Retail*: food sold at supermarkets, meat or produce markets, convenience stores, and other food purchased for consumption at home; and
(2) *Foodservice*: food that is prepared and consumed away from home, including fast food, restaurants, caterers, and institutions.

---

10. The U.S. Department of Agriculture (Low et al., 2015, p. 51) reported that food waste is the largest source of municipal solid waste.

These marketing channels supplied $1.6 trillion in food in 2014 in the U.S. (USDA ERS, 2017c). More people are consuming food away from home (through foodservice) than ever before. The share of food consumed through foodservice has been steadily growing over the past half-century, and in 2014 it reached 50.1 percent, overtaking retail sales for the first time (USDA ERS, 2017b). Seventy-nine percent of foodservice expenditures were for full-service and fast-food restaurants (USDA ERS, 2017b), leaving 21 percent for institutions, hotels, catering, cafeterias, and other forms of foodservice.

Because the food originates from the same producer sources, the structure and governance dynamics of one distribution channel can influence price, quality, and supply in the other. In the retail segment of U.S. food stores, for example, over half of the market share was controlled by four firms in 2011 (Howard, 2016). In 2011, Walmart held 33 percent of the market share, followed by Kroger (9 percent), Safeway (5 percent), and Supervalu (4 percent) (Howard, 2016). Although institutional foodservice purchasers do not interact directly with food retailers, the concentration in the retail segment means that the retailers influence the prices, volumes, and quality of food available for purchase in foodservice markets. As Burch et al. (2013, p. 216) assert, "The retail sector has moved beyond its traditional responsibility for food *distribution*, and is now strongly influencing patterns of *production* and *consumption*."

KEY FEATURES OF INSTITUTIONAL FOODSERVICE VALUE CHAINS

As described herein, the retail and foodservice distribution channels in agro-food value chains are interrelated because they are dependent on the same food sources, and the stages of production are similar. Fig. 1 portrays a general depiction of the stages of production for agro-food products destined for the institutional foodservice markets and the types of institutional actors commonly involved in each stage.[11] At first glance, it may not look that different from agro-food value chains for retail or other foodservice market segments.

However, the value chain for institutional foodservice is distinguishable by a unique set of key features; which, in turn, shape the opportunities for and barriers to increasing values-based procurement. These key features are:

- *Large volumes and consistent supply*: Conventional institutional foodservice sourcing criteria tend to place an emphasis on price, reliability, low transaction costs (ease of ordering and receiving), and a consistently high volume of supply (Feenstra et al., 2011; Hardesty et al., 2010; Low et al., 2015; Vogt and Kaiser, 2008).

---

11. As stated earlier, the specific structure for a particular type of food product may vary considerably from this general model.

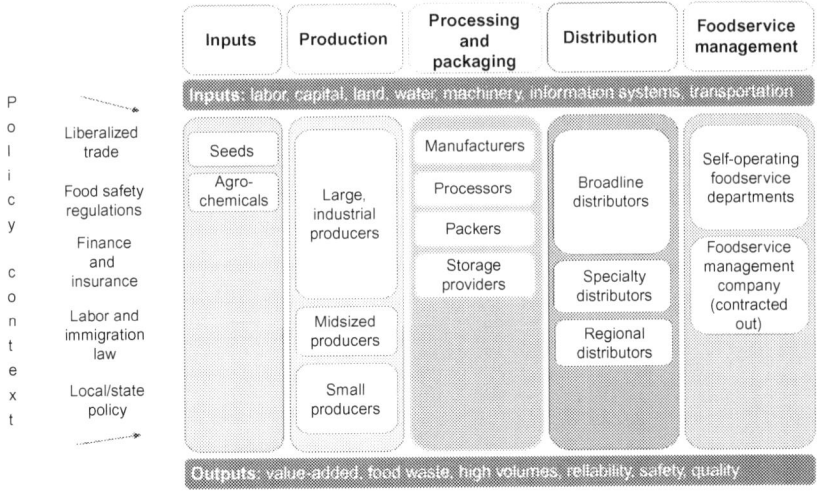

FIG. 1 Conventional institutional foodservice value chain.

- *The use of set menus, recipes, and products:* Because of the scale and need for coordinating large supplies and meeting their institution's food standards, institutions establish a combination of menus, recipes, and products. Making changes to one of the products or menu items will, consequently, change the upstream supply purchases or inputs and may also change the labor and skills involved in preparing and serving the new item.
- *High demand for processed or semiprocessed foods:* Partly because of the volume of food needed in a short period of time, many institutions purchase semiprocessed or fully processed foods rather than raw foods that then require more time and labor to prepare. For example, they may purchase precut frozen or canned vegetables instead of cutting fresh raw vegetables on-site, or they purchase frozen pizzas or chicken nuggets instead of making them from scratch.
- *Less branding and labeling:* Due to the way that institutional foodservice is typically prepared and served, end consumers have fewer opportunities to learn information about the food that they are selecting compared with the retail segment. This means that labeling and branding is more complex to engage as a strategy for building awareness in institutional foodservice, because information about the supply chain and production practices is more obscured and consumers have less access to information to evaluate their choices.
- *The use of contracting in procurement:* Many institutions and government agencies prefer multiyear, long-term contracts in order to keep transaction costs low and obtain rebates for committing to purchasing in large quantities over a long period. The contract provisions frequently specify price points, volumes, and other stipulations that allow them to accommodate specific requirements

or preferences. However, this structure also limits the ability of institutional foodservice purchasers to modify contract terms quickly and easily.
- *Highly differentiated sourcing criteria and purchasing practices:* The execution of institutional purchasing and foodservice management varies considerably across institutions. Some institutions have dedicated procurement departments or staff, while others—such as hospitals—purchase through group purchasing organizations outside the organization. Some institutions prepare and serve food in-house, while others contract it out to a private foodservice management company.[12] Many institutions are also subject to specific cost and regulatory constraints (especially public institutions). For example, hospitals and schools are required to meet certain nutritional and food safety standards (Low et al., 2015; Pullman and Wu, 2011). Values-based procurement increases this differentiation further.
- *A spectrum of public and private governance:* Some institutional foodservice is located within public institutions (such as public schools); others are located within private settings, such as corporate campuses or private universities; and some use a mixture of public and private funding. Publicly funded institutions have the advantage that they typically have governing boards (such as school boards) that are managing their legitimacy in the public eye, so they can be a potential lever for enacting values-based procurement. However, they also have the disadvantage that the governance process in public institutions can be more bureaucratic, highly scrutinized, and regulated and may be more complicated to change than private institutions.

While some of these features are hard to change because of the demands of institutional food consumption (e.g., high and consistent volume), the organizational approaches to meeting institutional demand for food stem from a set of long-established historical organizational practices, processes, and systems. These can be upgraded, adapted, or re-designed to incorporate the mission and values of the institution and end-consumer preferences.

Several of the preceding features represent barriers to entry for small and midsized producers and processors. In the absence of intermediary supporting organizations, such as cooperatives, associations, or food hubs; small and midsized producers and processors will not be able to meet the volume and regularity that institutions demand by themselves because of the high transaction costs (Matts et al., 2016). Moreover, most small and midsized farmers do not have capabilities in food processing, so they are more likely to provide foods that need more labor to prepare than an institution is set up to provide (e.g., see Chapter 14). Small and midsized farmers and processors are also likely to have a more difficult time meeting highly customized requirements stipulated in some long-term contracts. For example, certain food safety requirements

---

12. Declining funding and potential for cost-savings have contributed to the expansion of outsourcing (Pullman and Wu, 2011).

may require equipment, certifications, or processes that they do not yet have in place. In this case, supportive institutions such as agricultural extension services would be needed to help these producers build their capability to respond to highly specific and differentiated contract specifications.

Under the current market-led policy context, where the intermediaries and extension institutions noted herein are not consistently available or well-resourced, institutional foodservice will tend to purchase from large corporate suppliers. While institutions do purchase from multiple sources (distributors, producers, cooperatives, processors, etc.), it is most common for institutions to purchase food through distributors (Pullman and Wu, 2011; Low et al., 2015). Institutions purchase through broadline distributors most frequently—these are large, globally diversified companies that offer a wide variety of food items, including bulk and highly processed food items (Pullman and Wu, 2011). This makes institutional foodservice value chains more streamlined than agro-food supply chains generally, because this is how institutional purchasers currently minimize their transaction costs and achieve economies of scale in price and logistics (Feenstra et al., 2011; Pullman and Wu, 2011). For example, a school may have a multiyear contract with a large agro-food conglomerate for a specific type of food, such as Tyson for poultry products.

Ultimately, because most of the large distributors currently have few values-based purchasing food options (with a few exceptions), conventional institutional foodservice purchasing practices also tend to reproduce the worst problems of large-scale agro-food value chains (described in Chapter 2). These include contributing to environmental degradation and climate change, poor labor practices upstream in supply chains, favoring large-scale corporate agro-food businesses over smaller suppliers, toxic contamination, food waste, and low nutritional content (Berti and Mulligan, 2016; Clapp, 2016; Pullman and Wu, 2011). Consumers also have less choice over what baskets of food will be available for them to purchase when they eat from institutional foodservice; however, students and civil society organizations have had some success in mobilizing for changing menus and dining options in universities in recent years (e.g., through the work of Real Food Challenge, featured in the Preface).

SHIFTING INSTITUTIONAL FOODSERVICE TO VALUES-BASED PROCUREMENT

Although the key features noted herein in many ways constrain efforts to shift to values-based procurement, they also provide some opportunities and leverage points for change. This section provides an overview of how values-based procurement initiatives have been attempting to upgrade institutional foodservice value chains through some of these leverage points and identifies some other possibilities for thinking about food systems change in this context. The catch-all term "values-based procurement" is used as follows because the scope is to speak generally about these initiatives as a movement, but the limitations and trade-offs related to specific types of values-based purchasing are important to account for as well. For example, a values-based purchasing effort that is focused on procuring locally from small and midsized suppliers may still be reinforcing the use of exploitative labor, overuse of pesticides, or overconsumption of meat. Likewise, an effort focused on procuring organic or plant-based food through existing large distributors may improve the environment, but have less potential for increasing access to markets for small and midsized suppliers.

There has been steady growth in "values-based" purchasing in institutional foodservice value chains (Low et al., 2015; Feenstra et al., 2011). The increases in values-based purchasing at institutions have largely been driven by increased consumer demand, pressure from food activist campaigns, and changes in policy, such as the development of the Farm to School Program from the U.S. Department of Agriculture (USDA, 2011). In response to these pressures to do more values-based purchasing, institutional foodservice value chains are upgrading in several ways:

- *Product upgrading through menu upgrades and product innovations:* Many institutions are changing their menu options and working with upstream food manufacturers and distributors to create healthier products. In some cases, these upgrading efforts are motivated by policy changes (such as school food nutrition standards), while in other cases they are coming out of consumer activist campaigns.
- *Process upgrading by changing contractual stipulations:* Some initiatives are working to change contract provisions to increase the share of values-based or local/regional food purchasing, changing contractors to select one that can offer more values-based product offerings, or working to establish and improve standards within purchasing through efforts to share sample procurement models or educate decision makers about how to implement values-based procurement.
- *Functional upgrading in the conventional foodservice value chain:* Existing distributors and processors are adding new product offerings such as "organic" and "local" food. For example, Low et al. (2015) showed that more than 60 percent of schools engaged in farm-to-school initiatives were purchasing local food through a distributor rather than a manufacturer, producer, or cooperative. This suggests that the existing distributors are adapting to incorporate local food value chains within their set of offerings. However, due to the concentrated structure of the foodservice management market,[13] the effect of a large distributor adding a values-based food line can have repercussions for others in values-based agro-food chains, because it can constrain supply and increase prices in the short run.

---

13. The private market for foodservice management is highly concentrated, with four firms controlling 60 percent of the market share in 2017. The top firms were Compass Group PLC (21.5 percent market share), Sodexo (16.6 percent), Aramark (16.2 percent), and Delaware North Companies, Inc. (5.1 percent) (IBISWorld, 2017).

- *Chain upgrading—institutions are purchasing from new upstream suppliers:* Institutions engaged in values-based purchasing initiatives have increased purchasing directly from local producers, from new intermediaries (such as regional food hubs or new specialty or local/regional distributors), and with technical assistance from farm-to-institution intermediaries (Feenstra et al., 2011; Low et al., 2015). Several chapters in this volume exemplify this.

Growth of Agro-Food Intermediaries That Enhance Market Access, Information Sharing, and Capacity

Because these forms of social, environmental, and economic upgrading are happening through multiple avenues at once, the input-output structure for values-based institutional foodservice is becoming more diversified and complex—with more entities involved and a higher need for coordination, infrastructure, and information management. The growth of these and other forms of upgrading in values-based procurement in institutional foodservice represents a remarkable organizational innovation in agro-food value chains in the form of the rise of new intermediaries and improved regional infrastructure. The growth in intermediaries includes non-profits striving to establish procurement standards and build awareness of new purchasing models; agricultural extension services and research institutes that are building the capacity of growers and manufacturers to connect with institutional markets; regional food hubs and aggregators that are generating vendor lists and connecting buyers and suppliers that otherwise were not aware of each other; and consumer activist groups that are raising awareness of the consequences of conventional agro-food value chain practices for the environment, labor, health, and animal welfare. Although the proliferation of the new regional infrastructure is geographically uneven—highly concentrated on the West Coast and East Coast of the United States (Low et al., 2015), it suggests a gradual shift in agro-food value chains away from market-led sourcing practices.

Intermediaries are important for changing agro-food value chains because small and medium-sized producers and processors often have limited access to the best infrastructure, services/information, and technology. For example, access to finance and insurance is important for small and midsized producers and processors, who bear considerable upfront costs and risks if there are disruptions such as drought, flooding, bacterial outbreaks, workplace injuries, or pest infestations. At the processing and packaging stage of the value chain, packaging and cold storing technologies are critical for preserving the freshness and timeliness of delivered products (Pullman and Wu, 2011). Small and midsized suppliers have more limited access to advanced processing, packaging, and cold storing technologies compared with what conventional industrial agribusiness uses. Inventory tracking systems, which facilitate traceability and efficiency in distribution, are also a vital part of the infrastructure that makes agro-food value chains function well (Pullman and Wu, 2011). Small and midsized suppliers may lack the capability to purchase and maintain sophisticated inventory tracking systems, and this can be a barrier to entry for niche value chains, such as organics, that require a higher standard of traceability.

Regional food hubs represent an important organizational innovation (Berti and Mulligan, 2016). Food hubs are aggregators that act as bridging intermediaries between suppliers and buyers to resolve some of the structural barriers to re-territorializing food systems at a local or regional scale. The number of regional food hubs in the United States grew 288 percent between 2006 and 2014, operating with a mix of business models: 40 percent private business, 30 percent non-profit, and 20 percent cooperatives (Low et al., 2015). Notably, many regional food hubs host virtual directories and inventory management systems that facilitate transactions between participants. Many also assist farmers with product marketing and other forms of technical assistance. Preliminary research suggests that regional food hubs have a positive impact on the local economy (Jablonski et al., 2015).

Capacity-building intermediaries, typically non-profits and research institutes (many of which are featured in this volume), are also growing in size and scope. For example, the National Good Food Network describes its mission as follows:

> *The National Good Food Network is bringing together people from all parts of the rapidly emerging good food system – producers, buyers, distributors, advocates, investors and funders – to create a community dedicated to scaling up good food sourcing and access.*[14]

The network has a searchable database with how-to resources and research that local and regional stakeholders can easily access.

Limitations of the Values-Based Procurement Movement

Despite the upgrading and organizational innovations noted herein, there are still many barriers to upgrading institutional foodservice value chains. For example, during times when public budgets are cut, such as in the aftermath of the Global Economic Recession of 2007–09, institutions tend to downgrade through cost-cutting or outsourcing to save resources. As shown in Chapter 6, not all healthier food items are more expensive to procure, but there are widespread perceptions in institutional foodservice that they are more expensive, and this may make it politically challenging to implement these initiatives, especially when the institution is managing funding cuts. In addition, during such times institutions may lay off staff or switch to contractors to save costs, which can reduce capacity and skills in-house for implementing new menus, food preparation processes, or recipes.

---

14. National Good Food Network website, "About." Available from: http://www.ngfn.org/about (Accessed 30 December 2017).

The geographic unevenness of production and regional food infrastructure also makes it more challenging to procure local and fresh foods of certain kinds in rural areas or regions with more limited access to intermediaries and infrastructure (Askelson et al., 2016). This means that the cost to comply with national policies and implement new standards will vary geographically. Although intermediaries such as regional food hubs are one solution to the structural challenges of connecting multiple small producers with institutional buyers, another strategy would be to focus more on engaging midsized producers, manufacturers, and distributors (see Chapter 10; Klein, 2015).

In addition, some may question the legitimacy of values-based procurement efforts if they do not sufficiently take into consideration the context of long histories of inequity and cultural imperialism. Persistent inequities embedded in regional economies and across institutions create the conditions for these initiatives to become another case of "pay-to-play" consumer activism (Askelson et al., 2016), where only those institutions in relatively well-resourced areas can afford to spend time and energy on values-based purchasing. Likewise, efforts to change menus or product offerings do not always incorporate input from end consumers in low-income communities, raising questions about who gets to decide what types of food are "healthy" and how to respect cultural differences in food preferences as part of these initiatives and the voice of end consumers in deciding what types of healthier food they are willing to eat.

Values-based procurement initiatives should also take new advancements in technologies of production and supply chain coordination into account as both an opportunity for enhancing their work and as a potential barrier to upgrading (if only large corporations have the capacity to use new technologies). For example, there are opportunities for process upgrading in the form of new "smart farming" techniques, such as precision livestock farming, robotic agriculture, and machine imaging through drones, which have begun to diffuse agro-food value chains (Aravind et al., 2017; Banhazi et al., 2012; Lee, 2017; Ryu et al., 2015; Oberti and Shapiro, 2016; World Economic Forum, 2015). New technologies for inventory and ordering management show promise for helping small and midsized suppliers more efficiently link up with food aggregators and institutional buyers (Berti and Mulligan, 2016).

Finally, the continued consolidation of agro-food value chains at multiple stages of production is a particularly vexing problem, because transnational corporations have become increasingly dominant and lobby policymakers to shape policy in their interest (Clapp and Fuchs, 2009; Howard, 2016). The oligopolistic structure of agro-food markets means that large corporations can affect the price, supply, and quality of food produced, and the sheer scale of their uneven bargaining power can erode any returns that small and midsized firms can possibly capture from their efforts to innovate and upgrade. In the commodity food processing market segment, for example, the top four firms controlled 80 percent of U.S. crush capacity in 2011 (Howard, 2016): Bunge, Archer Daniels Midland, Cargill, and Ag Processing. Agriculture input markets are similarly highly concentrated; for example, four seed firms control 58 percent of the market share in that segment. Under these conditions, small and medium-sized producers and processors are price takers, having little negotiating power for several of the inputs and services that they depend on to carry out the value-added activities at their stage of production.

Likewise, concentration in agro-food chains more broadly means that a dominant firm in another segment—such as Walmart in the retail grocery segment—can influence prices, supply, and purchasing processes in which institutions and their procurement agents compete. For example, if Walmart decides to start selling local organic apples, it can severely limit the supply of local organic apples in a specific region and increase the price considerably (in the short-run). In this way, institutional foodservice value chains need to manage the risk of volatility in price and supply that stem from activities outside their direct control.

CONCLUSION

This chapter began with an overview of the broader policy context and private governance dynamics that shape agro-food value chains. It reviewed how the dominant policy emphasis on market-led liberalism has resulted in growing market concentration and financialization in the food system, as well as multiple societal responses and attempts to re-regulate the market to address negative externalities such as environmental degradation, public health epidemics, and labor exploitation. The values-based procurement initiatives in institutional foodservice can be understood as an emerging terrain of struggle in this double movement that is moving beyond a fixation on the role of the individual consumer and engaging the collective power of institutions and policymakers at multiple scales.

The chapter aimed to examine how a value chain analysis can help render more visible the complex structures and dynamics of agro-food production and consumption that are often obscured from the perspective of a single value chain actor, such as an end consumer, a food manufacturer, or institutional foodservice purchasing director. In doing so, it can be a useful tool for reflecting on the accomplishments and limitations of the values-based procurement initiatives to date and for identifying new strategies, alliances, and targets for shifting to values-based procurement in the future.

In many respects, as the cases throughout this volume demonstrate, the values-based procurement is constrained by the unique structural features of institutional foodservice procurement. For example, values-based procurement initiatives have focused on changing contracting provisions, connecting small local suppliers with institutional buyers, and upgrading menus and creating healthier food products. The overarching purpose of this volume is to document the similarities and differences across these approaches, to recognize different forms of upgrading as making a valuable contribution to a larger movement toward values-based procurement, and to promote the exchange of ideas, strategies, and lessons learned across different types of institutions and geographically diversified initiatives. Although implementation of farm-to-institution, regional food system infrastructure, and other values-based procurement initiatives has been somewhat uneven, there appears to be an overall trend of growth and expansion over time (Low et al., 2015).

However, situating institutional foodservice in a broader analysis of agrofood value chains is critical for ensuring that the movement is adequately addressing the root problems of the food system and not reproducing uneven power dynamics by generating a set of institutional purchasing practices that are only accessible to privileged communities and fail to incorporate critical reflections on equity, inclusion, and engagement from food system consumers and workers at all levels. Doing so might mean, for example, focusing less on "local" food, per se, and more on building alliances to advocate for policies that challenge the uneven market power of large agribusiness corporations, retailers, and financial institutions. It may also mean exploring new cross-institutional forms of collaboration, such as building coalitions between schools and hospitals, or hospitals and the justice system, to strengthen the collective power of institutions as food system actors. Additionally, it could mean exploring ways that new technologies can provide an avenue for open-source, collective forms of promoting transparency in agro-food value chains, building on existing efforts to better communicate the value of values-based food products and engage end consumers, or investing further in building the capacity of small suppliers to meet the safety and volume requirements of institutional buyers.

This volume also represents an opportune moment to reflect on new questions about the ongoing jostling of corporate power and technological innovation and how they will affect the future path forward for values-based institutional purchasing. How will GVCs continue to shift with the onset of automated agricultural production, food commodity trading with blockchain currencies, and diffusion of new logistical technologies and methods, such as drones, robots, and self-driving trucks? How will the emergence of new actors in food retailing and processing shape the future dynamics of values-based procurement initiatives, such as Amazon purchasing Whole Foods or Chinese mega-retailers investing in the U.S. livestock industry? How can values-based procurement initiatives avoid overly romanticizing the "local" and the dangers of aligning itself with isolationism and authoritarian populism by building a more sophisticated engagement with uneven power dynamics and equity in the field? What new methods and data are needed to understand and measure these changes in agro-foods value chains and their relevance to institutional foodservice value chains? The cases in this volume and analytical frameworks such as GVC analysis can help inform some necessary existential reflection on where the values-based procurement movement in institutional foodservice has come from, what it has achieved, and where it is headed in the future.

REFERENCES

Ahmed, G., Hamrick, D., Morgan, J., Nahapetyan, S., 2017. Russian Wheat Value Chain and Global Food Security. Duke University, Center on Globalization, Governance & Competitiveness, Durham, NC. Available at: https://gvcc.duke.edu/cggclisting/russian-wheat-value-chain-and-global-food-security/. Accessed December 28, 2017.

Alkon, A.H., Guthman, J. (Eds.), 2017. The New Food Activism: Opposition, Cooperation, and Collective Action. University of California Press, Oakland, CA.

Araghi, F., 2000. The great global enclosure of our times: peasants and the agrarian question at the end of the twentieth century. In: Magdoff, F., Foster, J.B., Buttel, F.H. (Eds.), Hungry for Profit: The Agribusiness Threat to Farmers, Food, and the Environment. Monthly Review Press, New York, pp. 145–160.

Aravind, K.R., Raja, P., Pérez-Ruiz, M., 2017. Task-based agricultural mobile robots in arable farming: a review. Span. J. Agric. Res. 15 (1), e02R01. https://doi.org/10.5424/sjar/2017151-9573.

Askelson, N., Cornish, D.L., Golembiewski, E., 2016. Rural school food service director perceptions on voluntary school meal reforms. J. Agric. Food Syst. Community Dev. 6 (1), 65–75. Available at: https://www.foodsystemsjournal.org/index.php/fsj/article/view/397.

Bair, J., 2005. Global capitalism and commodity chains: looking back, going forward. Compet. Chang. 9 (2), 153–180.

Bair, J., 2008. Analysing economic organization: embedded networks and global chains compared. Econ. Soc. 37 (3), 339–364.

Bair, J., Werner, M., 2011. Commodity chains and the uneven geographies of global capitalism: a disarticulations perspective. Environ. Plan. A 43 (5), 988–997.

Banhazi, T.M., et al., 2012. Precision livestock farming: an international review of scientific and commercial aspects. Int. J. Agric. Biol. Eng. 5 (3), 1–9. Available at: https://ijabe.org/index.php/ijabe/article/view/599.

Barrientos, S., Gereffi, G., Rossi, A., 2011a. Economic and social upgrading in global production networks: a new paradigm for a changing world. Int. Labour Rev. 150 (3/4), 319–340. Available at: http://search.proquest.com/docview/933126623/142F231242664335BA9/1?accountid=10598.

Barrientos, S., Mayer, F., Pickles, J., Posthuma, A., 2011b. Decent work in global production networks: framing the policy debate. Int. Labour Rev. 150 (3/4), 299–317. Available at: http://search.proquest.com/docview/933126600/142F231242664335BA9/2?accountid=10598.

Barrientos, S., Gereffi, G., Rossi, A., 2012. Economic and social upgrading in global production networks: a new paradigm for a changing world. Int. Labour Rev. 150 (3–4), 319–340. Available at: http://onlinelibrary.wiley.com.libproxy.lib.unc.edu/doi/10.1111/j.1564-913X.2011.00119.x/abstract.

Bernhardt, T., Milberg, W., 2011. Economic and social upgrading in global value chains: analysis of horticulture, apparel, tourism and mobile telephones. Manchester, UK. Available at: http://www.capturingthegains.org/publications/workingpapers/wp_201106.htm.

Berti, G., Mulligan, C., 2016. Competitiveness of small farms and innovative food supply chains: The role of food hubs in creating sustainable regional and local food systems. Sustainability 8 (7), 616. Available at: http://www.mdpi.com/2071-1050/8/7/616.

Burch, D., Lawrence, G., 2013. Financialization in agri-food supply chains: private equity and the transformation of the retail sector. Agric. Hum. Values 30 (2), 247–258. Available at: https://link-springer-com.libproxy.berkeley.edu/article/10.1007/s10460-012-9413-7.

Burch, D., Dixon, J., Lawrence, G., 2013. Introduction to symposium on the changing role of supermarkets in global supply chains: from seedling to supermarket: agri-food supply chains in transition. Agric. Hum. Values 30 (2), 215–224. Available at: https://link-springer-com.libproxy.berkeley.edu/article/10.1007/s10460-012-9410-x.

Carswell, G., De Neve, G., 2013. Labouring for global markets: conceptualising labour agency in global production networks. Geoforum 44, 62–70. Available at: http://www.sciencedirect.com/science/article/pii/S0016718512001315.

Castells, M., 2000. The Rise of the Network Society. Blackwell Publishers, Oxford/Malden, MA.

Clapp, J., 2014. Financialization, distance and global food politics. J. Peasant Stud. 41 (5), 797–814. Available at: http://www-tandfonline-com.libproxy.berkeley.edu/doi/full/10.1080/03066150.2013.875536.

Clapp, J., 2016. Food, second ed. Polity Press, Cambridge, UK/Malden, MA.

Clapp, J., Fuchs, D.A., 2009. Corporate Power in Global Agrifood Governance. MIT Press, Cambridge, MA.

Clapp, J., Isakson, S.R., Visser, O., 2017. The complex dynamics of agriculture as a financial asset: introduction to symposium. Agric. Hum. Values 34 (1), 179–183. Available at: https://link-springer-com.libproxy.berkeley.edu/article/10.1007/s10460-016-9682-7.

Coe, N.M., Hess, M., Yeung, H.W., Dicken, P., Henderson, J., 2004. 'Globalizing' regional development: a global production networks perspective. Trans. Inst. Br. Geogr. 29 (4), 468–484.

Coe, N.M., Dicken, P., Hess, M., 2008. Global production networks: realizing the potential. J. Econ. Geogr. 8 (3), 271–295. Available at: https://econpapers.repec.org/article/oupjecgeo/v_3a8_3ay_3a2008_3ai_3a3_3ap_3a271-295.htm Accessed December 28, 2017.

de Marchi, V.D., Maria, E.D., Micelli, S., 2013. Environmental strategies, upgrading and competitive advantage in global value chains. In: Business Strategy and the Environment. Available at: http://onlinelibrary.wiley.com.libproxy.lib.unc.edu/doi/10.1002/bse.1738/abstract.

DeLind, L.B., 2011. Are local food and the local food movement taking us where we want to go? Or are we hitching our wagons to the wrong stars? Agric. Hum. Values 28 (2), 273–283. Available at: https://link-springer-com.libproxy.berkeley.edu/article/10.1007/s10460-010-9263-0.

Dicken, P., 2007. Global Shift: Mapping the Changing Contours of the World Economy, fifth ed. Guilford Press, New York/London.

Dolan, C., Humphrey, J., 2000. Governance and trade in fresh vegetables: the impact of UK supermarkets on the African horticulture industry. J. Dev. Stud. 37 (2), 147–176.

FAO, 2013. Climate Smart Agriculture Sourcebook. Food and Agriculture Organization of the United Nations, Rome. Available at: http://www.fao.org/3/a-i3325e.pdf.

Feenstra, G., Allen, P., Hardesty, S., Ohmart, J., Perez, J., 2011. Using a supply chain analysis to assess the sustainability of farm-to-institution programs. J. Agric. Food Syst. Community Dev. 1 (4), 69–84.

Fernandez-Stark, K., Bamber, P., Gereffi, G., 2011. The Fruit and Vegetables Value Chain: Economic Upgrading and Workforce Development. Duke University, Center on Globalization, Governance & Competitiveness, Durham, NC. Available at: http://www.cggc.duke.edu/pdfs/2011-11-10_CGGC_Fruit-and-Vegetables-Global-Value-Chain.pdf. Accessed December 28, 2017.

Freidberg, S., 2004. The ethical complex of corporate food power. Environ. Plan. D 22 (4), 513–531. Available at: http://www.envplan.com.libproxy.lib.unc.edu/abstract.cgi?id=d384.

Gereffi, G., 1994. The organization of buyer-driven global commodity chains: How U.S. retailers shape overseas production Networks. In: Gereffi, G., Korzeniewicz, M. (Eds.), Commodity Chains and Global Capitalism. Greenwood Press, Westport, Conn. xiv, 334 pp.

Gereffi, G., 1995. Global production systems and third world development. In: Stallings, B. (Ed.), Global Change, Regional Response: The New International Context of Development. Cambridge University Press, Cambridge, UK/New York, NY.

Gereffi, G., 1999. International trade and industrial upgrading in the apparel commodity chain. J. Int. Econ. 48 (1), 37–70. Available at: http://www.sciencedirect.com/science/article/pii/S0022199698000750.

Gereffi, G., Fernandez-Stark, K., 2016. Global Value Chain Analysis: A Primer, second ed. Duke CGGC (Center on Globalization, Governance & Competitiveness). Available at: https://dukespace.lib.duke.edu/dspace/handle/10161/12488 Accessed December 28, 2017.

Gereffi, G., Memedovic, O., 2003. The Global Apparel Value Chain: What Prospects for Upgrading by Developing Countries? UNIDO, Vienna, Austria.

Gereffi, G., Korzeniewicz, M., Korzeniewicz, R.P., 1994. Introduction: global commodity chains. In: Gereffi, G., Korzeniewicz, M. (Eds.), Commodity Chains and Global Capitalism. Greenwood Press, Westport, CT. xiv, 334 pp.

Gereffi, G., Humphrey, J., Sturgeon, T., 2005. The governance of global value chains. Rev. Int. Polit. Econ. 12 (1), 78–104.

Gibbon, P., Ponte, S., 2005. Trading Down: Africa, Value Chains, and the Global Economy. Temple University Press, Philadelphia, PA.

Gibbon, P., Bair, J., Ponte, S., 2008. Governing global value chains: an introduction. Econ. Soc. 37 (3), 315–338.

Goger, A., 2013a. Managing Global Guilt and Local Norms: Governance in the Sri Lankan Clothing Industry (Dissertation). University of North Carolina at Chapel Hill, Chapel Hill, NC. Available at: https://cdr.lib.unc.edu/indexablecontent/uuid:f340c513-ee54-4f81-88ac-b7e7ba6badf9.

Goger, A., 2013b. The making of a 'business case' for environmental upgrading: Sri Lanka's ecofactories. Geoforum 47, 73–83. Available at: http://www.sciencedirect.com/science/article/pii/S001671851300064X.

Goger, A., Bamber, P., Gereffi, G., 2014. The Tobacco Global Value Chain in Low Income Countries. Duke University, Center on Globalization, Governance & Competitiveness, Durham, NC. Available at: https://gvcc.duke.edu/wp-content/uploads/2014-02-05_Duke-CGGC_WHO-UNCTAD-Tobacco-GVC-Report.pdf. Accessed December 28, 2017.

Goodman, D., Watts, M., 1997. Globalising Food: Agrarian Questions and Global Restructuring. Routledge, London/New York.

Guthman, J., 2009. Unveiling the unveiling: commodity chains, commodity fetishism, and the 'value' of voluntary, ethical food labels. In: Bair, J. (Ed.), Frontiers of Commodity Chain Research. Stanford University Press, Palo Alto, CA.

Hardesty, S., et al., 2010. Institutional food distribution systems: bringing students, farmers and food service to the table. J. Food Distrib. Res. 40 (1), 58–63.

Harrison, J.L., Getz, C., 2015. Farm size and job quality: mixed-methods studies of hired farm work in California and Wisconsin. Agric. Hum. Values 32 (4), 617–634. Available at: https://link-springer-com.libproxy.berkeley.edu/article/10.1007/s10460-014-9575-6.

Harvey, D., 1989. The Condition of Postmodernity. Blackwell, Cambridge.

Henderson, J., Dicken, P., Hess, M., Coe, N.M., Yeung, H.W.-C., 2002. Global production networks and the analysis of economic development. Rev. Int. Polit. Econ. 9 (3), 436–464.

Herod, A., 2001. Labor Geographies: Workers and Landscapes of Capitalism. Guilford Press, New York.

Hess, M., 2008. Governance, value chains and networks: an afterword. Econ. Soc. 37 (3), 452–459. Available at: http://www.tandfonline.com/doi/abs/10.1080/03085140802172722.

Howard, P., 2016. Concentration and Power in the Food System: Who Controls What We Eat? Bloomsbury Academic, London, UK/New York, NY.

Humphrey, J., Schmitz, H., 2002. How does insertion in global value chains affect upgrading in industrial clusters? Reg. Stud. 36 (9), 1017–1027. Available at: http://www.tandfonline.com/doi/abs/10.1080/0034340022000022198.

IBISWorld, 2017. IBISWorld U.S. Industry Reports: Food Service Contractors. Available at: https://www.ibisworld.com/.

Isakson, S.R., 2014. Food and finance: the financial transformation of agro-food supply chains. J. Peasant Stud. 41 (5), 749–775. Available at: http://www-tandfonline-com.libproxy.berkeley.edu/doi/full/10.1080/03066150.2013.874340.

Jablonski, B.B.R., Schmit, T.M., Kay, D., 2015. Assessing the Economic Impacts of Food Hubs to Regional Economies: A Framework Including Opportunity Cost. Cornell University, Charles H. Dyson School of Applied Economics and Management, Ithaca, NY. Available at: http://publications.dyson.cornell.edu/research/researchpdf/wp/2015/Cornell-Dyson-wp1503.pdf.

Kaplinsky, R., 1998. Globalisation, industrialisation and sustainable growth: the pursuit of the nth rent. Institute of Development Studies. Discussion paper 365.

Klein, K., 2015. Values-based food procurement in hospitals: the role of health care group purchasing organizations. Agric. Hum. Values 32 (4), 635–648. Available at: https://link-springer-com.libproxy.berkeley.edu/article/10.1007/s10460-015-9586-y.

Lee, H.W., 2017. Agriculture 2.0: how the Internet of Things can revolutionize the farming sector. Information and Communications for Development. Available at: http://blogs.worldbank.org/ic4d/agriculture-20-how-internet-things-can-revolutionize-farming-sector.

Low, S.A., et al., 2015. Trends in U.S. local and regional food systems. U.S. Department of Agriculture, Economic Research Service. Available at: https://naldc.nal.usda.gov/download/60312/PDF.

Matts, C., Conner, D.S., Fisher, C., Tyler, S., Hamm, M.W., 2016. Farmer perspectives of Farm to Institution in Michigan: 2012 survey results of vegetable farmers. Renew. Agric. Food Syst. 31 (1), 60–71. Available at: https://www.cambridge.org/core/journals/renewable-agriculture-and-food-systems/article/farmer-perspectives-of-farm-to-institution-in-michigan-2012-survey-results-of-vegetable-farmers/BEC2128F1FC400D645C1F7BFAEE76109.

Mayer, F., Gereffi, G., 2010. Regulation and economic globalization: prospects and limits of private governance. Bus. Politics 12 (3). Available at: http://www.degruyter.com.libproxy.lib.unc.edu/view/j/bap.2010.12.3/bap.2010.12.3.1325/bap.2010.12.3.1325.xml.

McMichael, P., 2005. Global development and the corporate food regime. In: Buttel, F.H., McMichael, P. (Eds.), New Directions in the Sociology of Global Development. Research in Rural Sociology and Development. Elsevier, Amsterdam, pp. 265–299.

McMichael, P., 2007. Development and Social Change: A Global Perspective. Sage Pine Forge Press, Thousand Oaks, CA.

Mcmichael, P., 2013. Value-chain agriculture and debt relations: contradictory outcomes. Third World Q. 34 (4), 671–690. https://doi.org/10.1080/01436597.2013.786290.

Morgan, D., 1979. Merchants of Grain. Viking Press, New York.

Murphy, S., Burch, D., Clapp, J., 2012. Cereal Secrets: The world's Largest Grain Traders and Global Agriculture. Oxfam International, Oxford, UK. Available at: https://www.oxfamamerica.org/static/oa4/cereal-secrets.pdf.

Oberti, R., Shapiro, A., 2016. Advances in robotic agriculture for crops. Biosyst. Eng. 146, 1–2.

Peck, J., Tickell, A., 2002. Neoliberalizing space. Antipode 34 (3), 380–404.

Pickles, J., 2006. Trade liberalization, industrial upgrading, and regionalization in the global clothing industry. Environ. Plan. A 38 (12), 2201–2206.

Piore, M.J., Sabel, C.F., 1984. The Second Industrial Divide: Possibilities for Prosperity. Basic Books, New York.

Polanyi, K., 1957. The Great Transformation. Beacon Press, Boston, MA. 1st Beacon paperback.

Porter, M.E., 1990. The Competitive Advantage of Nations. Free Press, New York.

Pullman, M., Wu, Z., 2011. Food Supply Chain Management: Economic, Social and Environmental Perspectives. Routledge, New York.

Ryu, M., et al., 2015. In: Design and implementation of a connected farm for smart farming system. 2015 IEEE Sensors, pp. 1–4.

Salerno, T., 2017. Cargill's corporate growth in times of crises: how agro-commodity traders are increasing profits in the midst of volatility. Agric. Hum. Values 34 (1), 211–222. Available at: https://link-springer-com.libproxy.berkeley.edu/article/10.1007/s10460-016-9681-8.

Scoones, I., et al., 2018. Emancipatory rural politics: confronting authoritarian populism. J. Peasant Stud. 45 (1), 1–20. Available at: http://www-tandfonline-com.libproxy.berkeley.edu/doi/full/10.1080/03066150.2017.1339693.

Selwyn, B., 2011. Beyond firm-centrism: re-integrating labour and capitalism into global commodity chain analysis. J. Econ. Geogr. 12 (1), 205–226. Available at: https://academic.oup.com/joeg/article/12/1/205/1160211 Accessed December 28, 2017.

Smith, A., et al., 2002. Networks of value, commodities and regions: reworking divisions of labour in macro-regional economies. Prog. Hum. Geogr. 26 (1), 41–63.

Sturgeon, T.J., 2009. From commodity chains to value chains: interdisciplinary theory-building in an age of globalization. In: Bair, J. (Ed.), Frontiers of Commodity Chain Research. Stanford University Press, Palo Alto, CA.

Thompson, J.J., Brawner, A.J., Kaila, U., 2017. "You can't manage with your heart": risk and responsibility in farm to school food safety. Agric. Hum. Values 34 (3), 683–699. Available at: https://link-springer-com.libproxy.berkeley.edu/article/10.1007/s10460-016-9766-4.

Tokatli, N., 2012. Toward a better understanding of the apparel industry: a critique of the upgrading literature. J. Econ. Geogr. Available at: http://joeg.oxfordjournals.org.libproxy.lib.unc.edu/content/early/2012/11/20/jeg.lbs043.

USDA, 2011. USDA Farm to School Team, 2010 Summary Report. U.S. Department of Agriculture, Food and Nutrition Service, Washington, DC. Available at: https://www.fns.usda.gov/sites/default/files/2010_summary-report.pdf.

USDA ERS, 2016. Retailing & wholesaling: overview. Available at: https://www.ers.usda.gov/topics/food-markets-prices/retailing-wholesaling/.

USDA ERS, 2017a. Agricultural trade. Available at: https://www.ers.usda.gov/data-products/ag-and-food-statistics-charting-the-essentials/agricultural-trade/.

USDA ERS, 2017b. Ag and Food Statistics: Charting the Essentials, October 2017. United States Department of Agriculture Economic Research Service, Washington, DC. Available at: https://www.ers.usda.gov/webdocs/publications/85463/ap-078.pdf?v=43025.

USDA ERS, 2017c. Market segments. Available at: https://www.ers.usda.gov/topics/food-markets-prices/food-service-industry/market-segments/.

USDA ERS, 2018. U.S. export share of production, import share of consumption. Available at: https://www.ers.usda.gov/topics/international-markets-us-trade/us-agricultural-trade/data/.

Vogt, R.A., Kaiser, L.L., 2008. Still a time to act: a review of institutional marketing of regionally-grown food. Agric. Hum. Values 25 (2), 241–255. Available at: https://link.springer.com/article/10.1007/s10460-007-9106-9.

Werner, M., 2012. Beyond upgrading: gendered labor and the restructuring of firms in the Dominican Republic. Econ. Geogr. 88 (4), 403–422. Available at: http://onlinelibrary.wiley.com.libproxy.lib.unc.edu/doi/10.1111/j.1944-8287.2012.01163.x/abstract.

World Economic Forum, 2015. Industrial Internet of Things: Unleashing the Potential of Connected Products and Services. World Economic Forum and Accenture, Cologny/Geneva. Available at: http://www3.weforum.org/docs/WEFUSA_IndustrialInternet_Report2015.pdf.

Zepeda, L., Reznickova, A., 2017. Innovative millennial snails: the story of Slow Food University of Wisconsin. Agric. Hum. Values 34 (1), 167–178. Available at: https://link-springer-com.libproxy.berkeley.edu/article/10.1007/s10460-016-9701-8.

Annelies M. Goger is the co-editor of this volume. She investigated global value chains and global production networks in wheat and tobacco while employed at the Duke University Global Value Chains Center (formerly the Center on Globalization, Governance and Competitiveness). She also brings previous experience studying globalization, governance, and ethical trade initiatives in the global clothing industry through her dissertation research in Sri Lanka, Europe, and the United States. She is currently a researcher at Social Policy Research Associates in Oakland, California specializing in food assistance, job training, and economic development programs in the Unites States. She holds a PhD in geography from the University of North Carolina at Chapel Hill, a MCP from the University of California at Berkeley, and a BA in sociology from Brandeis University.

Part II

Setting Purchasing Standards Through Policy

Chapter 4

From Foodservice Management Contracts to U.S. Federal Legislation: Progress and Barriers in Values-Based Food Procurement Policies

Raychel E. Santo*, Claire M. Fitch[†,a]
*Johns Hopkins Center for a Livable Future, Department of Environmental Health and Engineering, Bloomberg School of Public Health, Baltimore, MD, United States, [†]Farm Forward, Portland, OR, United States

Chapter Outline

Introduction	78	Foodservice Management Contracts as a Barrier to Increasing Regional and Sustainable Food Procurement	88
Using Institutional Foodservice Policies to Address Food System Challenges in the United States	79		
Market Structure and Rural Economy Considerations	81	Foodservice Contracts and the Rebate Pricing System	89
Environmental Considerations	85	Reflections on Uncertainties in the Larger Political Context	93
Health Considerations	86		
Labor and Equity Considerations	87	Conclusion	95
Animal Welfare	88	References	96

---

a. We would like to thank Karen Bassarab, Carolyn Hricko, and Anne Palmer for their feedback on drafts of this chapter, and Michael Milli for designing Fig. 1. Kate Clancy, Stacia Clinton, Alexa Delwiche, Noel Isama, Brent Kim, Bob Lawrence, Bob Martin, Shawn McKenzie, and David Schwartz also deserve thanks for reviewing the report on which this chapter was based.

INTRODUCTION

Growing public interest in the structure and effects of the U.S. food system has catalyzed demand for regionally[1] and sustainably[2] produced food in recent decades (Starr, 2010). Concerned about the environmental, health, animal welfare, and social justice implications of their food choices, an increasing number of people want to eat in ways that satisfy their taste buds and reflect their values (Cone Communications, 2014). Many people have begun organizing to reform institutional food procurement practices in line with these values (hereafter referred to as "values-based procurement"), with the belief that large institutions can use their purchasing power to change how food is produced, priced, and distributed for consumption. However, structural factors at multiple stages of the value chain shape which foods are available at schools, hospitals, and worksites, and these factors may constrain ethical eating ambitions (Barnett et al., 2005). This chapter focuses specifically on the structure of one stage of institutional agro-food value chains, foodservice management operations, and how their contractual agreements and policies can make it difficult for individuals—and even institutional decision makers—to understand and influence their institution's food procurement policies.

The structure of foodservice at any given institution shapes the possibilities for implementing values-based procurement. Some institutions manage their own foodservice operations ("self-operated"), but others choose to contract out to a third-party foodservice management company (Porter, 2006). Foodservice management companies provide services such as food procurement and preparation, menu development, price negotiation with food suppliers, staffing and management, regulatory compliance, and maintenance of space and infrastructure. Contracting with a foodservice management company can offer advantages and relieve institutions of the potential financial and administrative burden of self-operated foodservice management, but can also restrict an institution's control over foodservice, and may limit its procurement of food from regional, sustainable, or alternative suppliers.

Institutional foodservice management contracting is big business. The largest three companies (Compass Group, Aramark, and Sodexo) alone generated more than $37 billion[3] in revenue in North America in 2017, up by about 150

---

1. "Local" and "regional" are often used interchangeably, and are defined differently by different food system actors. We use the term "regional" with the understanding that it is inclusive of "local" and signifies that various scales and geographies are levied to supply a significant portion of the food needs of a geographical region (Clancy and Ruhf, 2010). We use "local" only where it is specifically used in the corresponding resource. Likewise, we use "community-based" where it is used in the corresponding resource.
2. The use of "sustainability" throughout this chapter encompasses social, environmental, and economic factors.
3. British pounds and Euros were converted to U.S. dollars based on average 2017 exchange rates (Compass Group, 2017a, p. 36).

percent since 2004[4] (Aramark, 2017a; Compass Group, 2017a; Sodexo, 2017). The total North American foodservice demand is now $83 billion annually, and about 60 percent of that business is outsourced to a foodservice management company compared to 51 percent in 2004 (Compass Group, 2017b; Compass Group PLC, 2004). These multinational corporations operate in various U.S. settings in the education, business and industry, health and elder care, sports, corrections, and leisure sectors. As private-sector enterprises, foodservice management companies have necessarily prioritized increased revenue, lower expenses, and customer satisfaction.

This chapter provides an overview of institutional foodservice structure and the role of policy, discusses existing barriers to the adoption of values-based procurement practices (particularly the way contracts are currently structured with foodservice management companies and rebate pricing systems), and reflects on the federal political context influencing institutional food procurement practices.[5] Using data from literature, our own interviews, and our work with other organizations in this sector, we provide a foundation for understanding institutional foodservice policy and politics, with the goal of contextualizing the case studies described throughout the rest of this volume. We argue that policies, from the institutional to the federal level, are instrumental in creating a favorable environment for values-based procurement by institutions.

USING INSTITUTIONAL FOODSERVICE POLICIES TO ADDRESS FOOD SYSTEM CHALLENGES IN THE UNITED STATES

The industrialization of U.S. agriculture, largely over the latter half of the 20 century, was characterized in part by specialization, mechanization, standardization, consolidation, and a greater reliance on off-farm inputs, with the majority of farmers abandoning diversified farming systems (Ikerd, 2008). Large corporations began to finance and operate industrialized food production facilities, acquiring small businesses and merging with other corporations to control multiple stages of the value chain (Howard, 2016). This concentration and vertical integration along food supply chains is credited with improving efficiency, reducing costs, and lowering consumer prices; however, it is also implicated in an array of negative environmental, labor, and economic impacts (FCWA and SRC, 2016; Ikerd, 2008; IPES-Food, 2017).

The markets for foodservice management and distribution have become similarly concentrated. The top three management companies operate foodservice in an estimated 45 percent of all North American institutional foodservice outlets (Fig. 1). Food distribution is perhaps even more concentrated.

---

4. The collective North American revenue of these three companies for fiscal year 2004 was $25.4 billion (Aramark Corporation, 2004; Compass Group PLC, 2004; Sodexho Alliance, 2005), after adjusting for inflation (Smart Asset, n.d.) and converting to U.S. dollars based on average 2004 exchange rates (Compass Group PLC, 2004, p. 13).
5. This chapter updates and expands upon a 2016 report for practitioners (Fitch and Santo, 2016).

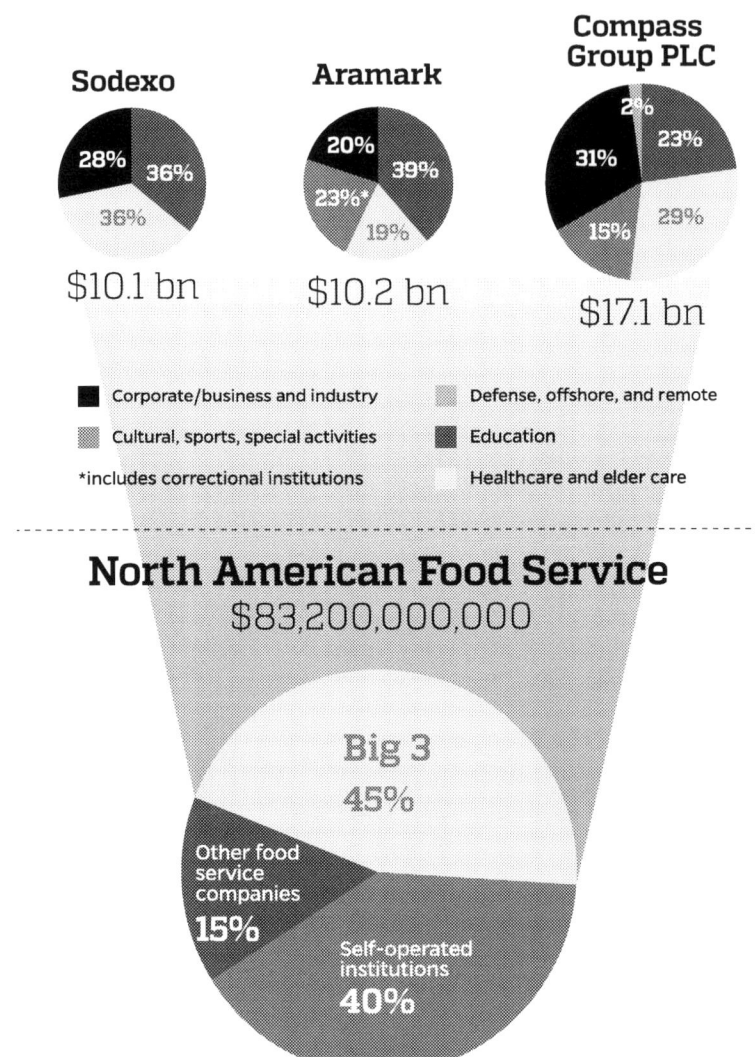

FIG. 1 Size and composition of the North American food service market. Note that Compass Group includes the following subsidiary companies: Bon Appétit Management Company, Canteen, Chartwells Higher Education and Schools Dining Services, Culinart Group, ESS Support Services, Eurest Dining Services, FLIK Hospitality Group, Gourmet Dining LLC, Levy Restaurants, Morrison Healthcare Food Services, Morrison Community Living, Restaurant Associates, TouchPoint Support Services, and Wolfgang Puck Catering. *(Data from Aramark (2017a,b), Compass Group (2017a,b), and Sodexo (2017).)*

The Federal Trade Commission (2015) estimates that two companies comprise 75 percent of the broadline food distribution market.[6]

Given these trends and their negative consequences; schools, hospitals, universities, government agencies, and other institutions have expressed growing interest in values-based procurement. Several key organizations—including Health Care Without Harm, Real Food Challenge, School Food Focus, and the Center for Good Food Purchasing—are working with specific institutions and foodservice management companies to advocate for comprehensive standards that encompass values-based criteria within institutional foodservice purchasing policies. Table 1 illustrates the growth and scope of these national procurement initiatives.

Institutional food policies are guidelines that direct the internal foodservice and procurement practices of a public or private organization, institution, or corporation. Ranging from nutrition standards to values-based procurement criteria, internal food policies are often included in bid solicitations for foodservice agreements and commodity contracts. In the case of public institutions, they may be enacted by legislative ordinances, executive orders, agency regulations, nonbinding resolutions, and informal agency guidance (FOE, 2017). Private institutions may enact similar food policies through various means, including executive orders, boards of trustees' decisions, or strategic plans. Private foodservice management companies may also have their own internal policies that direct their foodservice operations across all of their client institutions. For example, Bon Appétit Management Company's (a subsidiary of Compass Group) "Fish to Fork" preferred purchasing program guides company chefs to purchase seafood that meets specific traceability, size, distance, and species preferences (BAMCO, n.d.).

The following section provides more detail about the ways in which institutions are using food policies to address different socioeconomic, environmental, health, and animal welfare challenges facing the food system through their procurement policies. It also highlights how several non-profits—and in some cases, foodservice management companies themselves—have been instrumental in advocating for the incorporation of these issues into institutional values-based procurement policies.

Market Structure and Rural Economy Considerations

Amidst the accumulating market power of a handful of globalized agribusiness operations over the past few decades (Howard, 2016), consolidation of the value chain has threatened the continued existence of many small and midsized producers (Hullinger and Tanaka, 2015; Kirschenmann et al., 2008).

---

6. Sysco Corporation and US Foods, Inc.—the two largest distributors that contract with institutional foodservice—brought in nearly $68 billion in combined annual revenue in 2016 (Sysco, 2017; US Foods, Inc., 2017).

TABLE 1 Institutional Foodservice Procurement Policies Across Sectors in the United States

Name	Year Started	Criteria	Participating Institutions	Average Sustainable Food Spending Per Year
Farm Forward Leadership Circle	2017	Animal welfare	Colleges and universities: Bon Appétit Management Company (>1000 sites), plus five additional institutions	$10 million (projected)
Farm-to-school programs	1996–97	Healthy, regional	K-12 schools: 5,254 districts comprising 42,587 schools (42 percent of public schools in United States) in 2013–14 school year	$789 million (11 percent of all food expenditures)
Federal Food Service Guidelines (for concessions and vending operations)	2012, revised in 2017	Nutrition, regional, sustainable	Government agencies: Voluntary implementation of standard criteria is expected for all federal foodservice venues where food is sold	n/a
Global Animal Partnership (GAP) 1 broiler standards (for poultry purchases)	2016	Animal welfare	Foodservice management companies: All Aramark, Compass Group Sodexo facilities and select retail food outlets	Estimated >$1 billion
Good Food Purchasing Policy	2015 (national)	Local economies, valued workforce, animal welfare, nutrition, environmental sustainability	Government agencies and public K-12 schools: Policies passed in four cities and one county; campaigns underway in at least eight additional cities	Projected $500 million in active cities

Program	Year	Values	Scope	Dollar amount
Health Care Without Harm's Healthy Food in Health Care Pledge	2006	Healthy (antibiotic free), regional, sustainable	Healthcare facilities: 580 pledged; 850 in network (33 percent of hospitals in United States)	$147–216 million
Real Food Challenge Campus Commitment	2011	Local/community based, ecologically sound, fair, humane	Colleges and universities: 43 signatory institutions, plus 40 campuses with system-wide Real Food purchasing policies[a]	$85 million committed (by 2020)
School Food Focus	2007	Healthy, antibiotic free	K-12 schools: 45 districts (7855 schools)	$600 million
STARS Sustainable Dining Tracking	n/a	Local/community based, ecologically sound, fair, humane	Colleges and universities: 76 campuses have sustainable food policy; 286 campuses self-report spending average of 18 percent of food expenditures on sustainable food[b]	
Wasted food reduction commitments	2016-17	Environmental sustainability	Foodservice management companies: Aramark, Compass Group, Sodexo	n/a

[a] An additional 20 campuses are tracking their food purchases with the Real Food Calculator and several more have active student groups in the network.
[b] The question about whether campuses have a published sustainable food policy is a newer question in the STARS tracking system, so only campuses that have completed STARS version 2.1 have reported on this. However, STARS had been tracking self-reported percentage of expenditures on sustainable food in previous versions. Additionally, there is some cross-over with Real Food Challenge signatories.

Note: This table only highlights nationwide initiatives. Food procurement reform is also occurring in single institutions, as well as other foodservice sectors, including leisure (Hanson, 2015; NRDC, 2015) and prisons (Bulger, 2015).

Data from Aramark (2016, 2017b), AASHE (2017), CFGFP (2017), Compass Group USA (2017), CAP (2016), HCWH (2018), HHS (2017), Leadership Circle (2017), Mukherji and Weinronk (2017), School Food Focus (2017), Sodexo (2016a,b), and USDA (2017).

While there is not a binary distinction between small-scale and industrial producers, between "family" and corporate farms, or between "alternative" and "conventional" producers, many food producers in the middle of each spectrum (i.e., "agriculture of the middle") have more flexibility than industrial producers to transition to more socially, economically, and environmentally sustainable systems while also meeting a more significant portion of regional food demand compared with small-scale niche producers (Kirschenmann et al., 2008). The shifting national landscape from numerous small and midsized independent farms to few large-scale industrial farms has also been implicated in the decline of rural communities throughout the United States (Stofferahn, 2006). With their steady and significant demand, institutions are well positioned to support and expand these essential "agriculture of the middle" markets.

It follows that "supporting local farmers" or "aiding the local economy" are commonly cited reasons for engaging in farm-to-institution programs (Izumi et al., 2010a,b; Ng et al., 2010; Vogt and Kaiser, 2008). While evidence suggests that institutional food procurement has a negligible impact on participating farmers' incomes, it helps them diversify their markets, increase off-season sales, and gain an outlet for surplus and less desirable foods, such as "ugly" produce or less popular cuts of meat (Conner et al., 2012; Izumi et al., 2010c). Institutional purchasing may also help small and midsized farmers retain autonomy in face of the growing competitive threats from large-scale agribusiness (Izumi et al., 2010c).

Demonstrating the impact of institutional food procurement on the larger regional economy is challenging and contested. Some studies calculate the economic sales or jobs created within a region associated with the current or potential institutional procurement market (Becot et al., 2017; Mulik, 2016). Other studies translate these regional economic impacts into a multiplier effect—that is, the number of times one U.S. dollar cycles through a regional economy. A higher number means more money stays in a regional economy. Recent estimates suggest that farm-to-school sales generate economic multipliers of 1.03–2.4 (with a median multiplier of approximately 1.5) (Becot et al., 2017; Christensen et al., 2017).

Other experts argue that economic impact models, and their associated economic multiplier calculations, do not accurately capture the factors unique to alternative food value chains, such as distribution channels, processing facilities, farm inputs, labor needs, or production practices (Lynch et al., 2015). While farm-to-institution procurement does appear to create slightly more jobs in the regional economy, evidence suggests that there are also other socioeconomic benefits coming from regionalized food economies, such as the formation of relationships that strengthen social capital and networks in areas of impact (Buckley et al., 2013; Lynch et al., 2015). Real Food Challenge is one of the national initiatives that has specific criteria for "local and community based" food procurement (with specific ownership and size limitations for producers), and it explicitly aims to support small and midsized independent or cooperative producers and to challenge consolidation in the food industry (RFC, 2016).

Environmental Considerations

The current U.S. food system exacerbates many ecological crises, including soil, water, and air degradation; freshwater, fossil fuel, fishery, and phosphate rock depletion; and biodiversity loss (Shannon et al., 2015). It also lacks resilience to foster food security in the face of climate change, population pressures, and resource depletion (Shannon et al., 2015). No simple solution exists to minimize environmental externalities and maximize social good in our food system. For example, while many institutions have prioritized "local" food to reduce "food miles"—the distance the food travels from where it is produced to where it is consumed (Vogt and Kaiser, 2008)—and the greenhouse gas emissions associated with such transportation, in most cases, reducing food miles is far less effective for climate change mitigation than other dietary shifts such as eating less meat and dairy (Hoolohan et al., 2013; Weber and Matthews, 2008). Nevertheless, food procurement policies can lessen environmental degradation by encouraging more seasonal and sustainable food choices, reducing wasted food, and incentivizing sourcing from producers and processors that employ ecologically sound practices.

Shifting agricultural practices and the types of foods we consume are central to fostering a more ecologically sound food system. Organic and pasture-based systems can foster biodiversity, promote soil health, and reduce fossil fuel, synthetic fertilizer, and pesticide use, but they do not necessarily improve ecological burdens in terms of land use, fertilizer runoff, and climate change (Clark and Tilman, 2017; Dumont et al., 2013; Garnett et al., 2017). Sourcing organic and pasture-based foods can be valuable, but these shifts could be paired with dramatic reductions in animal product intake to be more environmentally beneficial (Clark and Tilman, 2017; Hoolohan et al., 2013; Garnett et al., 2017). Several foodservice operations have begun changing their menus to encourage more plant-centric eating by reducing meat portion sizes (including through blended meat offerings) and providing ample and enticing plant-based options (HCWH, 2018; Meatless Monday, 2017; Webster, 2015). Purchasing fewer animal products can also save money (FOE, 2017), which some institutions may use to buy pasture-based, organic, or humanely raised alternatives. Health Care Without Harm's "less meat, better meat" approach encourages such practices and offers resources to help institutions replace meat on their menus (HCWH, 2017).

Additionally, institutional food policies may include strategies to reduce and compost wasted food. Thirty-one to 40 percent of the U.S. food supply is wasted, which has myriad ethical and environmental implications (Buzby et al., 2014; Hall et al., 2009). Waste-sensitive procurement strategies can improve environmental and economic outcomes while shifting broader social conventions toward reducing the impact of wasted food (Sonnino and McWilliam,

2011). All three major foodservice management companies have set wasted food reduction goals for the next decade (Aramark, 2017b; Compass Group USA, 2017; Sodexo, 2016a).

Health Considerations

The standard American dietary pattern, characterized by a high intake of sugary desserts and drinks, processed and red meat, refined grains, and high-fat dairy, is associated with chronic diseases including Type 2 diabetes, cardiovascular disease, and some types of cancer (Shannon et al., 2015). Evidence suggests it also contributes to poor mental health outcomes including dementia and depression (Jacka et al., 2014).

A growing body of research indicates that interventions in retail food stores, cafeterias, and restaurants aimed at increasing the consumption of healthy foods may prove more effective and influential than interventions targeting individuals (Larson and Story, 2009; Seymour et al., 2004). School cafeterias are also particularly promising sites to encourage children to adopt the lifelong habit, and associated health benefits, of eating more fruits and vegetables (Knai et al., 2006; Yoder et al., 2014). Institutional policies that emphasize the procurement and preparation of high-quality vegetables, fruits, legumes, whole grains, and moderate amounts of animal products (including seafood) create food environments that may encourage healthier food consumption. The Good Food Purchasing Policy's nutrition standards support such objectives by focusing on prioritizing the purchase of whole or minimally processed foods, rather than moderately or ultra-processed products, and encouraging the reduction of red and processed meats purchases (CFGFP, 2017).

There are also broader public health implications of industrialized food production. Significant concerns have been raised over food safety and the contribution of widespread subtherapeutic antibiotic use for growth promotion and disease prevention in industrial food animal production operations and aquaculture facilities to the growing antibiotic resistance crisis (Silbergeld et al., 2008). Industrial food animal feeding practices can also result in the presence of bacterial pathogens, prions, metals, mycotoxins, and dioxins in animal feed and animal-based food products, which poses public health risks (Sapkota et al., 2007). Experts are also concerned about the generation and spread of novel infectious diseases, such as Nipah virus, SARS, and influenza, which may originate or transfer to humans in industrial food animal production operations (Leibler et al., 2009).

Thus far, institutional procurement policies have concentrated on reducing antibiotic use in poultry production. School Food Focus, Health Care Without Harm, the Center for Good Food Purchasing, and the Urban School Food Alliance have all led campaigns to shift institutional poultry procurement to antibiotic-free varieties. As part of its Good Food Purchasing Policy enactment, the Los Angeles Unified School District became the first of the country's large school districts to serve exclusively antibiotic-free chicken in 2016 (LAUSD, 2016).

Labor and Equity Considerations

Food system workers and their families, communities of color, and low-income communities disproportionately bear the adverse health and social impacts associated with U.S. food production and processing. Food production and processing workers perform strenuous labor with dangerous equipment, often in extreme temperatures and other severe environmental conditions, which contributes to their high rates of occupational injuries and disease (Fitch et al., 2017; Myers, 2010). Food system workers, a majority of whom are immigrant and migratory workers, also experience elevated rates of food insecurity, stress, and depression (Hiott et al., 2008; Minkoff-Zern, 2014). A growing emphasis on reducing these health risks, along with improving the poor working conditions, low wages, substandard housing, and inadequate labor rights that U.S. food system workers face has fueled interest in domestic Fair Trade schemes (Brown and Getz, 2008). The first domestic Fair Trade label, Food Justice Certified, was launched in 2010 to ensure safe working conditions, fair and equitable contracts, clean and safe farmworker housing, freedom of association and collective bargaining, among many other rights (Agricultural Justice Project, 2012).

Residents living in proximity to industrial crop and food animal operations face many similar physical, mental, and social health risks as workers (Casey et al., 2015; Harrison, 2011). Environmental justice concerns abound because these farms are concentrated primarily in communities that lack the socioeconomic and political power to prevent, mitigate, or adapt to these environmental inequities (Harrison, 2011; Nicole, 2013).

The Real Food Challenge and Good Food Purchasing Policy both include international and domestic Fair Trade sourcing as a central priority in their procurement standards.[7] Given the poverty-level wages associated with food system jobs (FCWA and SRC, 2016), some institutions may also consider additional policy changes (e.g., minimum wage increases, paid sick leave, workers compensation) to support not just farmworkers and producers, but also workers serving in their own cafeterias. Union contracts play an essential role in improving worker safety, health, and well-being by providing many of these benefits (Navarro and Muntaner, 2004), though they currently represent <2 percent of private-sector employees in agriculture, foodservice, and related industries

---

7. International Fair Trade certifications (e.g., Fair Trade USA, Fairtrade America, Equal Exchange) focus primarily on providing fair prices for small-scale, independent, and ecologically sustainable producers with the goal of challenging the inequities that arise from trading in global capitalist markets (Jaffee et al., 2004).

(Nesheim et al., 2015). Moreover, providing quality, nutritious food in institutional settings, most notably in schools and prisons where consumers do not have much—if any—choice of alternative food sources, may also be considered a social justice issue itself. Institutions can strive toward socially just procurement policies that provide food that supports the well-being of both consumers and producers.

Animal Welfare

Agricultural industrialization has profoundly altered livestock production. Since the 1950s, specialization and intensification have led to the widespread transition from diversified farms to industrial food animal production operations (Pew Commission, 2008a). The vast majority of food animals produced in the United States are raised in confined operations, with nearly half produced in the largest-sized operations (occupying <5 percent of the land used for animal production) (Graham and Nachman, 2010). Designed to produce abundant amounts of meat, eggs, or milk rapidly and cheaply, most industrial food animal production operations raise animals with limited to no outdoor access or the ability to exhibit their natural behaviors (Pew Commission, 2008a). To control for animals' aggressive behavior when they are under extreme stress, painful bodily alteration—such as de-beaking chickens, tail removal of pigs and dairy cattle, and dehorning beef cattle—is conducted, often without pain relief (Mason and Finelli, 2006). Animal production on small-scale, organic, or pasture-based farms does not necessarily ensure well-being either; farms should be audited before presuming higher animal welfare standards (Pew Commission, 2008b).

Consumers are increasingly expressing their demand for the humane treatment of farm animals, such as by purchasing animal products produced with certified higher welfare standards (Napolitano et al., 2010). As federal regulation regarding the treatment of farm animals remains limited, many states and a few local jurisdictions have passed animal cruelty legislation to outlaw or regulate certain practices (Rumley, n.d.). Some retailers and restaurants have also been demanding minimum animal welfare standards from their suppliers, initiating influential changes throughout the industry (Mench, 2008). For example, the big three foodservice companies and several fast food chains have committed to buying 100 percent of their chicken from sources that meet Global Animal Partnership (GAP) Step 1 Standards (i.e., slower growing) by 2024 (GAP, 2016; Sodexo, 2016b; Aramark, 2016).

FOODSERVICE MANAGEMENT CONTRACTS AS A BARRIER TO INCREASING REGIONAL AND SUSTAINABLE FOOD PROCUREMENT

Despite the progress made by several organizations and institutions, several hurdles continue to deter foodservice administrators, farmers, and distributors from implementing or expanding values-based procurement practices throughout the agro-food value chain. Commonly cited barriers include the cost of farmers' liability insurance and good agriculture practice (GAP) certification, anticipated variability in produce prices, year-round availability limitations, a lack of food processing infrastructure, and the additional time required for producers and purchasers to participate (Low et al., 2015; Thompson et al., 2014; Vilme et al., 2015). One of the primary barriers that has not been addressed as extensively in other literature is the question of how contracts with foodservice management companies influence the ability of institutions to adopt or implement values-based procurement policies.

To answer this question, we conducted semistructured, open-ended interviews with fourteen key informants who work in various capacities in different foodservice sectors over the summer of 2015. These informants included foodservice staff from the K-12 schools, higher-education, healthcare, and corporate sectors and non-profit staff who work on foodservice policies in the K-12, higher-education, and healthcare sectors. We asked informants to describe their involvement with institutional foodservice procurement practices, perceptions of barriers to regional and sustainable procurement, and opinions about what has been or could be successful in efforts to support such policies. We recruited these informants—who remain anonymous in order to protect their privacy and employment—through snowball sampling. Lack of publically available data and confidentiality restrictions limited the number of informants who would speak with us. However, despite these transparency challenges, we still gained useful insights about how the rebate pricing system may be shaping and constraining procurement policies.

Foodservice Contracts and the Rebate Pricing System

Institutions may choose to contract with ("outsource to") foodservice management companies because these companies often present appealing contracts that address budgetary and management concerns. Employees at one non-profit that works primarily with school districts that we interviewed explained that these proposals have created a perennial problem of school districts having to "fight the threat of foodservice management companies coming in," because while management companies can meet a low budget, the institutions that they work with have reported lower food quality and less flexible procurement options after outsourcing foodservice to a management company.

When institutions outsource their foodservice to a management company, they may encounter several different types of contracts, each with their own influence on the institution's finances and procurement options. As the consumer varies, so does the payer—in correctional institutions and school districts, for example, at least part of the cost is covered by the government. Under some contracts, an institution may set the price that consumers are charged, leaving the management company responsible for all operation costs—and consequently retaining all income and bearing all risk of increased costs (Obadia, 2015). Under other contracts, the institution may pay a fee to a management company and also cover the company's food and labor costs, but no menu or price point is set, theoretically allowing on-site foodservice providers more flexibility in menu design and procurement (Obadia, 2015).

There are opportunities to select regional and sustainably produced food under all types of contracts, depending on how they are written. Still, institutions outsourcing their foodservice may face unique barriers to regional food procurement. Due to their size, foodservice management companies are able to negotiate low prices through the formation of group purchasing organizations, in which they use their collective buying power to obtain volume discounts from large food producers, integrators, and distributors (Obadia, 2015). While institutions may join group purchasing organizations that are independent of foodservice management companies, the pricing models are often similar to those obtained by management companies. In group purchasing organizations, institutions' food purchases are often controlled and restricted to certain vendors in order to maintain group purchasing power and discounted pricing. Most independent group purchasing organizations require their clients to buy at least 80 percent of their products through preapproved vendors ("buying on contract"), while those connected with foodservice management companies often require on-contract purchases as high as 100 percent (Obadia, 2015). Foodservice managers have reportedly received incentives or bonuses from management companies for buying an increased percentage of their total product from approved vendors (Obadia, 2015).

Foodservice management companies do not generate all of their revenue from management fees or revenue from individual institutions; they also profit from off-invoice volume discount allowances (VDAs)—or rebates—from food suppliers. Management companies usually charge relatively low management fees, presumably in order to secure contracts with institutions, but may form agreements with distributors or directly with manufacturers and producers for rebates on their institutions' purchases. In these cases, management companies will ask for a rebate on a certain percentage of the sales of a product, and suppliers will mark up the price by that amount so that the client—the institution—pays an inflated price, and the difference goes to the management company (Bruske, 2011). For example, a distributor's sale of milk to an institution would be marked up so that a negotiated percentage of the sale would go to the foodservice management company, which has leverage due to its scale and volume of demand. In some cases, a percentage of the rebate that the foodservice management company receives is returned to the institution, but how much of the company's profit is returned is often undisclosed.

Group purchasing organizations independent of foodservice management companies may also secure discounts or rebates, which may be fully or partially redistributed to institutions (e.g., group purchasing organizations may withhold a percentage of the rebate as a fee). For example, the Urban School Food

Alliance formed in 2012 to coordinate bulk item purchases and menu creation among the country's urban public school districts to keep costs competitive (USFA, 2017). The lack of transparency surrounding the amount or percentage value of rebates and discounts makes it difficult to investigate the percentage of savings—in the form of rebates or discounts—that foodservice management companies and group purchasing organizations return to institutions.

These rebates have been the topic of recent investigations and growing public concern. The District of Columbia recently investigated Chartwells—owned by Compass Group USA—for claims that it fraudulently kept rebates from food manufacturers and distributors that the company was supposed to return to public school districts (Office of the Attorney General, D.C., 2015). In June 2015, Chartwells paid a $19.4 million settlement to a Washington, DC school district (Office of the Attorney General, D.C., 2015). New York State also received a $20 million settlement from Sodexo in 2010 (Bruske, 2011), and an $18 million dollar one from Compass Group in 2012 (Office of the Attorney General, New York, 2012) for similar claims that the foodservice companies were overcharging school districts and public universities.

As John F. Carroll—the Assistant Attorney General of New York during the 2010 settlement—explained in a speech, prior to 2007, it was not necessarily unlawful for foodservice management companies to retain rebates and discounts. Rules and regulations from the United States Department of Agriculture (USDA) did not require that they be credited back to public school districts (Bruske, 2011). Rather, institutions could negotiate, through the terms and conditions of their contracts with foodservice management companies, whether the discounts received by companies would be credited to the school district or retained by the company. In 2007, however, the USDA promulgated a final rule that included provisions whereby all rebates and discounts were to be credited against the cost of the food, requiring foodservice management companies to return all rebates received back to school districts (Bruske, 2011). Importantly, this rule only applies to public school districts, which receive federal funding for food and nutrition programs.

Carroll explained that rebates and discounts were not a significant economic factor for foodservice management companies prior to 2000. But since then, rebates have been a large economic factor in the foodservice business, and while some products' rebate payments are <5 percent, other products earn rebates of >50 percent of the purchase cost for individual items. Carroll estimated that the largest foodservice companies earn hundreds of millions of dollars in rebates, a fact not readily apparent from companies' or institutions' publicly available financial statements (Bruske, 2011). Carroll's investigation "found that most school participants had very limited understanding and knowledge about what the rebates were." After 2007, the USDA even provided an implementation timeline to help schools rebid their foodservice contracts in phases to ensure that all discounts and rebates would eventually be returned, but a 2009 USDA memo to all regional and state directors of Child Nutrition Programs states that, "several instances have been brought to our attention in which the rule's implementation timeline seems to have been misapplied, or in some cases, ignored" (USDA, 2009).

Carroll also described foodservice management companies' restrictions on the number of sources local site managers—employees working in schools—can use to purchase foods, as companies "tend to restrict purchases to the larger, industrial food producing companies," instead of smaller, regional food producers, which are less likely to have the economic power to enter into rebate agreements with foodservice companies (Bruske, 2011). He explained, "vendors that do not pay rebates...rarely appear on the list of approved vendors of foodservice companies," and concluded that rebates have an "inherent potential to create conflict of interest":

> *In at least one instance, I observed that a local produce wholesaler had to increase the prices it charged to the local school district for fresh produce – including locally grown produce — so that it could pay the foodservice company a rebate... I also observed that the local site manager found it difficult to meet 'buy local' requirements, which many foodservice contracts contain, and still comply with the foodservice company requirement that vendors pay rebates.*

(Bruske, 2011)

Following this description, the implicit expectation of rebate payments to foodservice management companies may encourage independent producers to increase their prices in order to enter the institutional foodservice market, or—if regional producers are unwilling or unable to raise their prices and offer rebates—may prohibit site managers from being able to purchase from regional farms.

In our interview with a former employee of a foodservice management company responsible for procuring food for a private college, the employee argued that there was "absolutely no transparency" around the percentage of food payments that were paid back to the foodservice company as a rebate. The former employee explained that staff only heard about these "kickbacks" when the chefs complained about them, but that it was hard to learn more because sharing invoices or any information about the preferred vendors list could cost staff members their jobs, and "their own paychecks were part of the system." The former employee argued that despite increased transparency on regional food purchases, all other institutional food purchases remain vague; ultimately, students do not, and cannot, know how much money is going toward their food and how much is relegated back to foodservice companies.

A director of a private university's institutional dining program expressed similar distrust of the current pricing system, despite positive advances in transparency. Asked about the rebates paid back to foodservice companies, he said, "Welcome to the gray world...they keep the keys to that safehouse so guarded that no one will ever get in." He explained that large group purchasing organizations are "the only ones that will ever truly know the value" of kickbacks received by foodservice companies. He described receiving "blind invoices" that do not disclose the actual price of food products retained by producers but, rather, the price paid, which presumably includes a built-in rebate for the foodservice company. According to this director, negotiated contracts should allow for increased purchasing from smaller, regional producers where "there is no kick back," and "you can reduce that rebate dollar." He argued that consumers should understand the implications of rebates in this context and how they can pose a barrier to buying from regional producers. He emphasized that students—and other consumers in different institutional settings—should understand where their money is going.

Employees at a non-profit focused on school districts reiterated that when schools have committed to buying more food from regional or sustainable producers, those districts that have contracts with foodservice management companies tend to be much less able to fulfill their procurement goals than schools with self-operated foodservices. We spoke with one foodservice director of a self-operated K-12 school district that has been operating a decade-long unlimited fruit and vegetable program (wherein students can take as much produce—some of which is sourced regionally—as they wish). Without preferred vendor lists or other general restrictions on sourcing, the district has found that its procurement challenges (which include finding regional farmers who are GAP-certified and can meet the district's high-volume needs) have become manageable and are "shrinking with time." The program's success—in terms of receptivity, participation, and growth—supports the idea that institutions with self-operated foodservice may have more flexibility in modifying their food procurement than those contracting with foodservice management companies.

As suggested by these informants, it is nearly impossible to legally access the financial records that show the percentage of food prices that is paid back to foodservice management companies and later paid as salaries and bonuses for site staff. Despite the lack of transparency, governmental programs, such as the USDA's Know Your Farmer, Know Your Food, and the aforementioned procurement advocacy organizations, are enabling more institutions to source directly from small and midsized producers in their regions and those who meet other sustainability, animal welfare, and social justice criteria. While these initiatives do not necessarily or automatically eliminate kickbacks or reform the rebate pricing system, they do foster transparency and may lead to broader efforts to create a more equitable pricing system.

REFLECTIONS ON UNCERTAINTIES IN THE LARGER POLITICAL CONTEXT

Policy at multiple scales (international, national, state, and local) can both help and hinder further progress on values-based institutional procurement.

The rapid growth in institutional procurement policy reform has come amid an era of growing partisanship and food lobbying. While the institutional values-based procurement movement has shown promising growth, several political uncertainties around international free trade policies, preemption, federal support for local and regional food systems, and regulatory authority could undermine the progress that has been made and halt future growth.

The protracted and fluctuating free trade agreement negotiations over the past five years could potentially result in more restrictions on values-based procurement policies in public institutions, because free-trade advocates feel that state or local sourcing requirements are a form of protectionism (FOE, 2016; Treat, 2016). While the Trump administration campaigned on an anti-free trade platform, it is not a guarantee that renegotiation of existing agreements such as the North American Free Trade Agreement (NAFTA) will support values-based procurement, nor that the restrictive provisions in other recently attempted agreements, including the Trans-Pacific Partnership (TPP), Trans-Atlantic Trade and Investment Partnership (TTIP), and Trade in Services Agreement (TiSA), will be off the table in the future (Hansen Kuhn, 2017). By impeding local citizens and institutions from setting priorities and values in their procurement schemes, trade agreements could represent a serious challenge to the rise of institutional values-based procurement.

At the federal-level, industry lobbyists representing large and powerful agribusiness corporations oppose regulatory uncertainty and restrictive or exclusionary policies at the state and local levels (Pertschuk et al., 2013). Public health scholars have recently highlighted how dramatic increases in state preemption bills and amendments threaten many local regulations, including those related to food and nutrition (Pomeranz and Pertschuk, 2017). While, to our knowledge, institutional food procurement policies have not been targeted by preemption laws thus far, Preemption Watch includes farm-to-institution as a key issue in its database filters (Grassroots Change, 2017). As values-based procurement policies continue to gain interest, they could become a target of lobbying efforts and legislation.

Changes in federal legislative and administrative priorities could also affect institutional values-based procurement. Several programs that support local and regional food systems or environmentally friendly farming practices face threats to funding, and possibly legislative authority (NSAC, 2017a,b). Reduced support for these programs might affect the price differential between conventional and independent, small and midsized, or sustainable producers—or reduce the availability of food from such alternative producers—which could discourage sustainable food procurement. Initiatives such as the Foodservice Guidelines Collaborative, a coalition of governmental (including the Centers for Disease Control and Prevention, Department of Defense, and USDA) and nongovernmental organizations aimed at increasing the availability of healthier foods and beverages and improving the environmental sustainability of federal foodservice facilities, may be vulnerable to changes in administration priorities.

In contrast, proposed legislation such as the 2017 Local Food and Regional Market Supply Act might enable procurement reform by improving infrastructure to connect producers and institutions (NSAC, 2017c).

Federal legislation may also shape whether institutions and other food system actors can increase transparency in the food procurement system. Transparency is essential for continued progress in institutional values-based procurement, because insufficient data and a lack of access to information is a commonly cited barrier to diversifying and changing the food system. Some informants even suggested that institutional food procurement is "purposefully opaque" in order to preserve the market power of the dominant corporations in various stages of the value chain. The proposed Regulatory Accountability Act of 2017 in Congress could effectively prevent future regulatory actions such as the one that the USDA took to prohibit federal funds from being used to cover rebates from vendors to foodservice management companies (Darrow et al., 2017). Whether or not the lack of transparency is intentional, political support to increase public transparency and traceability in food procurement is critical.

Lastly, state and municipal legislation has increasingly enabled and supported values-based procurement reform in many parts of the country, often through the justification that it will boost local and regional economies. At least 37 states have passed "Geographic Preference" legislation to allow local and state agencies to prioritize purchasing in-state agricultural products (Condra et al., 2015). Between 2012 and 2014 alone, eighteen states passed 28 bills supporting farm-to-school programs (Essex et al., 2015). While states have mostly concentrated on regional procurement values, several cities and counties—often in partnership with local food policy councils—have begun passing more comprehensive food procurement reforms (FOE, 2017), including, but not limited to, Good Food Purchasing Policy adoptees (CFGFP, 2017). This suggests that states and municipalities are a critical ally for the institutional values-based foodservice movement.

CONCLUSION

A large-scale shift among hospitals, schools, universities, and other institutions toward values-based food procurement could change the U.S. food system substantially. Institutional food policies not only have a significant potential to directly affect the economic and physical infrastructure of the food system; accurately and widely marketed efforts may also generate broader awareness and discussion about food system reform. The vast purchasing power and educational opportunity provided by institutions for assisting in this transition is only beginning to be tapped.

The prevailing model of institutional foodservice—marked by contracts with group purchasing organizations and foodservice management companies, rigid lists of preferred vendors, and little transparency about the origins of food products or the conditions under which they were produced—is deeply entrenched and difficult to change, posing challenges to values-based procurement, as our interviews unearthed. However, this model is relatively young and is being contested. Recent litigation has shed light on the continued problem of kickbacks—or rebates that are kept by foodservice management companies—and will likely encourage consumers, non-profit organizations, and institutions themselves to keep a close eye on contracts, and companies to re-visit their pricing structure. Attention to the federal, state, and local policy landscape is also merited to ensure procurement reform is not impeded, but enabled.

Across institutions and sectors, advocates are challenging the prevailing food procurement model to implement food procurement policies that benefit the environment, regional economies, human health, animal welfare, and social justice. As these initiatives expand, they are demonstrating that regional and sustainable food systems have the capacity to supply a large proportion of consumer demand while fostering natural resource sustainability, economic development, and social responsibility.

REFERENCES

Agricultural Justice Project, 2012. Social stewardship standards in organic and sustainable agriculture: standards document. Available from: https://www.agriculturaljusticeproject.org/media/uploads/2016/08/02/AJP_Standards_Document_9412.pdf. Accessed December 20, 2017.

Aramark, 2016. Aramark becomes first major food service company to ask suppliers to commit to humane conditions for broiler chickens by 2024 (press release). Available from: http://www.aramark.com/about-us/news/aramark-general/broiler-chicken-commitment. Accessed December 20, 2017.

Aramark, 2017a. Aramark reports fourth quarter and full year 2017 earnings. Available from: http://phx.corporate-ir.net/phoenix.zhtml?c=130030&p=irol-IRHome.

Aramark, 2017b. Aramark demonstrates progress minimizing environmental footprint (press release). Available from: http://www.aramark.com/about-us/news/aramark-general/green-thread-food-waste-progress?i=Environmental+Sustainability&s=Waste+Minimization. Accessed December 20, 2017.

Aramark Corporation, 2004. Form 10-K. Available from: http://getfilings.com/o0001193125-04-210901.html.

Association for the Advancement of Sustainability in Higher Education (AASHE) 2017. Sustainability Tracking, Assessment & Rating System (STARS) OP-7: food and beverage purchasing and OP-8 sustainable dining (Accessed 22 December 2017).

Barnett, C., Clarke, N., Cloke, P., Malpass, A., 2005. The political ethics of consumerism. Consum. Policy Rev. 15 (2), 45–51.

Becot, F., Kolodinsky, J.M., Roche, E., Zipparo, A.E., Berlin, L., Buckwalter, E., Mclaughlin, J., 2017. Do Farm-to-School programs create local economic impacts? Choices 32 (1), 1–8.

Bon Appétit Management Company (BAMCO) n.d. Fish to fork. Available from: http://www.bamco.com/timeline/fish-fork/ (Accessed 8 January 2018).

Brown, S., Getz, C., 2008. Towards domestic fair trade? Farm labor, food localism, and the 'family scale' farm. GeoJournal 73 (1), 11–22.

Bruske, E., 2011. John F. Carroll, Jr. speaks at SNA's legislative action conference (video file). Available from: https://vimeo.com/22455035. Accessed December 20, 2017.

Buckley, J., Conner, D.S., Matts, C., Hamm, M.W., 2013. Social relationships and farm-to-institution initiatives: complexity and scale in local food systems. J. Hunger Environ. Nutr. 8 (4), 397–412.

Bulger, M., 2015. Six U.S. correctional facilities with 'Farm to Prison' local food sourcing programs. Available from: http://seedstock.com/2015/01/04/six-u-s-correctional-facilities-with-farm-to-prison-local-food-sourcing-programs/. Accessed December 20, 2017.

Buzby, J.C., Wells, H.F., Hyman, J., 2014. The Estimated Amount, Value, and Calories of Postharvest Food Losses at the Retail and Consumer Levels in the United States, Economic Information Bulletin, ii-33. USDA, Washington, DC.

Casey, J.A., Kim, B.F., Larsen, J., Price, L.B., Nachman, K.E., 2015. Industrial food animal production and community health. Curr. Environ. Health Rep. 2 (3), 259–271.

Center for Good Food Purchasing (CFGFP), 2017. Good food purchasing standards 2.0—September 2017 update. Available from: https://goodfoodpurchasing.org/good-food-purchasing-standards-update-september-2017/#structure-scoring-updates.

Christensen, L.O., Jablonski, B.B.R., Stephens, L., Joshi, A., 2017. Economic Impacts of Farm to School: Case Studies and Assessment Tools. National Farm to School Network.

Clancy, K., Ruhf, K., 2010. Is local enough? Some arguments for regional food systems. Choices 25 (1), 123–125.

Clark, M., Tilman, D., 2017. Comparative analysis of environmental impacts of agricultural production systems, agricultural input efficiency, and food choice. Environ. Res. Lett. 12 (6), 064016.

Compass Group, 2017a. Full year results: press release. Available from: https://www.compass-group.com/en/investors/results-and-presentations.category1.year2017.html.

Compass Group, 2017b. The global leader in food services (factsheet). Available from: https://www.compass-group.com/content/dam/compass-group/corporate/Who-we-are/Compass%20Group_Factsheet_211117_FINAL.pdf.downloadasset.pdf.

Compass Group PLC, 2004. Annual report 2004. Available from: https://www.unglobalcompact.org/system/attachments/7515/original/Compass_Group_Annual_Report_2004.pdf?1282019197.

Compass Group USA, 2017. Compass group USA announces landmark commitment to reduce food waste by 25% by 2020 (press release). Available from: http://www.compass-usa.com/compass-group-usa-announces-landmark-commitment-reduce-food-waste-25-2020. Accessed December 20, 2017.

Condra, A., Broad-Leib, E., Schwartz Becker, E., et al., 2015. Increasing Local Food Procurement by Massachusetts State Agencies. Harvard Food Law and Policy Clinic, Boston, MA.

Cone Communications, 2014. Three-quarters of Americans say sustainability is a priority when making food purchasing decisions, according to new cone communications research. Available from: http://www.conecomm.com/2014-food-issues. Accessed December 20, 2017.

Conner, D., King, B., Kolodinsky, J., Roche, E., Koliba, C., Trubek, A., 2012. You can know your school and feed it too: Vermont farmers' motivations and distribution practices in direct sales to school food services. Agric. Hum. Values 29, 321–332.

Darrow, J.J., Brown, E.C., Kesselheim, A.S., 2017. The Regulatory Accountability Act of 2017—implications for FDA regulation and public health. N. Engl. J. Med. https://doi.org/10.1056/NEJMp1711643.

Dumont, B., Fortun-Lamothe, L., Jouven, M., Thomas, M., Tichit, M., 2013. Prospects from agroecology and industrial ecology for animal production in the 21st century. Animal 7 (6), 1028–1043.

Essex, A., Shinkle, D., Bridges, M., 2015. Harvesting Healthier Options: State Legislative Trends in Local Foods 2012–2014. National Conference of State Legislatures, Denver, CO.

Federal Trade Commission, 2015. FTC challenges proposed merger of Sysco and US Foods (press release). Available from: https://www.ftc.gov/news-events/press-releases/2015/02/ftc-challenges-proposed-merger-sysco-us-foods.

Fitch, C., Santo, R., 2016. Instituting Change: An Overview of Institutional Food Procurement and Recommendations for Improvement. Johns Hopkins Center for a Livable Future, Baltimore, MD.

Fitch, C., Hricko, C., Martin, R., 2017. Public Health, Immigration Reform and Food System Change. Johns Hopkins Center for a Livable Future, Baltimore, MD.

Food Chain Workers Alliance (FCWA) and Solidarity Research Cooperative (SRC), 2016. No Piece of the Pie: U.S. Food Workers in 2016. FCWA and SRC, Los Angeles, CA.

Friends of the Earth (FOE), 2016. Dangerous Liaisons: The New Trade Trio. Friends of the Earth International.

Friends of the Earth (FOE), 2017. Meat of the Matter: A Municipal Guide to Climate-Friendly Food Purchasing. Friends of the Earth U.S. and the Responsible Purchasing Network.

Garnett, T., Godde, C., Muller, A., Röös, E., Smith, P., de Boer, I....van Zanten, H., 2017. Grazed and Confused? Ruminating on Cattle, Grazing Systems, Methane, Nitrous Oxide, the Soil Carbon Sequestration Question—And What it all Means for Greenhouse Gas Emissions. Food Climate Research Network, Oxford.

Global Animal Partnership (GAP), 2016. Compass Group USA becomes first food service company to commit to 100% healthier, slower growing chicken by 2024 (press release). Available from: https://globalanimalpartnership.org/about/news/post/compass-group-usa-becomes-first-food-service-company-to-commit-to-100-healthier-slower-growing-chicken-by-2024/. Accessed December 20, 2017.

Graham, J., Nachman, K., 2010. Managing waste from confined animal feeding operations in the United States: the need for sanitary reform. J. Water Health 8 (4), 646–670.

Grassroots Change, 2017. Issues: food & nutrition. Available from: https://grassrootschange.net/category/farm-to-institution/.

Hall, K.D., Guo, J., Dore, M., Chow, C.C., 2009. The progressive increase of food waste in America and its environmental impact. PLoS One 4, e7940.

Hansen Kuhn, K., 2017. Trump's new trade policy: risks for North American food and farms. J. Food Law Policy 13 (1), 146–153.

Hanson, J.H., 2015. Critical factors for sustainable food procurement in zoological collections. Zoo Biol. 34 (5), 483–491.

Harrison, J.L., 2011. Pesticide Drift and the Pursuit of Environmental Justice. MIT Press, Cambridge, MA.

Health Care Without Harm (HCWH), 2017. Redefining Protein: Adjusting Diets to Protect Public Health and Conserve Resources. HCWH, Reston, VA.

Health Care Without Harm (HCWH), 2018. Menu of change: Healthy Food in Health Care—2017 program report with highlights and survey results (Accessed 8 December 2017).

Hiott, A.E., Grzywacz, J.G., Davis, S.W., Quandt, S.A., Arcury, T.A., 2008. Migrant farmworker stress: mental health implications. J. Rural. Health 24 (1), 32–39.

Hoolohan, C., Berners-Lee, M., McKinstry-West, J., Hewitt, C.N., 2013. Mitigating the greenhouse gas emissions embodied in food through realistic consumer choices. Energy Policy 63, 1065–1074.

Howard, P.H., 2016. Concentration and Power in the Food System: Who Controls What We Eat? vol. 3. Bloomsbury Publishing, London, UK.

Hullinger, A.M., Tanaka, K., 2015. Agriculture of the middle participation in state branding campaigns: the case of Kentucky. J. Agric. Food Syst. Community Dev. 6 (1), 107–120.

Ikerd, J.E., 2008. Crisis & Opportunity: Sustainability in American Agriculture. University of Nebraska Press, Lincoln, NE.

IPES-Food, 2017. Unravelling the Food–Health Nexus: Addressing Practices, Political Economy, and Power Relations to Build Healthier Food Systems. The Global Alliance for the Future of Food and IPES-Food.

Izumi, B.T., Alaimo, K., Hamm, M.W., 2010a. Farm-to-school programs: perspectives of school food service professionals. J. Nutr. Educ. Behav. 42 (2), 83–91.

Izumi, B.T., Wright, D.W., Hamm, M.W., 2010b. Farm to school programs: exploring the role of regionally-based food distributors in alternative agrifood networks. Agric. Hum. Values 27 (3), 335–350.

Izumi, B.T., Wright, D.W., Hamm, M.W., 2010c. Market diversification and social benefits: motivations of farmers participating in farm to school programs. J. Rural. Stud. 26 (4), 374–382.

Jacka, F.N., Sacks, G., Berk, M., Allender, S., 2014. Food policies for physical and mental health. BMC Psychiatry 14 (1), 132.

Jaffee, D., Kloppenburg, J.R., Monroy, M.B., 2004. Bringing the moral charge home: fair trade within the north and within the south. Rural. Sociol. 69 (2), 169–196.

Kirschenmann, F., Stevenson, S., Buttel, F., Lyson, T., Duffy, M., 2008. Why worry about the agriculture of the middle? In: Lyson, T.A., Stevenson, G.W., Welsh, R. (Eds.), Food and the Mid-Level Farm: Renewing Agriculture of the Middle. MIT Press, Cambridge, MA, pp. 3–22.

Knai, C., Pomerleau, J., Lock, K., McKee, M., 2006. Getting children to eat more fruit and vegetables: a systematic review. Prev. Med. 42 (2), 85–95.

Larson, N., Story, M., 2009. A review of environmental influences on food choices. Ann. Behav. Med. 38 (1), 56–73.

Leadership Circle, 2017. Members. Available from: http://farmforward.herokuapp.com/leadershipcircle/. Accessed December 20, 2017.

Leibler, J.H., Otte, J., Roland-Holst, D., Pfeiffer, D.U., Magalhaes, R.S., Rushton, J., Graham, J.P., Silbergeld, E.K., 2009. Industrial food animal production and global health risks: exploring the ecosystems and economics of avian influenza. EcoHealth 6 (1), 58–70.

Los Angeles Unified School District (LAUSD), 2016. School board approves chicken free of antibiotics and hormones L. A. unified first large district to switch to healthier option. Available from: https://home.lausd.net/apps/news/article/553037.

Low, S.A., Adalja, A., Beaulieu, E., Key, N., Martinez, S., Melton, A., Perez, A., Ralston, K., Stewart, H., Suttles, S., Vogel, S., Jablonski, B.B.R., 2015. Trends in U.S. Local and Regional Food Systems, AP-068. U.S. Department of Agriculture, Economic Research Service.

Lynch, J., Meter, K., Robles-Schrader, G., Goldenberg, M.P., Bassler, E., Chusid, S., Jansen Austin, C., 2015. Exploring Economic and Health Impacts of Local Food Procurement. Illinois Public Health Institute, Chicago, IL.

Mason, J., Finelli, M., 2006. Brave new farm? In: Singer, P. (Ed.), In Defense of Animals: The Second Wave. Blackwell Publishing Inc., Oxford, England.

Meatless Monday, 2017. Meatless Monday: foodservice implementation guide. Available from: http://www.meatlessmonday.com/images/photos/2017/07/meatless-monday-guide-foodservice-generic.pdf.

Mench, J.A., 2008. Farm animal welfare in the USA: farming practices, research, education, regulation, and assurance programs. Appl. Anim. Behav. Sci. 113 (4), 298–312.

Minkoff-Zern, L.A., 2014. Hunger amidst plenty: farmworker food insecurity and coping strategies in California. Local Environ. 19 (2), 204–219.

Mukherji, N & Weinronk, H 2017, Pers. comm 21 December 2017.

Mulik, K., 2016. Growing Economies: Connecting Local Farmers and Large-Scale Food Buyers to Create Jobs and Revitalize America's Heartland (Policy Brief). Union of Concerned Scientists, Washington, DC.

Myers, M.L., 2010. Review of occupational hazards associated with aquaculture. J. Agromed. 15 (4), 412–426.

Napolitano, F., Girolami, A., Braghieri, A., 2010. Consumer liking and willingness to pay for high welfare animal-based products. Trends Food Sci. Technol. 21 (11), 537–543.

National Sustainable Agriculture Coalition (NSAC), 2017a. House Budget Committee's spending proposal could derail Farm Bill (blog post). Available from: http://sustainableagriculture.net/blog/house-budget-hits-farm-bill-hard/.

National Sustainable Agriculture Coalition (NSAC), 2017b. Senate appropriators prioritize sustainable agriculture and rural development (blog post). Available from: http://sustainableagriculture.net/blog/senate-appropriations-markup-2017/.

National Sustainable Agriculture Coalition (NSAC), 2017c. The Local Farms Act. Available from: http://sustainableagriculture.net/our-work/campaigns/fbcampaign/localregional-food/local-farms-act/.

Navarro, V., Muntaner, C. (Eds.), 2004. Political and Economic Determinants of Population Health and Well-Being: Controversies and Developments. Baywood Publishing Company Inc., Amityville, NY.

Nesheim, M.C., Oria, M., Tsai Yih, P. (Eds.), 2015. A Framework for Assessing Effects of the Food System. Institute of Medicine and National Research Council, National Academies Press, Washington, DC.

Ng, S.L., Bednar, C.M., Longley, C., 2010. Challenges, benefits and strategies of implementing a farm-to-cafeteria program in college and university foodservice operations. J. Foodserv. Manag. Educ. 4 (1), 22–27.

Nicole, W., 2013. CAFOs and environmental justice: the case of North Carolina. Environ. Health Perspect. 121 (6), a182.

NRDC, 2015. Report: U.S. pro sports shifting to more sustainable game day food. Available from: https://www.nrdc.org/media/2015/150630-0. Accessed December 20, 2017.

Obadia, J., 2015. Food Service Management Companies in New England: Phase 1 Research Findings—Barriers and Opportunities for Local Food Procurement. Farm to Institution New England.

Office of the Attorney General, D.C., 2015. District reaches $19.4 million settlement with food-service contractors for DC public schools. Available from: http://oag.dc.gov/release/district-reaches-194-million-settlement-food-service-contractors-dc-public-schools. Accessed December 20, 2017.

Office of the Attorney General, New York, 2012. A.G. Schneiderman announces $18 Million settlement with compass group USA for overcharging NYS school lunch programs (press release). Available from: https://ag.ny.gov/press-release/ag-schneiderman-announces-18-million-settlement-compass-group-usa-overcharging-nys.

Pertschuk, M., Pomeranz, J.L., Aoki, J.R., Larkin, M.A., Paloma, M., 2013. Assessing the impact of federal and state preemption in public health: a framework for decision makers. J. Public Health Manag. Pract. 19 (3), 213–219.

Pew Commission on Industrial Animal Farm Production, 2008a. Putting Meat on the Table: Industrial Farm Animal Production in America. Johns Hopkins Bloomberg School of Public Health, Baltimore, MD. Available from: http://www.pcifapia.org/reports/.

Pew Commission on Industrial Animal Farm Production, 2008b. Animal well-being. Available from: http://www.pcifapia.org/_images/212-7_PCIFAP_AmlWlBng_FINAL_REVISED_7-14-08.pdf.

Pomeranz, J.L., Pertschuk, M., 2017. State preemption: a significant and quiet threat to public health in the United States. Am. J. Public Health 107 (6), 900–902.

Porter, D., 2006. Self-op vs. contract: what's right for your campus? Strategic Thinking. Available from: http://www.porterkhouwconsulting.com/articles/documents/SelfOpVsContract_000.pdf.

Real Food Challenge (RFC), 2016. The real food guide version 2.0. Available from: http://realfoodchallenge.org/sites/default/files/RealFoodGuide2.0.pdf.

Rumley, ER n.d. States' animal cruelty statutes (webpage). National Agricultural Law Center. Available from: http://nationalaglawcenter.org/state-compilations/animal-cruelty/ (Accessed 21 December 2017).

Sapkota, A.R., Lefferts, L.Y., McKenzie, S., Walker, P., 2007. What do we feed to food-production animals? A review of animal feed ingredients and their potential impacts on human health. Environ. Health Perspect. 115 (5), 663–670.

School Food Focus, 2017. We're reinventing school meals from coast to coast (webpage). Available from: http://www.schoolfoodfocus.org/. (Accessed December 20, 2017).

Seymour, J.D., Yaroch, A.L., Serdula, M., Blanck, H.M., Khan, L.K., 2004. Impact of nutrition environmental interventions on point-of-purchase behavior in adults: a review. Prev. Med. 39, 108–136.

Shannon, K.L., Kim, B.F., McKenzie, S.E., Lawrence, R.S., 2015. Food system policy public health, and human rights in the United States. Annu. Rev. Public Health 36, 151–173.

Silbergeld, E.K., Graham, J., Price, L.B., 2008. Industrial food animal production, antimicrobial resistance, and human health. Annu. Rev. Public Health 29, 151–169.

Smart Asset n.d. Inflation calculator. Available from: https://smartasset.com/investing/inflation-calculator#CrKrY6ghg4 (Accessed 8 January 2018).

Sodexho Alliance, 2005. Making every day a better day: reference document 2004–2005. Available from: http://www.sodexo.com/files/live/sites/sdxcom-global/files/020_Global_Content_Master/Building_Blocks/GLOBAL/Multimedia/PDF/Finance/Reference_Document/Sodexo-FY2005-Reference-Document.pdf.

Sodexo, 2016a. Sodexo commits to zero food waste to landfills (press release). Available from: http://www.sodexousa.com/home/media/news-releases/newsListArea/news-releases/zero-food-waste.html. Accessed December 20, 2017.

Sodexo, 2016b. Sodexo reinforces an already robust commitment to animal welfare by working with U.S. suppliers to improve conditions of broiler chickens (press release). Available from: https://www.prnewswire.com/news-releases/sodexo-reinforces-an-already-robust-commitment-to-animal-welfare-by-working-with-us-suppliers-to-improve-conditions-of-broiler-chickens-30038199-3.html. Accessed December 20, 2017.

Sodexo, 2017. Fiscal 2017 registration document. Available from: http://www.sodexo.com/home/finance/presentations-and-publications/financial-results.html.

Sonnino, R., McWilliam, S., 2011. Food waste, catering practices and public procurement: a case study of hospital food systems in Wales. Food Policy 36 (6), 823–829.

Starr, A., 2010. Local food: a social movement? Cult. Stud. Crit. Methodol. 10 (6), 479–490.

Stofferahn, C.W., Industrialized farming and its relationship to community well-being: an update of a 2000 report by Linda Lobao. Department of Sociology, University of North Dakota. Available from: http://citeseerx.ist.psu.edu/viewdoc/download?doi=10.1.1.578.6208&rep=rep1&type=pdf.

Sysco, 2017. 2017 annual report. Available from: http://investors.sysco.com/annual-reports-and-sec-filings/annual-reports.

Thompson, O.M., Twomey, M.P., Hemphill, M.A., Keene, K., Seibert, N., Harrison, D.J., Stewart, K.B., 2014. Farm to school program participation: an emerging market for small or limited-resource farmers? J. of Hunger & Environmental Nutrition 9 (1), 33–47.

Treat, S.A., 2016. Local governments could be required to abandon buy-local requirements, (blog post). Available from: https://www.iatp.org/blog/201606/local-governments-could-be-required-to-abandon-buy-local-requirements.

U.S. Department of Agriculture (USDA), Food and Nutrition Service, 2009. Memo code SP 07-2009. Available from: https://childnutrition.ncpublicschools.gov/regulations-policies/usda-policy-memos/2009/sp-07-2009.pdf.

U.S. Department of Agriculture (USDA), 2017. 2015 Farm to School census. Available from: https://farmtoschoolcensus.fns.usda.gov/.

U.S. Department of Health and Human Services (HHS), Food Service Guidelines Federal Workgroup, 2017. Food Service Guidelines for Federal Facilities. HHS, Washington, DC.

Urban School Food Alliance (USFA), 2017. Our mission. Available from: https://www.urbanschoolfoodalliance.org/#section-working-together.

US Foods, Inc., 2017. 2016 annual report. Available from: https://ir.usfoods.com/investors/financial-information/annual-reports/default.aspx.

Vilme, H., Lopez, I.A., Walters, L., Suther, S., Perry Brown, C., Dutton, M., Barber, J., 2015. Perspectives of stakeholders on implementing a farm-to-university program at an HBCU. Am. J. Health Behav. 39 (4), 529–539.

Vogt, R.A., Kaiser, L.L., 2008. Still a time to act: a review of institutional marketing of regionally-grown food. Agric. Hum. Values 25 (2), 241–255.

Weber, C.L., Matthews, H.S., 2008. Food-miles and the relative climate impacts of food choices in the United States. Environ. Sci. Technol. 42 (10), 3508–3513.

Webster, M., 2015. Menus of change: changing consumer behaviors and attitudes (powerpoint slides). Available from: http://www.menusofchange.org/images/uploads/pdf/MenusofChange15_GS3b_MWebster.pdf.

Yoder, A.B.B., Liebhart, J.L., McCarty, D.J., Meinen, A., Schoeller, D., Vargas, C., LaRowe, T., 2014. Farm to elementary school programming increases access to fruits and vegetables and increases their consumption among those with low intake. J. Nutr. Educ. Behav. 46 (5), 341–349.

Raychel Santo is a senior research program coordinator at the Johns Hopkins Center for a Livable Future, where she works on research projects related to local and regional food governance, urban agriculture, institutional food procurement, and the relationship between diet and climate change. Raychel earned her Master's degree in Food, Space & Society from Cardiff University School of Geography & Planning and her BA in Public Health and Environmental Change & Sustainability from Johns Hopkins University.

Claire Fitch is the director of Outreach for Farm Forward, where she manages the Leadership Circle program to align institutions' values around animal welfare with their purchasing decisions. She was previously the program manager of the Food System Policy Program at the Johns Hopkins Center for a Livable Future. Claire holds a Master's of Science in Public Health, with a focus in Human Nutrition, from the Johns Hopkins Bloomberg School of Public Health. She was a 2014 U.S. Borlaug Fellow in Global Food Security.

Chapter 5

The Good Food Purchasing Program: A Policy Tool for Promoting Supply Chain Transparency and Food System Change

Lindsey Day Farnsworth[*], Alexa Delwiche[†], Colleen McKinney[†]
*Center for Integrated Agricultural Systems, University of Wisconsin-Madison, Madison, WI, United States, †Center for Good Food Purchasing, Berkeley, CA, United States

Chapter Outline

Introduction	104	Creating Opportunities for Nontraditional Suppliers Through Requests for Proposals (RFPs) and Contracts	117
Using Policy to Leverage Public Spending for Healthier, Fairer Food	105		
Defining "Good Food"	106		
Implementation of the Good Food Purchasing Policy	107	Challenges and Lessons: The Crucial Roles of Transparency and Multilevel Advocacy in Institutionalizing the Good Food Purchasing Program	121
Nuts and Bolts: How the Good Food Purchasing Program Works	108		
Implementation and Data Collection	110	Misguided Perceptions of Present Sourcing Practices	121
Case Studies: Institutions Adopting the Good Food Purchasing Program for Food System Change	111	Data Quality and Availability	121
		Difficulties With Contracts and Bidding Processes	122
Evaluating and Institutionalizing Values-Based, Data-Driven Purchasing Practices: Oakland, CA	112	Future Directions	122
		Conclusion	123
		References	124
		Further Reading	125
Supply Chain Innovations and Growing Pains in Good Food Purchasing in Los Angeles	114		

Institutions as Conscious Food Consumers. https://doi.org/10.1016/B978-0-12-813617-1.00005-8
© 2019 Elsevier Inc. All rights reserved.

INTRODUCTION

Every year, public institutions across the United States—from school districts to city governments—spend billions of dollars on food. The U.S. Department of Agriculture's (UDSA) National School Lunch Program alone spends more than eleven billion dollars in public funds to serve lunches for more than 30 million children annually. These billions of dollars are spent with virtually no oversight or awareness of the conditions under which these foods were produced. Through first-hand experiences reviewing purchasing data from dozens of public institutions, the Center for Good Food Purchasing has found that without strong accountability tools in place, food companies that routinely cut corners along the supply chain (resulting in human rights violations, inhumane animal treatment, and environmental degradation) continue to receive substantial public contracts.

On October 24, 2012, the Mayor of Los Angeles, Antonio Villaraigosa, issued an Executive Directive officially ushering the *Good Food Purchasing Policy* into law (Villaraigosa, 2012). That same morning, the Los Angeles City Council unanimously adopted the *Good Food Purchasing Policy* through a City Council motion, affirming its legitimacy and expanding its reach. In essence, the policy mandated that all City of Los Angeles departments with food purchases greater than $10,000 annually adopt the Good Food Purchasing Program (hereafter the "Program") and initiate plans to incorporate its five foundational values—local economies, valued workforce, animal welfare, health, and environmental sustainability—into "all new contracts for food purchases" (Villaraigosa, 2012). Three weeks later, in November 2012, the board of the Los Angeles Unified School District (LAUSD) also adopted the *Good Food Purchasing Policy* through a school board resolution.

These policy victories were largely a result of the Los Angeles Food Policy Council's ongoing work with the City of Los Angeles and LAUSD to adopt better food purchasing practices, and they marked a watershed moment for institutional food purchasing. As Robert Gottlieb, Director of the Urban and Environmental Policy Institute at Occidental College commented, "the breakthrough of this [policy] is that it moves beyond 'local' and embraces a much deeper value system. This…can change the discourse nationally" (c.f. Delwiche, n.d.). To implement the policy, a multistakeholder working group of the Los Angeles Food Policy Council developed the Program, and Los Angeles Food Policy Council staff members guided participating regional institutional foodservice operators through baseline data collection and other implementation processes. This programmatic support proved crucial to enhancing the impact of the Policy and its enforceability because many of the operational dimensions of the Policy were unfamiliar to institutional foodservice operators.

This chapter tells the story of the Good Food Purchasing Policy by providing an overview of the foundational values that informed the development and implementation of the policy, and two case studies that show how it was put in to practice.[1] We underscore the dynamic interplay between policy and the implementation of values-based changes in foodservice purchasing. To make our arguments, we draw on four interviews conducted in 2016 and 2017, internal documents from the Center, policy statements by institutions that have adopted the Policy, media coverage, and participatory observations from working at the Center for Good Food Purchasing and implementing the Good Food Purchasing Program.

Ultimately, we argue that, while originally conceived as an implementation framework designed to accompany the 2012 Los Angeles Good Food Purchasing Policy, the Good Food Purchasing Program has expanded to become a national model for helping institutions, public interest groups, city food policy managers, and others nationally to bridge policy and implementation activities around foodservice and values-based food procurement. The Program serves two central functions. It remains an implementation framework for city-level policies, such as the formal Good Food Purchasing legislation passed by the City of Los Angeles. It also functions as an organizational level policy, for example, when it is adopted by school district foodservice directors, as an internal framework to guide food purchasing decisions.

The chapter begins with an overview of the values that informed the Good Food Purchasing Policy and background on how the Program was implemented. It then describes how the Program standards work, in practice, and describes the implementation process. We then present two case studies to highlight how various businesses, public interest groups, and institutional foodservice directors have used the Program to make changes in the sourcing and supply chain practices in large school districts in Oakland and Los Angeles, California. Subsequently, we assess the challenges that arose in the implementation process and identify actions that have contributed to the success of implementation. The chapter closes with a brief discussion of possible future directions of the Good Food Purchasing Program, including updating the Good Food Purchasing Standards, developing comparative benchmarking tools for institutions, and linking Good Food Purchasing initiatives within regions and nationally.

USING POLICY TO LEVERAGE PUBLIC SPENDING FOR HEALTHIER, FAIRER FOOD

This section outlines the conceptual and values-based underpinnings that drove "good food" purchasing in the *Good Food Purchasing Policy*. Then it provides a short history of how the Los Angeles Food Policy Council and its partners implemented the Program and then expanded it through the Center for Good Food Purchasing.

---

1. Two of the authors, Alexa Delwiche and Colleen McKinney, are staff members of the Center for Good Food Purchasing.

Defining "Good Food"

The Center and its partners believe that public institutions have significant responsibility and purchasing power to ensure that their decisions about spending and contracts use taxpayer dollars to support food production, processing, and distribution businesses that uphold values around issues such as labor and the environment. Across the country, public institutions at the federal, state, and municipal levels are already increasingly incorporating values into their purchasing decisions. Examples include the Sweatfree Purchasing Consortium, which is made up of cities and states that want to discourage the purchase of products sourced from sweatshops (Sweatfree Purchasing Consortium, n.d.), as well as the Environmental Protection Agency's Environmentally Preferable Purchasing Program, which provides guidance to institutions around purchasing more environmentally preferred products through bidding language and contracts (Environmental Protection Agency, 2017). In the same way, food procurement policies have become a popular tool for supporting local, healthy, and sustainable food systems.

However, through its experience working with institutions that aim to apply the Good Food Purchasing Program framework, the Center has encountered a number of issues that pose significant barriers to purchasing better food. These include the food system's complexity and lack of transparency, limited time and expertise at the institutional level to do the type of analysis required to make more informed food purchasing decisions, and a lack of consensus around which values, or "credence attributes," to prioritize and how to define and measure them (Pullman and Wu, 2012). In a sense, these barriers require action on three fronts:

(1) Cultivating the political will necessary to implement Good Food Purchasing policies across the supply chain;
(2) Developing new processes to incorporate the Program's values into organizational behavior through bidding practices, labor policies, menu design, and more; and
(3) Developing organizational capacity to maintain good record-keeping to ensure transparency and enable meaningful programmatic benchmarking.

While technological innovation can assist with some of these activities, many of them are political in nature and/or require behavioral change, making political advocacy and process-oriented technical assistance crucial to successful implementation in most places.

Credence attributes are product characteristics that result from production practices or nutritional fortification that add value to a product, but do not necessarily affect product appearance, such as "humanely raised" (Pullman and Wu, 2012, p. 124). According to a scan of existing procurement policies conducted by the Los Angeles Food Policy Council staff during the development of the Good Food Purchasing Program, many food procurement initiatives focus narrowly on a single credence attribute, such as meat raised without the routine use of antibiotics or environmentally sustainable production practices. While health and sustainability are important attributes of good food, single-issue campaigns risk pitting equally important environmental, health, social, and animal welfare concerns against one another, as if in a zero-sum game, thus limiting the potential for transformative change (Connelly et al., 2011, p. 321).

The Good Food Purchasing Program addresses this gap in values-driven procurement programs by providing a way to integrate multiple values into an actionable and measurable approach to good food procurement. Moreover, it became clear early on that for this work to have an impact, it was also crucial for implementing institutions to have mechanisms in place to assess programmatic progress and impact. The Good Food Purchasing Program provides a framework and toolkit to address these challenges and unite single-issue advocates under a common, mutually beneficial framework.

These five values are connected by the key theme of "transparency." In food supply chains, *transparency* refers to the open disclosure of labor and production practices, ingredients, and product origin (i.e., where a product was grown and processed). Transparency is as crucial a component of the Good Food Purchasing Program as the values themselves because it makes it possible to see the progress that producers, distributors, and institutional buyers are making in relation to the values at the core of the Program. Without transparency, it is virtually impossible to know what ingredients food contains, how it was produced, and whether people, animals, and the environment were treated fairly and sustainably in the process. As noted earlier, transparency requires political will, buy-in from all supply chain partners, and the development of organizational processes and practices, such as new bidding language and better record-keeping. While technological improvements can assist with the latter, much of this work requires behavioral changes that cannot be remedied by technology.

Implementation of the Good Food Purchasing Policy

Before the Los Angeles mayor issued his policy directive, the Los Angeles Food Policy Council formed a multistakeholder working group that spent two years developing the Good Food Purchasing Policy, a model procurement policy. The working group was comprised of approximately a dozen members representing non-profit organizations with expertise in labor, environmental sustainability, public health, and animal welfare. The group also consisted of staff from the Los Angeles County Department of Public Health, local farmers, distributors, value-added processors, and a few major foodservice providers. Because of the collaborative nature of the working group and the different interests represented, the group agreed that a good food procurement strategy should address the entire food system if transformation was the goal—none of the values represented in the working group could be left behind without creating divisions that would detract from common goals.

On the strength of this diverse working group, the mayor issued his directive, and the policy was later adopted by the City of Los Angeles and Los Angeles Unified School District in 2012. By that time, the policy had been vetted by more than one hundred local, state, and national public, private, and non-profit organizations. The Los Angeles Food Policy Council then built out the Program in detail, to guide data collection and implementation. It provided programmatic support in areas such as record-keeping, menu design, bidding processes, and assessing suppliers' adherence to the five values. The Program work underway in Los Angeles generated so much interest throughout California and nationally, that in 2015, a new non-profit organization, the Center for Good Food Purchasing, was created to oversee the expansion of the Good Food Purchasing Program to public institutions throughout California and beyond.

The Center for Good Food Purchasing now works with national partners, local food policy councils, and regional grassroots coalitions, administrators, and elected officials across the country to transfer, scale, and network the Good Food Purchasing Program. Seven participating city agencies (including departments of aging, parks departments, and event facility concessionaires) and four school districts have already formally adopted the Program in Los Angeles, San Francisco, Oakland, and Chicago. Together, these institutions spend more than $280 million on food every year, according to information about purchasing shared by the institutions with the Center. The cities of Austin; Cincinnati; Denver; New York; St. Paul; Minneapolis; Washington, DC; and others are in various stages of adopting the Program within their local institutions. This growing and aggregated purchasing power within and across cities has the potential to create a very clear demand for a food system that aligns with the five values of the Good Food Purchasing Program. Networking across regions and among institutions with similar values also has tremendous potential to change markets and transform supply chains.

NUTS AND BOLTS: HOW THE GOOD FOOD PURCHASING PROGRAM WORKS

The Good Food Purchasing Standards are a central component of the Program. The Standards provide institutions with a roadmap for purchasing more sustainable and equitable food. An institution is expected to meet a minimum or "baseline" in each value category by sourcing a certain amount of food from producers that reflect each of the five Program values. The standards establish a floor in each value category, but incentivize institutions to exceed the floor and improve their "score" across these criteria by awarding higher institutional performers with a higher star rating. Specifically, institutions are encouraged to make changes in the types of products they purchase to meet or exceed the baseline for each value (Figs. 1 and 2). Key aspects of the Good Food Purchasing Program are listed in Table 1.

Star rating	Number of points needed
★	5–9
★★	10–14
★★★	15–19
★★★★	20–24
★★★★★	25+

FIG. 1 Good Food Purchasing Program star rating system.

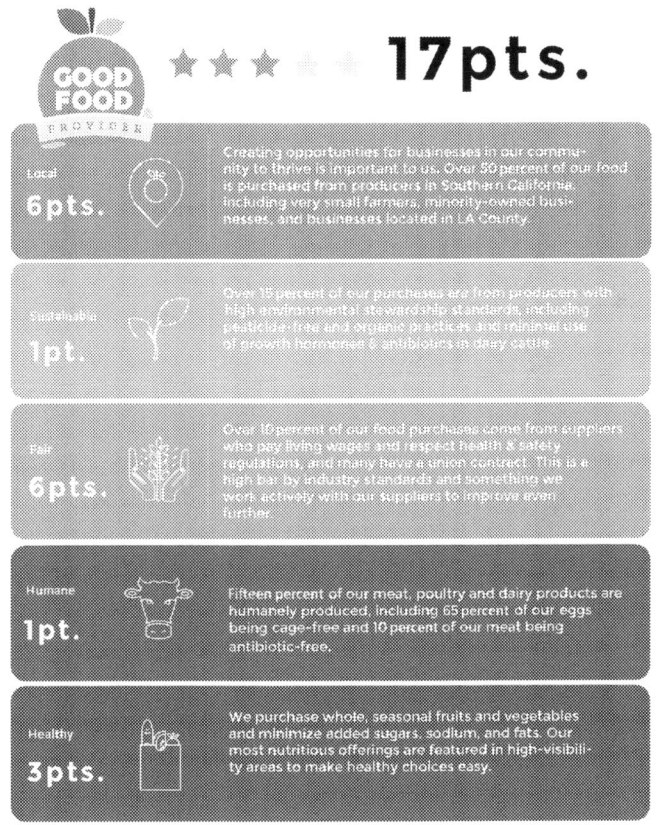

FIG. 2 Good Food Provider scoring example.

TABLE 1 An Overview of the Good Food Purchasing Program

The Good Food Purchasing Program aims to transform the way public institutions purchase food in five areas:
- **Local economies**: Support diverse, family and cooperatively owned, small and midsized agricultural and food processing operations within the local area or region.
- **Health**: Promote health and well-being by offering generous portions of vegetables, fruit, whole grains, and minimally processed foods, while reducing salt, added sugars, saturated fats, and red meat consumption and eliminating artificial additives. Improving equity, affordability, accessibility, and consumption of high quality culturally relevant food in all communities is central to advancing food purchasing practices.
- **Valued workforce**: Source from producers and vendors that provide safe and healthy working conditions and fair compensation for all food chain workers and producers from production to consumption.
- **Animal welfare**: Source from producers that provide healthy and humane conditions for farm animals.
- **Environmental sustainability**: Source from producers that employ sustainable production systems to reduce or eliminate synthetic pesticides and fertilizers; avoid the use of hormones, routine antibiotics and genetic engineering; conserve and regenerate soil and water; protect and enhance wildlife habitats and biodiversity; and reduce on-farm energy and water consumption, food waste and greenhouse gas emissions. Reduce menu items that have high carbon and water footprints, such as through plant forward menus that feature smaller portions of animal proteins in a supporting role.

The Good Food Purchasing Standards include a list of practices and performance indicators that correspond with each value category. At a minimum, participating institutions must meet the baseline standard in each category. Doing so will earn enough points to achieve one star in the Program—an accomplishment that indicates the institution is achieving above and beyond standard industry practices. To achieve a higher star rating, institutions can choose to earn a greater number of points in any or all of the value categories. This provides some flexibility in the implementation process for institutions, while holding all institutions to a common set of performance standards at the baseline level.

To support successful implementation over time, the Center works with participating institutions to establish supply chain transparency from farm to fork, evaluate how current purchasing practices align with the Good Food Purchasing Standards, assist with goal setting, measure progress annually, and celebrate institutional successes in shifting toward a values-based purchasing model. The Center issues a Good Food Provider verification seal to participating institutions that meet baseline requirements across the five value categories.

Implementation and Data Collection

Data collection is fundamental to program implementation and monitoring. The Center uses a multistep process in which Center staff members orient procurement professionals at partnering institutions (e.g., foodservice directors) to the types of data they will need to collect from each of their food distributors. Center staff members have developed knowledge of the types of data that are necessary to measure an institution's performance within each of the value categories. Types of data typically collected include the names of an institution's suppliers, the farms where the food was produced, and information about volume, price, and any known sustainability attributes (e.g., organic or animal welfare certifications).

The Center then cross-references suppliers identified by an institution's distributors with an internal database developed and maintained by the Center that includes detailed data on company size, production sites, labor violations, third-party certifications, and other information pertinent to evaluating supplier and product alignment with the five Program values, as outlined by criteria included in the Good Food Purchasing Standards. Although much of the data in the database is audit-based (i.e., not self-reported) public information, the Center has aggregated and augmented the information to support its program-specific analytic needs. The Center is not an auditing body (i.e., does not perform site visits), but instead relies on certifications based on third-party audits as the basis for its criteria and evaluation. The data maintained by the Center are available to institutions that work with the Center to assess and improve good food purchasing at their facilities. This database is the basis for the Center's analysis of individual institutions' performance on the Program's five core values and the Center's subsequent recommendations regarding strategy design and implementation.

Once the Center has analyzed a given institution's baseline data, it provides the foodservice director with a detailed assessment about how the current budget is allocated, how the institution is performing in each of the five value categories, and potential strategies to improve its Good Food Purchasing Program rating. For example, within the Local Economies category, if 10 percent of an institution's overall food is sourced within 250 miles, but 99 percent of that is coming from large farms, an institution might choose to focus on procuring more local product from small and midsized farms before it tries to further expand its total percentage of locally sourced products. By working with foodservice directors to better understand how their budgets are currently apportioned relative to different products and goals, the Center can help them strategize about how to reallocate energy and resources to achieve their Program goals across each value area.

CASE STUDIES: INSTITUTIONS ADOPTING THE GOOD FOOD PURCHASING PROGRAM FOR FOOD SYSTEM CHANGE

While the Good Food Purchasing Program continues to serve its original function as a benchmarking and strategic procurement planning tool for institutional foodservice operators, it has also evolved to serve other functions as stakeholders have adapted the program to respond to local challenges, opportunities, and priorities. For example, it has been employed as a policy tool for public officials to institutionalize fair, sustainable food purchasing; as a strategic planning tool for municipal food policy directors and food policy council staff; as an innovation tool for food businesses to spur new product development and sourcing practices; and as an organizing and accountability tool for local multisector coalitions to ensure that taxpayer dollars are spent on food contracts that truly serve the public.

This section highlights specific examples of challenges and innovations from cases in Oakland and Los Angeles. The example of Oakland Unified School District (OUSD) focuses on how the Program augmented preexisting farm to school efforts and codified these practices through policy. The Los Angeles-based examples (which include the Los Angeles Unified School District) demonstrate how the Program began with a policy mandate that has since been leveraged to develop new sustainable food products, improve labor conditions for the drivers of a local food distributor, and create opportunities for a wider range of suppliers to compete in the LAUSD commodity bidding process.

Evaluating and Institutionalizing Values-Based, Data-Driven Purchasing Practices: Oakland, CA

The case of the Oakland Unified School District (OUSD) provides an example of how the Good Food Purchasing Program, by providing valuable conceptual scaffolding and analytical tools, helped an institution that was already making strides to advance, expand, and ultimately formalize its sustainable food procurement work.

OUSD's Nutrition Services, which runs school food in the City of Oakland's public schools, began its transition toward healthier sourcing and meal planning around 2008 when it launched its Farm to School Program in partnership with the Community Alliance with Family Farmers (CAFF), a state-based nonprofit that promotes "sustainable food and farming systems through local and statewide policy advocacy and ... program[ming] ... to initiate institutionalized change" (CAFF, 2017). By 2010, healthy food procurement had become a central tenet of OUSD under the leadership of former superintendent Tony Smith and Nutrition Services Director Jennifer LeBarre. Smith viewed quality nutrition as integral to education reform. As he explained: "School food reform is not separate from school reform; it's part of the basic work we have to do in order to correct systemic justice, pursue equity, and give our children the best future possible" (Center for Ecoliteracy, 2011–2012).

OUSD started to transform its procurement practices by focusing on local sourcing, promoting more nutritious offerings, and improving the quality of meat served in school meals. For example, OUSD banned unhealthy products and ingredients such as soda and trans fats, and started purchasing larger amounts of chicken raised without antibiotics and organic, grass-fed beef.

To afford higher quality meat, foodservice redesigned its recipes to feature a combination of legume-based and animal-based proteins. OUSD's Nutrition Services also built linkages between K-12 students and regional farms by bringing farmers' markets directly to public school campuses and by implementing a "California Thursdays" program, in partnership with the Berkeley-based non-profit, the Center for Ecoliteracy, which features a weekly lunch sourced entirely from California-grown ingredients. By the fall of 2013, foodservice had made considerable strides toward healthier school food programming and its efforts attracted state and national acclaim and earned OUSD a "School Food Innovation Award" from the Center for Ecoliteracy (Grady, 2013).

Still, OUSD's foodservices' Director Jennifer LeBarre saw room for improvement. While at a School Food Focus conference in 2014,[2] she learned about the Los Angeles Food Policy Council's Good Food Purchasing Program through Joann Lo, the co-chair of the Los Angeles Food Policy Council working group that developed the original Good Food Purchasing Standards.[3] LeBarre saw potential for OUSD to strengthen its food procurement work by building on the Good Food Purchasing Program model. In particular, LeBarre hoped to improve OUSD's performance on issues such as labor and animal welfare, which had not been major focal points of foodservices' past procurement work. She also sought assistance with analyzing the impact of OUSD's shift in purchasing practices. Following the conference, Lo facilitated an introduction between LeBarre and Los Angeles Food Policy Council staff to explore how the Program could be used to help assess and expand OUSD's good food purchasing practices.

In early 2015, the newly minted Center for Good Food Purchasing began working with LeBarre and OUSD's Farm-to-School Supervisor, Alex Emmott, to conduct Program assessments of the 2012–13 and 2014–15 school years to examine the evolution of OUSD's healthy, sustainable food procurement work. A follow-up was conducted for the 2015–16 school year. Between the 2012–13 and 2015–16 school years, OUSD more than doubled humane and sustainable purchases as a result of implementing California Thursdays. OUSD also procured more than 30 percent of its food products locally, within 250 miles of Oakland. Although local purchases stayed the same between the two periods evaluated, OUSD increased its food purchases from small and midsized farmers (including produce, seafood, meat, and grain products), as defined by the Center's Good Food Purchasing Standards, from less than $5000 to more than $250,000. Findings from these two evaluation periods helped to justify OUSD Nutrition Service's investments in its farm to school initiatives to the Board, as well as identify priorities for the future. OUSD Nutrition Services also found

---

2. This conference, which took place in Oakland, California, and which was hosted by the organization School Food Focus, brought together school foodservice personnel from across the country to network and discuss values-based food procurement.
3. Lo is also the co-director of the Food Chain Workers Alliance.

that the Good Food Purchasing Program helped it evaluate related initiatives, such as its California Thursdays Program, by focusing on specific goals and indicators (Real Food Media, 2016).

As such, the assessment enabled LeBarre and Emmott to shift foodservice's thoughtful, but still relatively ad hoc approach toward a more systematic strategy by comparing the district's performance to date to the criteria for a five-star Good Food Purchasing Program rating. With guidance from the Center, they began to develop a strategy map based on a planning tool designed to help OUSD Nutrition Services identify specific actions that it can take to meet its Program goals while staying within its forecasted budget. OUSD Nutrition Services also developed metrics to track and measure progress toward their goals to help the district achieve a five-star rating within ten years.

As OUSD's good food procurement program evolved, LeBarre and Emmott wanted to ensure that their investments in evaluation metrics and Good Food Purchasing strategy design did not depend on the presence of specific champions, and would not be jeopardized by changes in the district's leadership or priorities. To codify their good food procurement work, they began collaborating with school board members to draft a food purchasing policy that would effectively institutionalize the OUSD Good Food Purchasing Program strategy map. In November 2016, the OUSD School Board voted in favor of a resolution that affirmed OUSD's participation in the Good Food Purchasing Program. At the time that this chapter was written, OUSD Nutrition Services anticipated that the school board would vote on a more detailed procurement policy and administrative regulations in late 2017, which would include specific goals, metrics, and benchmarks for meeting Good Food Purchasing targets over the short, medium and long term (Real Food Media, 2016).

The OUSD case demonstrates how the Program has been used to promote systematic progress toward all five value categories at the institutional level. In addition to using the Program as an internal evaluation and planning tool, this case also shows how the Good Food Purchasing framework can be used to codify district-wide procurement practices through a formal policy formation process. This marks a noteworthy departure from the policy-driven implementation of the Good Food Purchasing Program in Los Angeles and illustrates how the Program can both be used by foodservice operators to implement policy mandates (as was the case in Los Angeles), as well as by decision-making bodies, such as school boards, to institutionalize better food purchasing practices that have been operationalized, but are not yet codified. In other words, while Los Angeles represents a top-down case, Oakland is an example of how the Good Food Purchasing Program can also be used from the bottom up.

Supply Chain Innovations and Growing Pains in Good Food Purchasing in Los Angeles

In the Los Angeles Unified School District (LAUSD), the Good Food Purchasing Policy paved the way for improvements across the supply chain from the development of sustainable new products, to better labor standards, to bidding and contract language that promotes more socially and environmentally responsible food procurement. However, even with the Good Food Purchasing Policy and Program in place, these improvements did not necessarily come easily. Particularly with respect to labor practices and contracts, Good Food Purchasing Program advocates—both inside and outside of LAUSD—continued to play a key role in holding the district and its suppliers accountable and ensuring that the policy translated into tangible changes on the ground. This section illustrates some of these advances and challenges, and it highlights the roles different groups played in contributing to product innovation, improved labor conditions, and the development of a more values-based bidding and contract process.

Fostering Innovation in Product Development

The Good Food Purchasing Program has also spurred innovation by creating demand for new types of products and supply chain partnerships. For example, LAUSD's adoption of the Good Food Purchasing Program created a learning curve for the district's suppliers, as many of them had never fully considered the labor practices and environmental sustainability of the supply chains in which they participated. Gold Star Foods, a major school food distributor in the Western United States, was an early proponent of the Good Food Purchasing Program and an active participant in the Los Angeles Food Policy Council Working Group where the Program originated. In fact, Gold Star Foods' visibility and enthusiasm for the Program played an important role in garnering support of key influencers in the Mayor's Office and school district who became critical allies in internally navigating and successfully advancing the policies (McKinney, 2017).

Gold Star Foods' now-CEO, Sean Leer, saw the school district's adoption of the Program as an opportunity to modify some of its product offerings to better meet the new nutritional and sustainability standards and to create a niche for itself in the institutional food procurement market. He soon began working with one of his manufacturers to develop a healthier, more sustainable bread product. The result was a whole-wheat bun made from wheat sourced from a cooperative of Food Alliance certified sustainable, family-owned farms, in the Central Valley of California and milled and baked in Los Angeles into 40 million servings of bread products for the students. Gold Star Foods created 65 regional food chain jobs as a result of this shift. It eliminated high fructose corn syrup and reduced sodium. LAUSD foodservice representatives reported that kids loved the buns, and that they (foodservice staff) loved that the prices stayed the same. The bread products not only made their way onto 650,000 LAUSD students' plates, but to hundreds of other districts outside of Los Angeles—showing the ripple effects that one district can have through its commitment (PolicyLink, 2016). In this way, the Good Food Purchasing Program and related policies can create new opportunities for existing suppliers by

Improving Labor Conditions

In 2015, a coalition of food chain worker advocates demonstrated how LAUSD's commitment to the five values of the Program could motivate suppliers that work with the district to bring their practices into alignment with the values. Making this case helped advocates secure a union contract between truck drivers organizing to join the International Brotherhood of the Teamsters (a union with a record of championing freight drivers and warehouse workers) and Gold Star Foods. As previously noted, Gold Star Foods was an early champion of the Good Food Purchasing Program and has made innovative changes to its product offerings to comply with the Program. Still, like many other suppliers, Gold Star Foods had room for improvement. In 2015, Gold Star Foods' drivers brought attention to some of the company's labor practices by seeking assistance from the Teamsters to address entry wages, safety concerns, and treatment of senior drivers (Teamsters Local Union No. 572, 2015).

In addition to the Good Food Purchasing Policy, LAUSD had official policies that helped secure better working conditions for Gold Star Foods drivers. For example, the district had a sweat-free policy that explicitly required vendors working with the district to sign a certification of compliance confirming that it adheres "to the provisions of the district's Sweat-Free Procurement Policy," which include "safe and healthy working conditions" and "non-poverty wages standard[s]," in alignment with the district's Good Food Purchasing Policy. In signing the certification, vendors acknowledge that the district's penalties for violations include contract cancellation and vendor debarment (LAUSD, 2015). This policy led the LAUSD School Board and Food Services Division to take seriously Gold Star Foods' drivers' grievances and helped galvanize support from food justice organizations, such as the Food Chain Workers' Alliance.[4]

Ultimately, the Teamsters and Gold Star came to agreement, and the drivers secured significant improvements in truck and route safety, as well as a seniority system for route selection. In addition, as a result of unionization, hourly wages for new hires increased from thirteen dollars an hour to nineteen dollars an hour. According to Shaun Martinez, a researcher with the Teamsters, the benefits of unionization are not limited to pay increases and safety improvements; unionization also changes the culture and power dynamics of a company in important

---

4. As previously mentioned, Food Chain Workers' Alliance's co-director, Joann Lo, also served as co-chair of the Los Angeles Food Policy Council's Good Food Procurement Working Group and now serves as a member of the Center for Good Food Purchasing's Governance Board. The Food Chain Workers Alliance is an important partner helping to support the national expansion of the Good Food Purchasing Program and coalition building efforts within each city.

ways by making it safer for workers who observe labor violations to come forward and call attention to them (Day Farnsworth, 2017, pers. comm., 4 April).

Today, the Teamsters and Gold Star have a positive relationship. Martinez reports that driver turnover is now lower thanks to higher wages and improved working conditions, and it saves the company costs (Day Farnsworth, 2017, pers. comm., 4 April). Moreover, recent evaluations of LAUSD's Good Food Purchasing Program performance indicate that Gold Star's willingness to work with the Union to improve its labor practices has substantially improved LAUSD's Program score (McKinney, 2017).

In general, working conditions along the food chain are exploitative and often unsafe: Eight out of ten of the lowest paying jobs in the United States are in the food system (United States Department of Bureau of Labor Statistics, 2012), and many major food companies demonstrate a pattern of labor law violations, injuries, and even fatalities. LAUSD's procurement policies are helping to transform the playing field by exposing violations and incentivizing positive working conditions. In this instance, they paved a way for an improved relationship between one of the district's vendor and its employees. This example provides a promising model for future efforts to improve working conditions and wages along the supply chain through an explicit focus on food chain workers' rights. The Program alone cannot bring an end to worker exploitation and injustice in the food system, but it can bring attention to potential problems in institutional supply chains that labor leaders and workers can solve with employers.

By promoting supply chain transparency and cultivating new policy and programmatic champions for nutrition, valued workforce, environmental sustainability, animal welfare, and local economies, the Good Food Purchasing Program has spurred novel opportunities for advocacy organizations and coalitions to advocate for supply chain improvements such as fair working conditions for food chain workers.

Creating Opportunities for Nontraditional Suppliers Through Requests for Proposals (RFPs) and Contracts

Requests for proposals (RFPs), contracts, and bidding processes are opportunities for businesses, such as food processors and distributors, to become sole or priority suppliers for institutions such as school districts. Securing contracts with large buyers can be hugely beneficial to suppliers because it usually provides a stable source of revenue for the duration of the contract. As noted herein, this was a motivating factor for the CEO of Gold Star Foods, who saw an opportunity to fill a niche not yet filled by other distributors. As such, proposals, contracts, and bidding processes are critical instruments for ensuring vendor compliance with institutions' good food purchasing goals and the values they promote. This case, also from LAUSD, underscores the importance of contractual language in creating a context that is conducive to advancing the Program's aims. It also demonstrates some of the ways that contracts and RFPs alone are insufficient to ensure good food procurement practices due to persisting pressures to compete on cost and volume as well as systemic factors that still overwhelmingly favor large suppliers.

The "Chicken Story" at LAUSD dates back to 2010, when a Los Angeles Unified school board member unsuccessfully attempted to block a five-year chicken contract with Tyson due to Tyson's poor record on worker health and safety (Blume, 2017). At that time, the district did not have written policies to help the board navigate such issues, so the contract was approved. LAUSD then adopted and implemented the Good Food Purchasing Policy in 2012. In 2015, the district released a new bid for its commercial and commodity food products. This was the first bidding process for a foodservice contract that the district released since the new policy had been adopted. As such, chicken became the district's test case for applying the Good Food Purchasing Policy in the contracting process.

In February of 2015, LAUSD released an RFP that included turkey, chicken, beef, potatoes, and cheese. The evaluation criteria set forth in the RFP, which totaled 210 points, included the following: (1) Price (weighted at 75/210 points), evaluated based on the lowest overall cost to the district, and (2) the Good Food Purchasing Plan (weighted at 40/210 points), which detailed how prospective vendors would assist the district in achieving a five-star Good Food Purchasing rating within the five-year contract period. Prior to this process, with support from LA Food Policy Council staff, LAUSD staff developed purchasing plans to include in the appendix of the RFP to provide clear guidance to the vendors. The criteria accounting for the remaining 95 points included product taste and quality, experience, delivery and implementation plan, social responsibility, safety plan, small/minority/women-owned utilization commitment, and work-based learning partnership commitment.

Also included within the Good Food Purchasing Commitment section of the RFP was compliance with the Urban School Food Alliance's "Chicken Standard," which required that "all chicken products must be produced under a USDA Process Verified Program that includes compliance with the following:

- No animal by-products in the feed,
- Raised on an all vegetarian diet,
- Humanely Raised as outlined in the National Chicken Council Animal Welfare Guidelines for Broilers dated January 2014, and
- No antibiotics ever. If your company cannot supply the full volume of "No Antibiotics Ever" chicken requested on our procurement, please provide a written plan of when your company will meet the above Standard. In the meantime, your company must have the capacity for USDA Process Verified (third party) for Therapeutic Use Only chicken as defined in the Natural Resources Defense Council's 'Support For Antibiotic Stewardship in Poultry Production' dated December 2013; OR School Food Focus/The Pew Charitable Trusts 'Purchasing Guidelines That

Minimize the use of Antibiotics in Poultry Production' dated December 5, 2014." (LAUSD, 2015)

The Urban School Food Alliance (the Alliance) is a non-profit coalition of the largest school districts in the country, including LAUSD, which "shares best practices and uses its purchasing power to continue to drive quality up and costs down while incorporating sound environmental practices" (Urban School Food Alliance, n.d.). As a new coalition, one of the first issues the Alliance decided to address was the widespread presence of antibiotics in food and the associated increase in antibiotic resistant bacteria, which the Centers for Disease Control has deemed one of the greatest public health threats of our time (Watanabe, 2014). The Alliance's Chicken Standard was developed in partnership with the Natural Resources Defense Council, a key partner in the development of the Good Food Purchasing Standards, providing expert guidance on the antibiotic language included within the Environmental Sustainability category of the Standards. The Alliance's Chicken Standard was included within the RFP as a complementary, synergistic requirement, which would help LAUSD meet its goals within the Good Food Purchasing Program.

Despite the district's efforts to incorporate robust Good Food Purchasing criteria into the RFP, which included the Alliance's Chicken Standard, price and volume initially prevailed as the determining factors, and the Procurement Division advised the School Board to award the multimillion dollar contract for chicken to Tyson Foods and Pilgrim's Pride because of their competitive prices and volumes, even though both companies had a record of noncompliance with several of the Program's value categories.

This recommendation spurred vocal opposition from a multistakeholder coalition of local, state, and national organizations, organized and led by the Food Chain Workers Alliance. The coalition argued that the RFP represented a crucial opportunity to create space for chicken producers and distributors that are better able to comply with the production criteria associated with the district's Good Food Purchasing Policy. The coalition's effective organizing, which included letter writing, visits to board members, and a call-in day, led the LAUSD's procurement division to withdraw its recommendations.

LAUSD later re-released the RFP with modified language that enabled broadline distributors to submit bids for commercial chicken. Because broadline distributors are not vertically integrated like large chicken processors (i.e., companies that have ownership and/or control over the entire supply chain from farm to fork), they have greater flexibility with regard to their sourcing practices. This creates crucial openings to develop new supply chain partnerships with producers and processors that meet the Program's values-based criteria. As a result of the revised RFP, Gold Star Foods submitted a bid and was awarded the contract to supply approximately 30 percent of the district's commercial chicken produced without routine antibiotics. The district's Procurement Division initially recommended awarding the remaining two-thirds of the contract to Tyson and Pilgrim's Pride. This again stirred opposition from labor, environmental, and animal rights advocates, and ultimately the two companies withdrew their bids, though both companies identified reasons other than the opposition for withdrawing their bids (Blume, 2017).

During 2016, LAUSD purchased a minimal amount of chicken from Gold Star Foods, largely substituting chicken with turkey from Jennie-O, a large national supplier of turkey items to school districts. In March 2017, through a new RFP process, the district recommended four additional vendors for a $50 million contract over five years to provide chicken produced without routine antibiotics. Companies included: A&R Food Distributors, Don Lee Farms, Perdue, and Somma Food Group. Again, the coalition mobilized to evaluate the companies, applying lessons learned from the past two years.

Many of the coalition members were disappointed with the vendor recommendations, which largely continued to privilege low costs over other values outlined in the RFP. However, because of the limited time between the public notice of the contract awards and the board vote, the coalition did not have adequate time to mobilize opposition, as they had in the past. As such, they worked with the school board to amend the contract. The school board ultimately removed a vendor (A&R Food Distributors) that was in conflict with labor advocates from the contract, and included key provisions to incorporate more accountability measures to monitor vendor compliance with the Good Food Purchasing Program mid-contract:

- Remove the Award to A&R Food Distributors, Inc. for Chicken Biscuits and Chicken Patties and direct the Superintendent to rebid these items to return to the Board with a new recommendation, if possible, before June 30, 2017.
- Amend the language of the Board Report to read that "an annual survey be conducted on Goodman Food Products, Inc. dba Don Lee Farms, Perdue Farms Inc., Somma Food Group, and its subcontractors, to analyze their progress and improvement on all aspects of the Good Food Purchasing Policy and that the survey shall be provided to the Board; and that the Board shall reserve the right to terminate the contract at any point in the five-year term if policies are not followed with fidelity."
- Amend the language to read that the Board strongly urges the remaining Awardees to use California based sourcing, processing and distribution to better meet the provisions of the Good Food Purchasing Policy. (Board of Education of the City of Los Angeles, 2017)

Ultimately the coalition in Los Angeles that emerged during the adoption of the Good Food Purchasing Policy proved as, or more, influential in subsequent years as it monitored ongoing implementation efforts and ensured that the district fulfilled its commitment to the Program's core values at key decision points. In this way, the coalition leveraged the Good Food Purchasing Policy and Program to shape RFP requirements that helped hold both the school district and its suppliers accountable. Had the coalition's efforts concluded with the passage of the policy, these opportunities to influence the bidding process, hold all parties accountable, and ultimately transform purchasing practices may well have been overlooked.

CHALLENGES AND LESSONS: THE CRUCIAL ROLES OF TRANSPARENCY AND MULTILEVEL ADVOCACY IN INSTITUTIONALIZING THE GOOD FOOD PURCHASING PROGRAM

Common challenges to implementing the Good Food Purchasing Program include misguided perceptions of present sourcing practices, limited data quality and availability, and, as noted, difficulties institutionalizing policy through contracts and bidding processes. Poor transparency across the food supply chain is the unifying theme that runs throughout these challenges, which are discussed in more detail as follows.

Misguided Perceptions of Present Sourcing Practices

The Center for Good Food Purchasing has found that it is not uncommon for institutional foodservice providers to think their present performance on indicators, such as local and sustainable sourcing, is better than it actually is due to an historical lack of supply chain transparency, less rigorous definitions around values, and/or unsubstantiated sourcing claims made by their distributors. By grounding institution-specific baseline evaluations in hard data, the Center sometimes reveals that an institution's performance vis-a-vis the Good Food Purchasing Standards may not be as good as foodservice directors anticipate. However, once institutions have a clear understanding of their baseline performance in all five core areas, they are in a position to make informed decisions about what to prioritize and how to best allocate their energy and resources to improve their procurement operations through incremental steps, as the OUSD example illustrated.

Data Quality and Availability

Many institutions and their supply chain partners are not accustomed to the detailed level of data collection that the Program requires. In particular, smaller distributors often have less sophisticated tracking systems because they lack the capital needed to invest in tracking technology and have difficulty providing quality data on past transactions, making ex post facto baseline data collection more challenging. The Center for Good Food Purchasing has also found that existing contracts between institutions and foodservice providers can present barriers to implementing the Good Food Purchasing Program, even when there is traceability capacity due to proprietary clauses included in contracts.

Despite these challenges, in general, the Center has seen dramatic improvements in traceability and supply chain transparency among early adopters, once they have fully understood what data are required for program participation. Where there is a commitment to the Program, the key is to clarify what types of traceability information are expected from program participants at the outset of program implementation and also support distributors as they navigate challenges associated with the initial learning curve. In instances in which contracts impede supply chain transparency, bidding processes represent crucial opportunities for institutions to redefine expectations with regard to traceability and the Program's core values.

Difficulties With Contracts and Bidding Processes

Contracts are powerful mechanisms through which to formalize an institution's commitment to sustainable, fair, humane, local, and environmentally-responsible purchasing. However, to be vehicles for change, contracts and bidding processes must also be changed, and the decision makers implementing them may require some guidance about how to balance competing needs for price and other values. In the case of LAUSD, translating the Good Food Purchasing Program into the district's bidding process required carefully crafting contractual language to create space for companies better able to fulfill Good Food Purchasing Program's standards.

Two years of organizing in response to the chicken contracts yielded valuable lessons for the Los Angeles coalition that can inform emerging organizing efforts in other cities. First, partnering with public interest groups in the bidding process will help ensure robust, meaningful review of supplier alignment with the Good Food Purchasing Program during the bidding process. In the future, assembling a community/expert review panel to evaluate bidders' Good Food Purchasing Plans may help identify compliant bidders early in the process. Second, a transparent, accountable contracting process is a critical provision, and should be included in good food purchasing policies whenever possible. Such a provision offers opportunities for the community to gain information about recommended contract awards and respond to any areas where companies may not align with the Program values.

As noted throughout the chapter, the success of this model depends on a commitment to all five core Good Food Purchasing values and to transparency across the food supply chain. It also requires buy-in from *both* institutional leadership *and* partnering foodservice providers and distributors. Finally, these efforts most often also necessitate outside engagement to ensure institutional accountability to the Program as demonstrated in the LAUSD chicken RFP and Teamsters examples.

FUTURE DIRECTIONS

The Center for Good Food Purchasing has ambitious expansion plans for the future, as well as updating the Good Food Purchasing Standards, developing new comparative benchmarking tools, and exploring ways to more strategically stitch together Good Food Purchasing initiatives at the regional and national scales.

The Center recently completed its first update of the Good Food Purchasing Standards since they were released in 2012. The Good Food Purchasing Standards 2.0 are the product of an extensive stakeholder engagement process that began in spring 2016 and culminated in summer 2017. It reflects the contributions and feedback from more than 80 national experts who provided issue-specific guidance on the five Good Food Purchasing Program values as well as on considerations for implementation at the institutional level. Modifications made to the standards were intended to help build greater alignment among local and national partners, adapt to changes in the institutional procurement field since the initial version of the standards, and incorporate a national and more geographically accountable scope into the Program, as it expands throughout the United States.

Some modifications resulting from the 2017 standards update include (1) an expansion of scoring options to give greater incentive to institutions to purchase at the highest level across the values; (2) the creation of additional pathways for achievement in categories where meeting the baseline have been particularly challenging across institution types (e.g., the inclusion of a carbon and water footprint reduction for animal products and food waste reduction as an option within the Environmental Sustainability category and the introduction of a meat reduction target as an option within the Animal Welfare category); and finally (3) to ensure that equity considerations are distributed across all of the value categories. Going forward, the Center plans to review and update the Good Food Purchasing Standards every five years. In addition, the Center has found that program participants would like to be able to compare themselves to peer institutions, so it is exploring the development of mechanisms that would enable institutions to benchmark their performance metrics against aggregate sectoral and national procurement data.

Finally, the Center sees tremendous potential for its institutional partners to engage in collective action at the regional and national levels. The Center's regional cohort-based model initially emerged as a strategy for supporting shifts in food procurement at the institutional level within a single region. Similarly, as the Center has expanded its national reach, it has also begun to engage in new opportunities for networking efforts across cohort groups and in conjunction with organizations working on related national campaigns such as the Real Food Challenge, School Food Focus, and Health Care Without Harm. Additionally, in partnership with the non-profit PolicyLink, the Center is in the early stages of developing a statewide policy strategy with national implications for helping to scale these regional efforts.

CONCLUSION

The Good Food Purchasing Program was created in 2012 to help local institutional foodservice operators implement the Los Angeles Good Food Purchasing

Policy. Since then, it has expanded to become a national model for promoting healthier, more humane, and more socially and environmentally responsible institutional food purchasing. The case studies featured in this chapter represent examples of how actors as diverse as foodservice operators, food businesses, public interest groups, and food policy directors are using the Program to:

- Systematically chart and holistically improve their procurement practices,
- Innovate and develop new products that meet the growing demand for good food by major purchasers,
- Institutionalize the Program through organizational and municipal policies and contracts, and
- Hold institutions and their suppliers accountable for upholding their commitments to the Program.

This chapter also highlights challenges associated with implementing the Good Food Purchasing Program. They include a prevailing lack of transparency across institutional food supply chains, limited implementation and procurement analytics capacity within many institutions, and reluctance by some institutions and their suppliers to fully actualize the Program by changing their labor practices, bidding processes, and contracts.

Finally, this chapter illustrates the constructive tension between policy and program implementation and the unique roles of internal and external Program champions. In Los Angeles, the Program was initiated from the top down, while in Oakland it evolved from the bottom up. Yet, in each instance, the Program's success required champions at both the implementation and policy levels. Further, as the LAUSD labor and contract examples demonstrate, longer-term program success also appears to benefit from vocal and well-organized public interest groups that are not afraid to use the Program (and associated policies) as an accountability tool.

In just five years, the Good Food Purchasing Program has spurred novel opportunities for diverse stakeholders to turn institutional food purchasing into a tool for good. The Center for Good Food Purchasing is rapidly expanding its network to support organizations across the United States in incorporating the values of nutrition, valued workforce, environmental sustainability, animal welfare, and local economies into their procurement practices. While the Center has an ambitious agenda for the future, if history is an indication, its partners are also likely to influence the shape and trajectory of the Good Food Purchasing Program in innovative, constructive, and unforeseeable ways.

REFERENCES

Blume, H., 2017. Chicken nearly disappeared from L.A. school lunches. Now, it's making a comeback. The Los Angeles Times. 13 April. Available from: www.latimes.com. Accessed April 26, 2017.

Board of Education of the City of Los Angeles, 2017. Regular meeting stamped order of business. Available from: https://boe.lausd.net/sites/default/files/03-14-17RegularBdOBSTAMPED.pdf.

Center for Ecoliteracy, 2011–2012. Rethinking school lunch: Oakland Unified School District Feasibility Study executive summary. Available from: http://www.ousd.org/cms/lib07/CA01001176/Centricity/Domain/118/RSL-OAK_FS_2pg-Summary_2012-03-14.pdf.

Community Alliance with Family Farmers, 2017. About. Available from: http://www.caff.org/about/.

Connelly, S., Markey, S., Roseland, M., 2011. Bridging sustainability and the social economy: achieving community transformation through local food initiatives. Crit. Soc. Policy 31 (2), 308–324.

Delwiche, A., n.d. Abridged historical account of GFPP process_AD. Internal document. Los Angeles Food Policy Council.

Environmental Protection Agency, 2017. About the Environmentally Preferable Purchasing Program. Available from: https://www.epa.gov/greenerproducts/about-environmentally-preferable-purchasing-program.

Grady, B., 2013. Fresh, locally sourced food in OUSD school meal program draws kudos. Oakland Local. 17 November. Available from: http://oaklandlocal.com/2013/11/fresh-locally-sourced-food-in-ousd-school-meal-program-draws-kudos/.

Los Angeles Unified School District Procurement Services, 2015. Strategically-sourced commercial & commodity chicken, p. 9.

McKinney, C., 2017. The Good Food Purchasing Program in Los Angeles Unified School District (2014–2015). Available from: http://laschoolboard.org/sites/default/files/01-24-17Tab1Good%20Food%20Purchasing%20Program%20Update.pdf.

PolicyLink, 2016. The Los Angeles Good Food Purchasing Program: changing local food systems, one school, supplier, and farmer at a time. Available from: http://www.policylink.org/sites/default/files/LA_GFFP_FINAL_0.pdf.

Pullman, M., Wu, Z., 2012. Food Supply Chain Management: Economic, Social and Environmental Perspectives. Routledge, New York.

Real Food Media, 2016. Interview with Jennifer LeBarre, April 4, 2016. Oakland Unified School District.

Sweatfree Purchasing Consortium n.d. Home page. Available from: http://buysweatfree.org/ (Accessed 7 November 2017).

Teamsters Local Union No. 572 2015. Teamsters 572 letter to LAUSD re Gold Star safety concerns. Available from: Teamsters Local Union No. 572.

United States Department of Bureau of Labor Statistics, 2012. Occupations with the lowest median wage, May 2012. Available from: https://www.bls.gov/oes/2012/may/high_low_paying.htm.

Urban School Food Alliance n.d., Home page. Available from: https://www.urbanschoolfoodalliance.org/ (Accessed 8 November 2017).

Villaraigosa, A., 2012. Executive directive no. 24. Subject: Good Food Purchasing Policy. Available from: http://lacity.cityofla.acsitefactory.com/sites/g/files/wph281/f/mayorvillaraigosa331283141_10242012.pdf.

Watanabe, T., 2014. L.A. unified to ban the purchase of chicken raised with antibiotics. The Los Angeles Times. 9 December. Available from: www.latimes.com.

FURTHER READING

Board of Education Oakland Unified School District, 2016. Resolution No. 1617-0079—District commitment—Good Food Purchasing Policy/Program. Available from: https://ousd.legistar.com/LegislationDetail.aspx?ID=2868168&GUID=08C30E57-C3B5-44EE-8D00-FA5039000551&Options=&Search=.

Center for Ecoliteracy, 2011. Rethinking school lunch: Oakland Unified School District Feasibility Study full report. Available from: http://www.ousd.org/cms/lib07/CA01001176/Centricity/Domain/95/OUSD%20FeasibilityStudy%20Full%20Report.pdf.

Center for Good Food Purchasing, 2017. Home page. Available from: http://goodfoodpurchasing.org/.

Food Chain Workers Alliance & Solidarity Research Cooperative, 2016. No Piece of the Pie: U.S. food workers in 2016. Available from: http://foodchainworkers.org/wp-content/uploads/2011/05/FCWA_NoPieceOfThePie_P.pdf.

Los Angeles Unified School District Procurement Services, 2011. Certification of compliance with the District's Sweat-Free Procurement Policy, pp. 4–5. Available from: https://psd.lausd.net/Vendors/Docs/851/C-984.pdf.

Oakland Unified School District, 2015a. Nutrition services. Available from: http://www.ousd.org/Page/934.

Oakland Unified School District, 2015b. Oakland Unified School District fast facts 2014–2015. Available from: http://www.ousd.org/cms/lib07/CA01001176/Centricity/Domain/1/OUSD%20District%20Docs/OUSD%20Fast%20Facts%202014-15.pdf.

Lindsey Day Farnsworth is postdoctoral fellow at the University of Wisconsin-Madison's Center for Integrated Agricultural Systems. Her work includes applied research on community and regional food distribution, policy, access, financing, and values-based food supply chains. She has also worked internationally as an evaluator on agricultural development projects in Latin America. At the local level, she serves on the Madison Food Policy Council and Public Market Development committee. Lindsey holds a PhD from the Nelson Institute for Environmental Studies, UW-Madison; an MA in Urban and Regional Planning, UW-Madison; and a BA in Sociology and Latin American Studies from Brandeis University.

Alexa Delwiche is the executive director of the Center for Good Food Purchasing. She previously served as managing director of the Los Angeles Food Policy Council (LAFPC) from 2013 to 2015 and Food Policy Coordinator from 2009 to 2013 for the LAFPC and the Los Angeles Food Policy Task Force. At LAFPC, she spearheaded the development, launch, and implementation of the Good Food Purchasing Program. Previously, Alexa worked for the United Farm Workers and the San Francisco Board of Supervisors. Alexa has a Master's of Public Policy from the University of California at Los Angeles.

Colleen McKinney is the associate director for the Center for Good Food Purchasing, where she leads Good Food Purchasing Program's expansion and implementation efforts in collaboration with national and local partners. She works with participating institutions to assess purchasing practices, set goals, monitor progress, and celebrate success. Previously, she was a Program Associate at the Los Angeles Food Policy Council, where she was instrumental in developing the Good Food Purchasing Program. She holds a Master's of Public Policy from the University of Southern California.

Chapter 6

From Navigating the Regulatory Environment to Designing a Good Food Supply for Institutions: Cases From Philadelphia

Jonathan Deutsch[*,†], Alexandra Zeitz[*], Benjamin Fulton[*], Brandy-Joe Milliron[†], Catherine Bartoli[‡]

[*]Center for Food and Hospitality Management, Drexel University, Philadelphia, PA, United States, [†]Department of Nutrition Sciences, Drexel University, Philadelphia, PA, United States, [‡]Get Healthy Philly, Philadelphia Department of Public Health, Philadelphia, PA, United States

Chapter Outline

Introduction	127	Project 2: Support for Compliance	135
Background: Philadelphia's Food Systems and Levers for Institutional Change	129	Project 3: Changing the Upstream Value Chain to Increase Compliance	138
Literature Review: Procurement Policy Implementation	130	Conclusion	142
		References	143
Project 1: Education About Compliance and Cost	131	Further Reading	145

INTRODUCTION

The health of the most vulnerable people in modern society is powerfully impacted by institutional food (Tsui et al., 2015). Consumers of institutional foods, such as prisoners, active duty military members, boarding school and university students, shelter residents, those in assisted-living, and individuals in other residential environments, may consume most or all of their food from an institutional foodservice provider, and their food choices are dictated by what is provided. Furthermore, some individuals who eat institutional food as part of their overall intake at hospitals, early childhood centers, and senior care centers, are particularly vulnerable from a health standpoint. In addition to limitations in food choice, consumers of institutional food are often also contending with economic hardship and unequal access to healthy environments and food. Consumption of safe, healthy, and wholesome food is critical to reversing health disparities (Tsui et al., 2013).

Several cities and counties across the United States, including large municipalities such as New York City and Los Angeles County (Karpilow, 2012), have adopted nutrition standards for foods purchased, served, or contracted through municipal agencies as a mechanism to create healthy food and beverage environments to curb lifestyle-related diseases, with a particular focus on vulnerable populations (Gardner et al., 2014; Robles et al., 2013). Such standards are shaped by the Dietary Guidelines, published every five years by the U.S. Department of Agriculture (USDA) and Health and Human Services (HHS), and supported federally through the Centers for Disease Control and Prevention (CDC), HHS, and the U.S. General Services Administration's guidelines created for federal agency worksites (CDC, 2012). In so doing, large municipalities serving food in a variety of venues become a strong market force in cultivating demand for healthy products required by comprehensive nutrition standards (CDC, 2011).

The City of Philadelphia ("City") implemented its own nutrition policy in 2014 with the authorship of Mayoral Executive Order ("Executive Order") 04-14, which recognized the potential for nutrition standards to support the health of Philadelphians served by City agencies. The resulting Philadelphia Nutrition Standards ("Standards") required minimum nutrition criteria for foods purchased, served, and sold by City agencies (City of Philadelphia, 2014), including recommendations related to:

- *Nutrition data*: for example, maximum sodium and/or sugar per serving,
- *Promoted ingredients*: for example, fresh fruits and vegetables; whole grains,
- *Prohibited ingredients:* for example, no partially hydrogenated oils in the ingredient list,
- *Food preparation:* for example, elimination of deep-frying, and
- *The foodservice environment:* for example, portion control; the availability of water.

The Standards propelled City departments toward the provision of healthy food options, and demonstrated the City's commitment to improving the health of those it serves. The Standards also demonstrated the City's leadership in promoting positive health outcomes through environmental change, and acted as a model for other institutions on creating healthy food environments.

This chapter illustrates how Philadelphia introduced comprehensive nutrition standards in City departmental foodservice. The City's foodservice work is a collaboration with several City agencies, including the Procurement and Finance Departments, which are charged with contract execution and oversight.

The effort is further supported through organizational collaborations such as the Drexel Food Lab (DFL) at Drexel University to assist City agencies and others in navigating and implementing nutrition standards. The authors of this paper are a part of that collaboration. After providing some background and a brief literature review, the group presents three brief case studies that describe the trajectory of DFL's work as it shifted upstream, from institutional education to supplier product development, to ultimately influencing marketplace improvements that support implementation of the Standards. The chapter will then conclude with a reflection on progress, a proposal for the potential of supply-side solutions, and future opportunities.

BACKGROUND: PHILADELPHIA'S FOOD SYSTEMS AND LEVERS FOR INSTITUTIONAL CHANGE

Philadelphia, with more than 1.5 million residents, has some of the highest rates of poverty, hypertension, diabetes, and deaths from heart disease among the nation's ten largest cities. Thirty-eight percent of Philadelphians have been diagnosed with hypertension, and 15 percent of residents have reported a diabetes diagnosis (City of Philadelphia, 2016). In 2010, the Philadelphia Department of Public Health established Get Healthy Philly (GHP) to address these lifestyle-related chronic diseases. GHP brings together government, community-based organizations, academia, and the private sector to create and enhance environments that promote healthy eating, active living, and are smoke free for all Philadelphians. For several years, GHP has worked to create healthy food and beverage environments in Philadelphia communities through the support of healthy retail and farmers' market initiatives (The Food Trust, 2014, 2016). The Executive order gave support to GHP for the creation of the Standards and their application to City-supported program food purchases.

The City of Philadelphia purchases and serves food in a variety of ways to distribute more than 14.5 million meals annually. Some departments purchase directly through a bid process for specific items such as bread, poultry, and milk, where program staff prepare and serve meals on site. Other departments contract foodservice operations to outside companies who manage food purchases and deliveries. The variety of avenues by which food enters into City programs impacts not only the array of products utilized, but also how easily items may be identified to fit the needs of a diverse recipient population. The Standards have been implemented through an ongoing collaboration of GHP, City agencies, and partner organizations, including DFL. GHP works with department administrators and staff to train individuals on content and the process for Standards implementation. The Standards are included in City food bids and service related contracts that include a food provision. Since incorporating the Standards, there has been a considerable increase in the number of products replaced with lower sodium alternatives as one example of how departments are changing their foodservice offerings. The Health Promotion Council of

Southeastern Pennsylvania, a local nonprofit, has conducted ongoing trainings and technical assistance for kitchen and contract staff to align menus and foodservice operations with the Standards. Through this technical assistance, departments relayed that certain products were challenging to procure, providing the impetus for the collaboration between GHP and DFL.

LITERATURE REVIEW: PROCUREMENT POLICY IMPLEMENTATION

U.S. cities have ushered in a variety of initiatives to improve food procurement, including healthy workplace procurement guidelines, institutional policies, and municipal regulations (cf. Gardner et al., 2014; Hyseni et al., 2016; Kimmons et al., 2012; MacLellan et al., 2009; Maes et al., 2011; Pitts et al., 2016). The efficacy of such policy and procurement initiatives can be evaluated by identifying how easily they are understood and adopted along the food supply chain. The success of potentially effective policies is predicated on clear language (understanding), staff training (turning understanding to adoption), along with access to products that meet policy requirements (adoption to full implementation) (Gardner et al., 2014).

An analysis of the path from understanding, to adoption, to implementation can assist practitioners in identifying barriers to successful policy implementation. Barriers to implementation of healthy foodservice guidelines in hospital and federal worksite cafeterias include supply-side barriers, such as vendors not having compliant food and beverage selections; and demand-side barriers, such as the lack of foodservice staff training or acceptability by clients (Pitts et al., 2016). School cafeterias experience similar challenges (MacLellan et al., 2009), including a lack of healthy food options compounded by resource limitations, a combination common among public institutions. Indeed, both supply- and demand-side barriers were evident in this project.

Along with the preceding technical hurdles, new regulations are often met with organizational and institutional resistance to change (Pache and Santos, 2010). Foodservice managers are busy professionals who are required to balance competing demands (MacLellan et al., 2009). Many are daunted by adding another layer of complexity to their extant responsibilities of food and labor cost-control, relationship management with stakeholders, equipment and facilities maintenance, and regulations compliance, while providing appetizing, safe, and wholesome food on time, at the proper temperature, that attempts to rise above (un)popular perceptions of "institutional" food (Payne-Palacio and Theis, 2015; Johns and Pine, 2002). While some foodservice operations have a centralized designated procurement person or department, most small operations, and even many large operations, do not (Payne-Palacio and Theis, 2015). In those cases, foodservice managers may lack procurement expertise and struggle to balance procurement among other priorities. Therefore, the addition of a new nutrition policy at the institutional, municipal, state, or federal level may be perceived as an unwelcome, unrealistic, or unaffordable hurdle by a foodservice manager saddled with procurement responsibilities (Lederer et al., 2014). Recognizing these challenges, and hearing some of them from foodservice managers, the City integrated a step-wise approach to implementing the Standards.

Despite the stresses new policies might put on individual operators of institutional foodservice, the implementation of institutional foodservice guidelines focused on health and sustainability can positively impact public health and the food system (Kimmons et al., 2012). Evidence suggests that a multitiered, multistakeholder approach is critical in facilitating such changes in institutional foodservice (Kimmons et al., 2012; MacLellan et al., 2009; Pitts et al., 2016), in which policies and regulations are supported by supply-side activities and demand-side restrictions. Such multicomponent interventions include reformulating products to make them healthier or policy-compliant (supply-side) or pricing unhealthy foods higher than healthy foods (demand-side) (Hyseni et al., 2016). Roberto et al. (2015) similarly suggest that health solutions related to food and obesity-prevention are not either/or scenarios, but complex interconnected relationships, including supply- and demand-side explanations and approaches that combine government regulation with market-based solutions and public education campaigns to create lasting change.

DFL and GHP worked together to identify the barriers to implementing the Standards by assessing understanding of the Standards as well as supply- and demand-side constraints. In each, identification of specific barriers—lack of understanding, technical skill, or appropriate products—informed interventions aimed at education, technical assistance for foodservice staff, or supply-side innovation. The three case studies that follow illustrate how DFL is mitigating supply-side issues alongside ongoing GHP trainings and technical assistance to city departments and hospitals, the combination of which address the integration of healthy options.

PROJECT 1: EDUCATION ABOUT COMPLIANCE AND COST

When DFL embarked on the collaboration with GHP, one of the initial goals was to help city agencies identify and procure foods that meet the City's Standards. Even after understanding the requirements of compliant products, interactions with foodservice managers revealed a belief that to purchase compliant foods would, in many cases, comprise a larger share of the food budget, or be cost prohibitive. For example, if frozen fried fish sticks were deemed noncompliant due to their preparation method (deep frying), a compliant alternative, such as breading fresh or frozen fish fillets in reduced-sodium whole grain bread crumbs and baking them, would be more expensive in both product costs and labor. This perception was consistent with consumer studies that have documented that making healthier choices in the supermarket tends to cost more (Parker-Pope, 2007).

Further complicating perceptions of the cost implications of foodservice pricing was a lack of transparency and sharing of prices common to the foodservice industry. Most foodservice vendors do not post transparent pricing information on their websites (Payne-Palacio and Theis, 2015). Prices may vary customer-to-customer based on a number of factors including the presence of a contract, the size of the account, affiliation with a group purchasing organization, credit/payment history, or simply the procurement professional's ability and skill in negotiating and monitoring prices invoice-to-invoice to avoid unexpected increases (Payne-Palacio and Theis, 2015).

From personal experience shopping as private consumers and purchasing for restaurants and institutions as foodservice professionals, the DFL could generate several examples of healthier foods, per the Standards, that either cost less, more, or the same amount as conventional foods. For example, skim milk (compliant) is typically cheaper than whole milk (noncompliant); reduced-sodium chicken broth (compliant) is often sold at the same price point as conventional chicken broth (noncompliant); and lean ground beef (compliant) typically costs more than beef with a higher percentage of fat (noncompliant). Therefore, DFL designed a simple study to test the market and clarify whether compliance with the Standards would be cost-prohibitive for City agencies. There was no intent to conduct a definitive or generalizable study showing a relationship between complying with the regulations and food expenditures; however, the DFL sought to collect data to better nuance the conversation with agencies.

Vendors serving Philadelphia institutions were given a list of 54 commonly ordered foods and asked to provide pricing information, framed as a potential sale. On this list of products were many matched-pairs of compliant and noncompliant products, though they were not presented to the vendor as such. For example, prices were sought for both Kix, a compliant cereal, and Raisin Bran Crunch a noncompliant cereal, or for whole-milk (noncompliant) and part-skim (compliant) mozzarella. The results yielded numerous examples of compliant food products priced lower or at the same amount as noncompliant products, as well as instances where compliant foods were priced higher than noncompliant foods.

As displayed in Table 1, many of the compliant products (in the white cells) were priced lower per unit (ounce) than the noncompliant alternative in the shaded cell.[1] Specifically, fat-free milk, whole wheat tortillas, and portioned haddock are priced lower than their noncompliant alternatives of 2 percent strawberry milk, (white) flour tortillas, and breaded haddock portions. In the case of milk and tortillas, the compliant product is simply less expensive—no additional labor cost or other hurdle to adoption is at play. In the case of the haddock portions, the cost is significantly less (one dollar per pound), but if the menu outcome is to be comparable (a breaded fish portion), the additional labor and ingredient cost for preparing and breading the item needs to be considered.

---

1. Comparing prices per unit rather than case is advisable, as pack size varies by product.

TABLE 1 Price Comparisons From Vendors Serving Philadelphia of Compliant and Noncompliant Products

Item	Pack	Price (Case)	Price (Ounces)
2% strawberry milk	18/8 oz.	$15.99	$0.11
Fat-free milk	18/8 oz.	$14.69	$0.10
Goya fajitas flour tortillas	24/8 oz.	$14.99	$0.08
Goya whole wheat tortillas	24/12 oz.	$19.85	$0.07
Haddock portions, breaded	10 lb.	$28.90	$0.18
Pacific haddock, portioned	10 lb.	$18.90	$0.12
Hellman's real mayonnaise	4/1 gal	$62.95	$0.12
Hellman's low fat mayonnaise	4/1 gal	$58.95	$0.12
Hunt's stewed tomatoes	6/105 oz.	$21.00	$0.03
Hunt's no salt added diced tomatoes	6/105 oz.	$21.00	$0.03
Kikkoman soy sauce	1/5 gal	$73.33	$0.11
Kikkoman reduced sodium soy sauce	1/5 gal	$70.83	$0.11
Kraft shredded whole milk mozzarella	1/5 lb.	$13.95	$0.17
Kraft part-skim shredded mozzarella	1/5 lb.	$13.95	$0.17
Pepperidge farm original white bread	12/24 oz.	$2.50	$0.01
Pepperidge farm stone ground 100% whole wheat bread	12/24 oz.	$2.50	$0.01
Planters honey roasted peanuts	6/46 oz.	$64.56	$0.23
Planters dry roasted unsalted peanuts	6/46 oz.	$64.56	$0.23
Yopliat low fat strawberry	6/32 oz.	$19.39	$0.10
Yopliat nonfat plain	6/32 oz.	$19.39	$0.10
Perdue chicken nuggets	5/5 lb.	$54.75	$0.14
Perdue chicken breast tenders	2/5 lb.	$26.50	$0.17
Perdue oven roasted turkey breast	2/10 lbs.	$73.80	$0.23
Perdue oven roasted turkey breast reduced sodium	2/10 lbs.	$85.00	$0.27
Spaghetti	20/1 lb.	$15.95	$0.05
Whole wheat spaghetti	20/1 lb.	$26.95	$0.08

Compliant cheaper
Same price
Noncompliant cheaper

In other instances, the cost for compliant and noncompliant products is identical or close enough that when calculated per ounce, the difference is within one cent per portion. Specifically, low-fat mayonnaise, no-salt-added canned tomatoes, reduced-sodium soy sauce, part-skim mozzarella, whole wheat bread, unsalted peanuts, and nonfat plain yogurt are identical, or nearly identical, in price to noncompliant alternatives such as conventional mayonnaise, stewed tomatoes, conventional soy sauce, white bread, honey roasted peanuts, and sweetened low-fat yogurt. In those cases as well, no additional labor hurdles exist in using the compliant product.

There are also instances of compliant products costing more. Compliant chicken breast tenders, which then need further preparation, are more expensive than chicken nuggets; reduced sodium deli meat is more expensive than its conventional counterpart; and whole wheat pasta costs more than conventional pasta made from white flour. Finally, the inclusion of more fresh produce to a menu incurs additional costs. In those cases, agencies can be advised to use strategies including off-setting higher priced foods with lower-priced foods to stay within budget, adjusting portion sizes of higher-priced foods, reviewing menu selections, or initiating conversations with vendors to explore more affordable selections of higher-priced items.

Armed with real-world data on the impact of the comprehensive food regulations on pricing of a selection of commonly purchased items, GHP communicated the results of this exploratory study at an interdepartmental meeting of City food procurement professionals. In addition, DFL collaborated with GHP to author a toolkit (Philadelphia Department of Public Health, 2017) which addressed the commonly held misconception that compliant procurement decisions will always cost more, as well as additional educational resources on the following topics:

- *Product selection.* Products including fresh produce, lean meats, plant-based proteins, and low-sodium canned goods that are generally compliant; products such as bread, meat, and dairy, which can be compliant if correctly selected; and products such as snacks, frozen foods, baked goods, broth, cereal, and beverages that will take extra work, such as reading nutrition labels or asking questions of vendors or manufacturers to ensure compliance.
- *General purchasing guidelines for compliant food.* For example, choosing items that contain zero grams of trans fats and no partially hydrogenated oils in the ingredient list; and choosing items labeled "low sodium" over other choices, and asking vendors for nutrition labels.
- *Guidance on specific product selection and what to look for on a label to ensure compliance.* For example, a nutrition facts label for compliant bread is presented in the toolkit, highlighting the areas that show compliance.
- *Recommendations for cooking methods.* For example: Easily compliant (steam, boil, bake, stew, braise, and poach), potentially compliant (stir-fry, saute, and roast), and noncompliant (pan-fry and deep fry) cooking methods.
- *Tips for healthy cooking.* For example:
    - Adding citrus juice or other acidic foods like a touch of vinegar to dressings, roast meat, and sautéed vegetables and sauces to enhance flavor without adding salt.
    - Adding fresh vegetables such as beets, carrots, and celery to a braised dish to add flavor and to increase the nutritional value of a dish.
    - Roasting meats and vegetables in a hot oven (400°F) to enhance browning and add to flavor.
    - Toasting spices in a dry pan until aromatic to help enhance flavor without adding more salt or fat.
- *Healthy cooking substitutions with recipes.* For example:
    - Moving from canned soup to broth-based soup made from water, chicken, and/or vegetable scraps with additional vegetables, with meat/fish and grains added.
    - Moving from white rice to 50:50 white rice/brown rice blend to "ease into" adding whole grains.
- *Recipes.* These include reformulated recipes detailed in the next section of this chapter.

To relay these findings, DFL and GHP provided education and hands-on trainings to foodservice managers, cooks, and purchasers to help dispel misconceptions about compliant food costing more and provided an opportunity to make suggestions for positive ways to incorporate compliant food. For example, blanket statements such as, "consumers will reject brown rice" might be tempered with a recipe incorporating a percentage of brown rice to "ease into" a whole grain dish. Challenges with sourcing a compliant barbecue sauce can be overcome with a cost-neutral homemade version. The combination of hands-on culinary training paired with education and resources, such as acceptable and accessible quantity recipes, is effective in making healthy changes. The provision of such information to City agency purchasers can help them in their decision-making process. Additionally, the data gathered supports purchases that align with the Standards and helps departments uphold policy without the extra cost burden.

PROJECT 2: SUPPORT FOR COMPLIANCE

DFL conducted observations at foodservice sites to determine where it could provide technical assistance in helping foodservice managers and staff follow the Standards. While the educational outreach described in the first case study can be beneficial to many constituencies, targeted support around specific issues may be warranted. Working with GHP and Philadelphia's Office of Homeless Services, DFL identified a shelter for the homeless that needed practical solutions and guidance to move from understanding the Standards to fully adopting them into their food procurement strategy.

The GHP- and DFL-identified shelter provides safe haven for more than 250 men and offers nearly 3000 meals each week to homeless and other food insecure men. The site also provides supportive infrastructure services including showers, emergency shelter, clean clothing, mailroom services, case management, medical assistance, and counseling. While the organization is an independent 501(c)3, Philadelphia's Office of Homeless Services contracts with the organization to provide food, and therefore the site must comply with the Standards.[2]

DFL observed that many items on the site's menu closely followed the procurement guidelines including fresh fruit, fat-free milk, and 100 percent juice. Yet some donated items, particularly casseroles prepared by community members, were noncompliant. Further discussion with the shelter revealed that the donated casseroles were a prominent part of the organization's identity, as stated on its website: "Donated Casseroles are the 'Main Course' of a complete, hot meal that we serve to hundreds of God's poor and homeless each day....Bread, a beverage and a dessert are also provided for each guest at every meal."

---

2. The editors have deliberalitely anonymized this information to protect subjects.

The site also lists nine sample recipes that are suggested models for donors. Donated main courses help the organization stretch its food budget and provide more food to more people. However, by accepting donated food rather than purchasing these items, the organization gives up the ability to control the nutritional quality of the meals they serve. The sample recipes are nearly absent of whole grains, fresh vegetables, and lean proteins, and exceed sodium recommendations. So while the shelter follows the Standards for foods that easily meet compliance requirements, the main entrees may not align with the nutrition Standards.

The recipe posted on the shelter's site for "Chicken au Gratin Casserole" serves as one example of these nutritional issues. The ingredients include two 10 ¾ ounce cans cream of celery soup; ½ pounds of elbow macaroni; two cups, eight ounces Kraft sharp cheddar cheese; and one cup milk. The instructions are simple and only include: combine ingredients, mix lightly; bake in a casserole pan in a 350 degree oven for 30 minutes; cool and freeze.

From a culinary standpoint, the recipe is problematic for a number of reasons. Most obviously, it does not call for chicken, despite the title. In addition, it does not specify par-cooked or raw macaroni, preheating the oven, and the quantity of cheese is unclear. From a nutrition standpoint, the recipe is absent whole grains, vegetables of any sort, and does not specify non- or reduced-fat dairy. While the yield is six to eight servings, even at the smaller serving size (1/8 recipe), according to DFL nutrition analysis using USDA Supertracker, the dish contains 17 g of fat, with 40 percent of the daily value of saturated fat, based on a 2000 cal daily diet; and 623 mg of sodium, representing nearly one-third of the daily value for sodium based on a 2000 cal diet. Other recipes are similarly problematic. For example, the Shepherd's Pie casserole provides more than 1200 mg of sodium, more than 50 percent of DV. The beef stroganoff casserole contains 25 g of fat at the smaller serving size, with 12 g of saturated fat, more than 50 percent of the daily value.

GHP and DFL saw an opportunity to encourage donated food to be compliant with the Standards in a manner that would preserve the relationship with donors, the culinary offerings of the site, and meet Standards. DFL worked with Drexel University students to reformulate the nine sample casseroles to make them compliant with the Standards and conducted acceptance testing with students, faculty, and visitors to show that they were more appealing in flavor, appearance, and overall acceptability when compared with the original formulations. An important note is that the essence of the recipe was preserved—DFL and GHP were not advocating taking a cheesy noodle bake, for example, and replacing it with lentil burgers or jerk rasta pasta. Rather, recipe development was focused on maintaining the cultural and culinary essence of the dish—which the team sees as the province of the site—while making it compliant with the Standards.[3] The revisions incorporate nutrient-dense ingredients including

---

3. A future project could be to develop culturally relevant foods that may be received with greater enthusiasm by the population served.

low-fat dairy, lean proteins, vegetables, and whole grains. In addition, the recipe instructions are clearer from a culinary perspective. The reformulated recipes were then provided to the organization for use on their website, included in the GHP foodservice toolkit to encourage their use by other charitable organizations, and recommended as potential menu items for casseroles cooked on site. Figs. 1 and 2 show the original and reformulated recipes, and include the nutrition analysis generated using the USDA's Supertracker.

In Project 2, spending time at the shelter helped DFL identify opportunities to improve the quality and nutrition of menu items beyond the Standards themselves. It was instructive to observe how sites work both with and around the guidelines. While no one intended to provide unhealthy food, by not revising long-standing practices to the new guidelines, the site was unintentionally serving less-than-optimal food from a nutrition perspective. By promoting healthier casserole ideas as models, donors can provide compliant meal solutions at no extra cost to the site. This work challenged the team to not only think about working with purchasers to make sensible and compliant purchases, but also to consider making advances on the supply side to make compliant products available. Particularly given that, in this situation, the vast majority of meals provided were in the form of donated casseroles, the focus on creating and soliciting healthier versions helps the shelter in serving meals that align with the Standards they strive to uphold.

Ingredients
4 cups chopped cooked meat (left over beef, lamb, etc.)
4 cups mashed potatoes
4 cups gravy (or canned gravy)
1 tsp salt,
½ tsp pepper
⅛ tsp paprika
2 large onions, chopped or minced
2 Tbs butter or margarine

Instructions
1. Combine meat, gravy and onions.
2. Line bottom of buttered casserole pan with well beaten mashed potatoes.
3. Add thick layer of meat mixture then cover with layer of potatoes. (A really thick layer of potatoes will act as a crust.)
4. Dot with pats of butter.
5. Bake at least 3/4 hour until potatoes are browned on top.
6. Freeze.

Nutrition facts

Servings per container = 8	
Serving size	1/8 casserole
Amount per serving	
Calories	**315**
	% Daily value
Total fat 12 g	
Saturated fat 6 g	25%
Monounsaturated fat 4 g	
Cholesterol 67 mg	22%
Sodium 1202 mg	52%
Total carbohydrate 28 g	21%
Dietary fiber 3 g	11%
Total sugars 3 g	
Added sugars 0 g	0%
Proteins 27 g	58%
Vitamin A	9%
Vitamin C	12%
Calcium	5%
Potassium	15%
Iron	17%

FIG. 1 Original recipe for Shepherd's Pie.

Ingredients
4 Idaho potatoes, chopped
1 small cauliflower, chopped (about 6 cups)
¼ cup milk
2 tablespoons olive oil
1 tablespoon butter
1 pound 90% lean ground beef
1 large onions chopped (about 2 cups)
4 cups diced carrots
2 cup diced parsnips
1 teaspoon dried rosemary
1 teaspoon dried thyme
2 tablespoons flour
3 cups beef broth
1, 10 ounce package frozen peas, thawed
1, 15.5 ounce can lentils, drained and rinsed
salt and pepper to taste

Instructions
1. Place potatoes and cauliflower in a large pot and cover with water. Place the pot on the stove over high heat. Bring the water to a boil and cook the vegetables for about 20 minutes or until soft. Drain the water and mash the vegetables with milk. Season with salt and pepper.
2. Heat a large sauté pan over medium heat and add olive oil and butter. Add ground beef and cook until starting to brown, about 10 minutes. Add onion, carrots and parsnips and cook for about 15 minutes, until they begin to soften. Season with rosemary, thyme, salt and pepper.
3. Add the flour to the meat and vegetables, stirring to coat. Add the beef broth, while stirring. Bring the broth to a simmer and cook for about 15 minutes, until the vegetables are soft and the sauce has thickened. Add the peas and lentils.
4. Spray an aluminum casserole pan with cooking spray. Pour the beef and vegetable mixture into the bottom of the pan. Top with the mashed potatoes, and spread into an even layer. Wrap with plastic wrap and freeze.

Nutrition facts

Servings per container = 8	
Serving size	1/8 casserole
Amount per serving	
Calories	**353**
	% Daily value
Total fat 11 g	
Saturated fat 1 g	16%
Monounsaturated fat 5 g	
Cholesterol 39 mg	13%
Sodium 173 mg	8%
Total carbohydrate 44 g	34%
Dietary fiber 10 g	41%
Total sugars 9 g	
Added sugars 0 g	0%
Proteins 22 g	49%
Vitamin A	84%
Vitamin C	70%
Calcium	9%
Potassium	28%
Iron	27%

FIG. 2 Reformulated recipe for Shepherd's Pie.

PROJECT 3: CHANGING THE UPSTREAM VALUE CHAIN TO INCREASE COMPLIANCE

While education and technical assistance can have positive outcomes for individual sites, much greater impact can be achieved when food manufacturers and distributors ensure that compliant food is available, of high quality, and reasonably priced. Having a supply of compliant food shifts the burden from foodservice managers, who are challenged with satisfying diverse stakeholder groups, navigating extensive procurement guidelines, and finding healthy food in a catalog of noncompliant, but potentially delicious, competitively priced (often subsidized), and aggressively marketed foods. Food processing manufacturers and distributors, on the other hand, have much to gain in sales and profits by formulating, producing, and distributing compliant products to a receptive marketplace.

Despite the benefits to manufacturers, GHP encountered companies who expressed reluctance to invest in the research and development (R&D) expense of reformulating food without assurances that the market would be receptive and of sufficient size to provide a return on their investment. As such, GHP partnered with DFL to assist manufacturers in producing compliant food by assuming the burden of initial R&D investment and then making completed formulas available to manufacturers for commercialization and marketing at scale. Further, by helping these companies navigate the City's procurement process, GHP and DFL directly connected them to opportunities providing a market-based incentive to formulate products compliant with the City of Philadelphia's standards.

GHP and DFL identified a number of products that City agencies and partner hospitals struggled to source through ongoing conversations and market scans. For example, it is fairly easy to source compliant sliced sandwich bread, which must be whole grain, with less than or equal to 180 mg of sodium per serving and greater than or equal to 2 g of fiber per serving. But sandwich bread does not satisfy Philadelphians' cultural or culinary demand for a cheesesteak or hoagie, and agencies have difficulty sourcing a compliant hoagie roll. As shown in Table 2, none of the existing vendors and major suppliers of hoagie rolls offered a product compliant with the comprehensive food standards guideline for sodium, which specifies less than or equal to 290 mg for a hoagie roll in the category that includes "All other grains/starches (buns, hoagie rolls, dinner rolls, muffins, bagels, tortillas, waffles, etc." In addition, the standards recommend purchasing whole grain-rich baked goods (City of Philadelphia, 2017).

DFL approached the development of a compliant hoagie roll with a two-pronged approach. First, it reached out to one of the largest hoagie roll producers in the region to assess the company's interest in producing a compliant alternative. The company was open to the development, especially as it would meet the needs of other institutional customers as well. Second, DFL developed a hoagie roll in its own facility that meets the Standards and satisfies sensory expectations of multiple audiences, including foodservice managers, university students, and industry advisors. With a solid formulation in hand, DFL is now in talks with the large hoagie roll company about testing the manufacture of the formulation in its at-scale facilities and addressing commercialization issues such as shelf-life and ingredient sourcing.

Another project takes on mainstays of institutional foodservice: prepared deli salads, such as potato salad, coleslaw, and macaroni salad. These are typically high in fat, sodium, and sugar, and low in fiber. Few, if any, compliant deli salads are available in the Philadelphia market, and while they can be made compliant when prepared from scratch, labor and facilities capacity may limit this possibility. Further, where compliant salads exist, they often conflict with cultural and consumer expectations. As shown in Table 3, an option such as the fire-roasted edamame salad may not be seen by many consumers as an acceptable substitution for potato salad, coleslaw, or macaroni salad.

Working with a regional fresh-cut produce supplier, DFL reformulated traditional deli salads into produce-forward, appealing kits that align with consumer

TABLE 2 Hoagie Rolls From Existing Vendors and Drexel Food Lab Compliant Formulation

Company	Product	Calories	Fat	Sodium	Protein	Fiber	Sugar	Carb
GHP Standards	Hoagie rolls, etc.			290				
Amoroso's Baking Company:	School Italian roll #495	210	2.5	440	9	4	4	40
Drexel Food Lab:	6" roll, 50 percent WW flour	209	2.5	290	7	2.5	2.5	42

TABLE 3 Deli Salads From Existing Vendors and Drexel Food Lab Compliant Formulation

Company	Product	Calories	Fat	Sodium	Protein	Fiber	Sugar	Carb
GHP Standards	Other			480				
Sandridge Food	Baked potato salad	320	27	700	3	1	3	18
	Fire roasted edamame salad	120	6	460	6	4	2	9
Boston Salads	Elbow macaroni salad	286	21	598	3	1	5	21
	Red cabbage slaw	180	13	670	1	2	11	14
Drexel Food Lab	Potato and celery salad	77	4	100	1	1	1	10
	Mixed bean salad	95	2	247	4	4	2	16
	Cous Cous salad w/ butternut squash	124	6	124	3	2	3	17
	Asian cabbage slaw	83	7	93	2	1	1	4

expectations for potato salads, slaws, and vegetable-forward salads with familiar ingredients. Rather than distribute salads in premixed buckets or tubs in the fashion of traditional wet potato and macaroni salads, the kits arrive in boxes and are mixed on-site for freshness, requiring no additional preparation other than mixing. These kits are available for use in institutions, which can buy directly from the manufacturer, as well as in prepared foods sections of supermarkets and in restaurants in Philadelphia and beyond.

Based on the progress of the hoagie roll and deli salad projects, DFL sees additional opportunities for R&D work on commonly purchased items for which compliant sourcing is a challenge. These projects are in various states of commercialization: a hamburger blended with fresh vegetables to make it compliant as well as sustainable and price-competitive; a compliant soup mix made from dried vegetable scraps; tofu-based desserts; and breakfast sausages that incorporate plant-based proteins. These products have significant potential for impact.

DFL continues to engage multiple manufacturers, an approach that presents challenges as well as strong potential. First, it is difficult to make inroads with a food manufacturer with whom no prior relationship exists. Because DFL is part of a university academic program, it leverages connections such as employers of interns and alumni, fellow members of organizations such as the Institute of Food Technologists (IFT) or Research Chefs Association (RCA), or vendors serving the Drexel culinary program or City agencies to identify and connect with potential partners. Second, DFL struggles with the pace of many manufacturers' schedules. Because DFL is strictly focused on R&D and it is unable to actually manufacture products for sale, it depends on commercial partners to bring ideas to market. Often these partners' focus is divided among competing priorities, unexpected distractions, or larger, more lucrative projects. Third, food product development is a slow, multiyear process from conception, through formulation, optimization, consumer and sensory testing, test marketing, and launch. Many new food products fail (Aramouni and Deschenes, 2015), and while DFL realizes success in formulation/conceptualization, time will tell whether products can achieve consumer acceptance, achieve financial sustainability in the marketplace, and support dietary improvements. Decisions about how to create and implement effective branding, labeling, and consumer education campaigns also factor into the success of these reformulated food products.

CONCLUSION

The Philadelphia Nutrition Standards are changing the way City departments buy, prepare, and serve food. GHP and DFL began working together to identify barriers to Standards implementation, and arrived at solutions that evolved in their complexity and moved upstream in the supply chain from demand-side issues, such as price and recipes, to supply-side issues such as the manufacture and distribution of compliant products. By working with manufacturers to assist them in formulating healthier products, DFL can fill the gap between the demand and supply, and point food suppliers toward new markets for compliant products. While there is much more work to be done, the evolution of work—from assisting foodservice providers through *educating* them about procurement guidelines, to *supporting* them with guidance and technical assistance, to working with manufacturers to *supply* market-driven solutions—is a model that might hold potential in other markets.

In addition, this work may help to assuage condemnation of food procurement regulations as harmful to agencies in restricting flexibility without providing additional resources to comply with the guidelines, and costly to manufacturers, by restricting the products they can sell. As the reform of school food in many municipalities has proven, improved nutrition through regulation may concurrently provide opportunities for manufacturers to reach new markets with new products, which would not have been introduced in the absence of those regulations (Morgan and Sonnino, 2013).

DFL sees potential through technology transfer of developing and licensing additional health-promoting products following the completion of this project as part of their long-term sustainability and commitment to health promotion. DFL aims to create a new generation of food science and culinary professionals who apply their skills to addressing food system problems, improve the food system from the inside by educating students about food policy, and giving them practical hands-on culinary and food science skills to address problems, and collaborate with manufacturers to bring new products to market in a way that allows for licensing intellectual property back to the university to allow for financial sustainability of these efforts.

Implementation of nutrition standards policy is multifaceted, and taking a step-wise approach has proven critical to the early successes in Philadelphia. The support provided to City agencies through direct technical assistance and resource sharing paired with manufacturer engagement addressed both demand and supply-side challenges. DFL's flexibility and openness to considering a wide variety of avenues to address agency needs has encouraged further engagement and elevated agency work in providing healthy options to constituents.

REFERENCES

Aramouni, F., Deschenes, K., 2015. Methods for Developing New Food Products: An Instructional Guide. DESTech, Lancaster, PA.

Centers for Disease Control and Prevention, 2011. Improving the Food Environment Through Nutrition Standards: A Guide for Government Procurement. U.S. Department of Health and Human Services, Centers for Disease Control and Prevention, National Center for Chronic Disease Prevention and Health Promotion, Division for Heart Disease and Stroke.

Centers for Disease Control and Prevention, 2012. Health and sustainability guidelines for federal concession and vending operations. https://www.gsa.gov/portal/getMediaData?mediaId=239667. Accessed April 14, 2017.

City of Philadelphia, 2014. Executive order no. 4-14. Available from: http://www.phila.gov/health/pdfs/Executive%20Order.pdf.

City of Philadelphia 2016, Community Health Assessment Available from: (Accessed 17 April 2017). http://www.phila.gov/health/pdfs/CommunityHealthAssessment2016.pdf.

City of Philadelphia, 2017. Chronic disease—citywide nutrition standards. Available from: http://www.phila.gov/health/chronicdisease/workplace.html.

Gardner, C.D., et al., 2014. Food-and-beverage environment and procurement policies for healthier work environments. Nutr. Rev. 72 (6), 390–410.

Hyseni, L., et al., 2016. The effects of policy actions to improve population dietary patterns and prevent diet-related non-communicable diseases: Scoping review. Eur. J. Clin. Nutr. 71 (6), 694–711.

Johns, N., Pine, R., 2002. Consumer behaviour in the foodservice industry: a review. Int. J. Hosp. Manag. 21 (2), 119–134.

Karpilow, K.A., 2012. Healthy Procurement in Public Agencies (A Nutrition Primer Module). Addendum to Understanding Nutrition: A Primer on Programs and Policies in California, third ed. California Center for Research on Women and Families, Public Health Institute, Sacramento, CA.

Kimmons, J., et al., 2012. Developing and implementing health and sustainability guidelines for institutional foodservice. Adv. Nutr. Int. Rev. J. 3 (3), 337–342.

Lederer, A., et al., 2014. Toward a healthier city. Am. J. Prev. Med. 46 (4), 423–428.

MacLellan, D., Taylor, J., Freeze, C., 2009. Developing school nutrition policies: enabling and barrier factors. Can. J. Diet. Pract. Res. 70 (4), 166–171.

Maes, L., et al., 2011. Effectiveness of workplace interventions in Europe promoting healthy eating: a systematic review. Eur. J. Pub. Health 22 (5), 677–683.

Morgan, K., Sonnino, R., 2013. The School Food Revolution: Public Food and the Challenge of Sustainable Development. Routledge, London.

Pache, A., Santos, F., 2010. When worlds collide: the internal dynamics of organizational responses to conflicting institutional demands. Acad. Manag. Rev. 35 (3), 455–476.

Parker-Pope, T., 2007. A high price for healthy food. 5 December 2007. Well. Available from: https://well.blogs.nytimes.com/2007/12/05/a-high-price-for-healthy-food/comment-page-17/.

Payne-Palacio, J., Theis, M., 2015. Foodservice Management: Principles and Practices, 13th ed. Pearson Education South Asia Pte Ltd, Singapore.

Philadelphia Department of Public Health, 2017. Foodservice toolkit, City of Philadelphia. Available from: https://drive.google.com/file/d/0B2GPbD7d2z78MnZ6aW8ya2FHX28/view.

Pitts, S.B.J., et al., 2016. Implementing healthier foodservice guidelines in hospital and federal worksite cafeterias: barriers, facilitators and keys to success. J. Hum. Nutr. Diet. 29 (6), 677–686.

Roberto, C.A., et al., 2015. Patchy progress on obesity prevention: emerging examples, entrenched barriers, and new thinking. Lancet 385, 2400–2409.

Robles, B., et al., 2013. Comparison of nutriton standards and other recommended procurement practices for improving institutional food offerings in Los Angeles County, 2010–2012. Adv. Nutr. 4, 101–202.

The Food Trust, 2014. Healthy corner store initiative: overview. Available from: http://thefoodtrust.org/uploads/media_items/healthy-corner-store-overview.original.pdf.

The Food Trust, 2016. Get Healthy Philly, Farmers Market and Philly Food Bucks Report 2016. Available from: http://thefoodtrust.org/uploads/media_items/2016-get-healthy-philly-farmers-market-and-philly-food-bucks-report.original.pdf.

Tsui, E.K., et al., 2013. Missed opportunities for improving nutrition through institutional food: the case for food worker training. Am. J. Public Health 103 (9), e14-e20.

Tsui, E.K., et al., 2015. Institutional food as a lever for improving health in cities: the case of New York City. Public Health 129 (4), 303–309.

FURTHER READING

Centers for Disease Control and Prevention, 2017. About the Sodium Reduction in Communities Program. Available from: https://www.cdc.gov/dhdsp/programs/about_srcp.htm.

Drexel University, 2017. Culinary arts and food science—the Drexel Food Lab. Available from: http://drexel.edu/hsm/academics/Culinary-Arts-Food-Science/food-lab/.

New York City Mayor's Office of Contract Services, 2015. New York State Food Purchasing Guidelines, City of New York. Available from: https://www1.nyc.gov/assets/mocs/downloads/ pdf/epp/New%20York%20State%20Food%20Purchasing%20Guidelines%203.pdf.

Philadelphia Department of Public Health, 2016. 2016 Community Health Assessment (CHA), City of Philadelphia. Available from: http://www.phila.gov/health/pdfs/CommunityHealth Assessment2016.pdf.

Jonathan Deutsch, PhD is a professor in the Center for Food and Hospitality Management and Department of Nutrition Sciences at Drexel University and the James Beard Foundation's Impact Fellow, leading a national curriculum effort to train the next generation of chefs in preventing food waste. He oversees the Drexel Food Lab, a product development and food innovation lab that solves real world problems. He is the author or editor of six books including Barbecue: A Global History (with Megan Elias), Culinary Improvisation, and Gastropolis: Food and Culture in New York City (with Annie Hauck-Lawson). When not in the kitchen, he can be found behind his tuba.

Alexandra Zeitz is the manager of the Drexel Food Lab, a culinary innovation and product development research lab housed in the Center for Food and Hospitality Management. Alexandra is a graduate of the Culinary Arts program at Drexel University. During her time at Drexel, she assisted in the founding of the Food Lab, and completed her Co-Op and later worked in the bread department of High Street on Market and James Beard nominated Fork Restaurant, both located in Philadelphia. Alexandra now works as a full-time staff researcher at Drexel, managing the Food Lab and its projects. She is also completing her Masters of Education at Drexel University in Creativity and Innovation.

Benjamin Fulton is a graduate student in the food science program offered by the Center for Food and Hospitality Management and Department of Nutrition Sciences of Drexel University in Philadelphia, Pennsylvania. He is also a part of the Drexel Food Lab team, which, through research and culinary innovation, aims to solve real world problems in areas such as sustainability, nutrition, and noncommunicable disease prevention.

Brandy-Joe Milliron is an assistant professor in the Department of Nutrition Sciences at Drexel University. As a community nutrition scientist, Brandy-Joe has a broad background in obesity prevention and management. Her current research explores the ways in which nutrition, wellness, and healing behaviors promote healthy cancer survivorship and caregiver health. In addition to cancer prevention and control, her research spans from global nutrition and maternal, infant, and young child feeding, to garden-based nutrition education, food-related decision making, and the evaluation of community food projects.

Catherine Bartoli is the Healthy Food Procurement Coordinator for Get Healthy Philly, Division of Chronic Disease Prevention, within the Philadelphia Department of Public Health. In this position, Catherine works with city agencies and hospitals to implement comprehensive nutrition and procurement standards. She was previously the Nutrition Coordinator for Alum Rock Union Elementary School District's Child Nutrition Services department. Catherine received her M.A. in Urban Studies from Temple University and M.S. in Nutritional Science with a focus on community nutrition education from San Jose State University. She is also a registered dietitian and licensed dietitian nutritionist.

Part III

Creating Shared Identity Using Place-Based Connections

Chapter 7

Making Local Sourcing Standard Practice: Lessons From Michigan

Kathryn Colasanti*, Colleen Matts*, Kaitlin K. Wojciak[†]
*Center for Regional Food Systems, Michigan State University, East Lansing, MI, United States
[†]Michigan State University Extension, East Lansing, MI, United States

Chapter Outline

Introduction	149	Case Studies	164
Growth in Farm-to-Institution Activity	151	Case Study 1: Baxter Child Development Center	165
History of Farm-to-Institution Organizations in Michigan	151	Case Study 2: Calumet Center	166
State Government Policy	158	Case Study 3: Bronson Methodist Hospital	168
Non-Profit and MSU Extension Facilitation	160	Conclusion	170
		References	171
Supply Chain Partnerships	163	Further Reading	173

INTRODUCTION

In response to challenges such as the loss of small farms, industry consolidation, rising obesity rates, and inequitable access to healthy food options, consumer support for local food in the United States has soared. This interest in farm-direct sourcing has driven a dramatic rise in farmers' markets and other direct market sales over the past 20 years (USDA, 2017), and it has matured into a movement converging with initiatives for public health, sustainability, and racial equity (Pirog et al., 2014). Over the same time period, a growing number of institutions have aligned with this local food movement and taken steps to use their foodservice programs to help local farm businesses, support their local economies, and provide fresh, healthy foods to their clients (Beery and Joshi, 2007; Hardesty et al., 2010; Vallianatos et al., 2004; Vogt and Kaiser, 2008). These efforts are broadly described as "farm-to-institution" activity, which can encompass a variety of initiatives, such as on-site gardens and classroom agricultural education, but centers around purchasing and serving foods sourced from farmers and producers in the "local" geographic region, which many institutions define by state boundaries (Izumi et al., 2010a; Thompson et al., 2016).

Beyond documenting the presence of farm-to-institution programs, a growing body of literature has helped to define the nature and potential of such programs, as well as the motivations and barriers practitioners experience. One study showed that school foodservice personnel were motivated to serve local food because of positive student reaction, competitive prices, and the sense of helping area farmers (Izumi et al., 2010c). Other research suggests that producers are interested in selling to institutions for both economic and social reasons (Izumi et al., 2010b), although thus far, their actual sales to institutions tend to be small (Beery and Joshi, 2007; Feenstra et al., 2011; Matts et al., 2015). Overall, the existing research indicates that there are barriers to implementing farm-to-institution models and that realizing the full potential of farm-to-institution programs will require changes in food system distribution, more education of all stakeholders, and improved information flow across the supply chain (Feenstra et al., 2011; Hardesty et al., 2010). Indeed, research from the United States Department of Agriculture (USDA) shows that supportive state policy, a high density of farmers' markets in the area, and an urban setting with affluent, highly-educated residents all help facilitate successful farm-to-school programs (Ralston et al., 2017).

Michigan has been on the leading edge of the national farm-to-institution trend, witnessing substantial increases in local food procurement activity over the past ten years (Colasanti and Matts, 2013). The agricultural diversity in the state, which is second only to California in the number of different crops cultivated commercially (Michigan Department of Agriculture and Rural Development, 2014), provides extensive opportunities to source Michigan foods. In our personal observations in Michigan, the most motivated and forward-thinking foodservice professionals take advantage of the agricultural diversity around them by initiating relationships with local growers and suppliers. Research focusing on Michigan shows that the K-12 schools engaged in local sourcing (approximately 54 percent), purchase through a variety of suppliers: 80 percent through broadline (full-service) distributors, 58 percent through federal food programs, 14 percent directly from local farmers, and 3 percent through food hubs (Thompson and Matts, 2015). However, it is only through nearly two decades of dedicated funding, outreach, technical assistance, and advocacy that Michigan is moving toward widespread adoption of local food sourcing. In this chapter, we examine the three arenas that have had the most impact in advancing farm-to-institution activity in the state: state government policy, non-profit and agriculture extension agency facilitation, and supply chain partnerships.

We start the chapter by sharing the history of farm-to-institution activity and organizations in Michigan. We then discuss the lessons evident in the role various actors have played in supporting farm-to-institution activity. We follow this with case studies of three different institutions: a child development center, a juvenile detention facility, and a hospital system. Each example highlights the ways in which personnel are making local sourcing a standard practice of their foodservice program and the organizational motivations underlying these changes. We pull from our own research and work, as participants of farm-to-institution activities in Michigan, to illustrate our points. We conclude by reflecting on how collective efforts are leading toward the institutionalization of farm-to-institution, in which local sourcing in institutional foodservice moves from a niche activity to a widespread standard practice.

GROWTH IN FARM-TO-INSTITUTION ACTIVITY

Surveys of farm-to-school activity in Michigan indicate that local sourcing practice and interest has grown substantially over the past decade (Table 1). In 2004, more than 10 percent of responding school foodservice directors reported purchasing foods from a local farmer (Izumi et al., 2006). Ten years later, 2014 data from the Michigan Department of Education showed that 54 percent of foodservice directors sourced local foods for their school food program in the 2014–15 school year (Thompson and Matts, 2015). In part because non-profit organizations and federal initiatives have focused their efforts on food in K-12 schools, they are ahead of other institutional sectors in farm-to-institution activity. However, interest in farm-to-institution is also evident in other sectors, including early childcare and education sites, as well as among Michigan vegetable farmers (Matts, 2013; Colasanti and Thompson, 2016; Matts et al., 2015).

This body of data shows that despite the challenges, a growing number of institutions are seeking out Michigan grown and produced foods, and a substantial share of local farmers are expressing interest in selling to institutions (even if they do not currently do so). The growth in farm-to-institution activity, however, has only occurred through a parallel growth in the networks of support and technical assistance. We now turn to an overview of the history of organizations supporting farm-to-institution activity in Michigan.

HISTORY OF FARM-TO-INSTITUTION ORGANIZATIONS IN MICHIGAN

Strong support organizations have been an important driving force in the growth of farm-to-institution activity in Michigan (Table 2). Drawing on national interest in farm-to-school, state agency and academic partners began gathering regularly in 2001 to discuss farm-to-school potential. Organizational support for farm-to-institution work solidified in the mid-2000s when two regional non-profit organizations dedicated staff time to facilitating farm-to-school programs and supporting school foodservice directors' efforts to purchase local foods. The Michigan Food Policy Council, created by State Executive Order in June 2005, also formed a task force dedicated to farm-to-institution. Staff at the Michigan State University Center for Regional Food Systems (CRFS) have served as co-ordinators of statewide farm-to-school activities since 2007, providing financial

TABLE 1 Survey Data Showing Interest in Local Sourcing in Michigan

Survey Year	Population Surveyed	Sample Size	Response Rate	Key Findings	Source
2004	Michigan K-12 school foodservice directors	383	58 percent	• 11 percent purchased local foods in the last year • 83 percent interested in purchasing local foods	Izumi et al. (2006)
2009	Michigan K-12 school foodservice directors	270	28 percent	• 41 percent purchased local foods in the last year • 70 percent interested in purchasing local foods	Colasanti et al. (2012)
2012	Metro Detroit K-12 schools and districts, hospitals, colleges and universities	80	24 percent	• 78 percent purchased local foods in the last year • Of those who had not purchased local foods, 67 percent were interested in doing so	Matts and Colasanti (2013)
2012	Michigan K-12 school foodservice directors	317	34 percent	• 54 percent purchased local foods in the last year • 89 percent interested in purchasing local foods	Matts and Smalley (2013)
2012	Great Start Readiness Programs	76	33 percent	• 32 percent engaged in farm-to-school activities • 69 percent interested in connecting with local farmer • 53 percent interested in starting a garden	Matts (2013)

Year	Group	N		Findings	Source
2012	Michigan vegetable farmers	311	38 percent	• 7 percent currently selling to institutions • 47 percent interested in selling to K-12 schools • 41 percent interested in selling to hospitals • 40 percent interested in selling to colleges or universities	Matts and Smalley (2014)
2013	Michigan K-12 school foodservice directors	354	38 percent	• 68 percent purchased local foods in the last year • 82 percent interested in purchasing local foods	Matts and Smalley (2014)
2013–14	National K-12 school foodservice directors	12,585	70 percent	• 43 percent participated in farm-to-school activities • 18 percent planned to start farm-to-school activities	United States Department of Agriculture (2016)
2014	Michigan K-12 school foodservice directors	878	100 percent[a]	• 54 percent currently purchasing local foods	Thompson and Matts (2015)

[a] In this dataset, "No" responses from this optional question were indistinguishable from answers left blank; 97 respondents who did not provide a "Yes" answer to the initial question but still marked a source of local foods in the follow-up question are excluded from this count. Therefore, the actual number of school food service directors purchasing local foods may be higher than our reported number.

TABLE 2 Overview of Farm-to-Institution Support Organizations in Michigan

Scope	Organization	Farm to Institution Role
Regional organizations	Groundwork Center for Resilient Communities (previously known as the Michigan Land Use Institute)	• Farm-to-school support • Participates in the Michigan Farm to Institution Network (MFIN) • Supporting Partner of the National Farm to School Network • Coordinated initial, regional 10 Cents Pilot • Provides communications support for the state-level 10 Cents Pilot
	Food System Economic Partnership, southeast Michigan (disbanded)	• Farm-to-school support
State-level organizations	Ecology Center	• Farm-to-hospital support • Co-lead of MFIN • Coordinator of Michigan's Healthy Food in Health Care program, part of the national Health Care Without Harm Campaign
	Michigan State University Center for Regional Food Systems	• Farm-to-school support and tracking • Co-convener of the Michigan Good Food Charter, including the Institutional Food Purchasing Workgroup • Co-lead of MFIN • Provides evaluation support for the state-level 10 Cents Pilot • State Lead/Core Partner of the National Farm to School Network
	Michigan State University Extension	• Community food systems, including farm-to-school and farm-to-institution, support • Management team member of MFIN • Supporting Partner of the National Farm to School Network

TABLE 2 Overview of Farm-to-Institution Support Organizations in Michigan—cont'd

Scope	Organization	Farm to Institution Role
State of Michigan	Michigan Food Policy Council (disbanded)	• Co-convener of the Michigan Good Food Charter • Coordinator of farm-to-institution task force
	Michigan Department of Education	• Farm-to-school and farm-to-institution support and tracking • Leadership team member of MFIN • Administers state-level 10 Cents Pilot • Supporting Partner of the National Farm to School Network
	Michigan Department of Agriculture and Rural Development	• Farm-to-school and farm-to-institution support • Leadership team member of MFIN • Provides support for the state-level 10 Cents Pilot • Supporting Partner of the National Farm to School Network

and technical assistance, training, resources, and policy education. Staff from the Ecology Center, a non-profit based in Ann Arbor, Michigan, began working with hospitals in the state on local and sustainable food procurement in 2008.

Farm-to-institution interest was on the rise just as the Michigan economy was sliding into the Great Recession of 2008–09. While states across the country suffered during this period, Michigan's woes were particularly severe and persistent, due to the state's economic dependence on the automobile industry, which experienced major losses during this time. Surveys of foodservice professionals at that time showed that many were motivated to engage in farm-to-institution activities because of the opportunity to support their local economy and provide nutritious food to vulnerable populations (Colasanti et al., 2012; Matts and Smalley, 2013). Although recessions can make local procurement more difficult by further restricting budgets, economic hardship may also make the potential benefits of local food sourcing more salient.

The presence of a broad, shared agenda for food system change in Michigan has also been a key catalyst in the rise of farm-to-institution activity. Since its release in 2010, the Michigan Good Food Charter (Colasanti et al., 2010a) has served as a shared vision for a food system that is "rooted in local communities and centered on 'good food' – food that is healthy, green, fair, and affordable." The organizations leading the Charter development process saw institutions as a critical part of "good food" because of the large number and size of institutional foodservice programs across the state. As a way to identify opportunities to leverage the collective buying power of institutions, these organizations formed an Institutional Food Purchasing Workgroup as one of five workgroups in the Charter development process. Ultimately, the statewide, participatory process of creating the Charter and the Charter's recognition of the value of farm-to-institution activity helped strengthen and diversify support for the role of institutions in a locally based food system.

The Institutional Food Purchasing Workgroup brought an opportunity to engage stakeholders in developing a common vision for institutional food purchasing and planted the seeds for the network that would emerge later. An institutional foodservice director, a graduate student, and a farm-to-institution specialist at the Center for Regional Food Systems (CRFS) co-led the workgroup of 26 members, including foodservice providers from various institution types, farmers, specialty and full-service distributors, state agency staff, and farm-to-institution facilitators from non-profit organizations. As the workgroup discussed an overarching goal for 2020, some members were concerned about being publicly held to any local food purchasing target. Others were hesitant to commit to locally sourcing more than 5 or 10 percent. Nonetheless, the group eventually agreed to put forward the goal that "Michigan institutions will source 20 percent of their food products from Michigan growers, producers and processors" by 2020 (Colasanti et al., 2010a, p. 8).

Beginning in 2010, in response to the Michigan Good Food Charter and as part of overall restructuring, a new Community Food System Work Group within Michigan State University (MSU) Extension began to include farm-to-institution support in their work. In 2012, the CRFS farm-to-school staff team began to reach out to various types of early childcare and education (ECE) programs, first by offering mini-grants to support local food sourcing, and then through training, technical assistance, and resource provision tailored to the unique needs of these programs.

Several years later, in 2014, the Charter's 20 percent local sourcing goal was the driving force behind the launch of the Michigan Farm to Institution Network (MFIN). The network, co-led by the MSU Center for Regional Food Systems and the Ecology Center with support from MSU Extension, provides a space for learning, sharing, and collaborating around the shared goal of getting more local food to institutional foodservice programs. By 2017, the more than 550 network members included institutional foodservice directors and buyers as well as farmers, food vendors and suppliers, advocates, supporters, and researchers. Among the MFIN members who identified their affiliations, approximately one-fourth are from institutions, primarily K-12 schools and hospitals. Non-profit and university supporters, advocates, and researchers (including over 20 MSU Extension educators), comprise another quarter of the network's members. Food producers and suppliers make up more than 15 percent of network members, including farmers, food hub managers, distributors, processors, and representatives from agricultural commodity groups. Staff from state, city, and county government as well as public health practitioners and consultants round out the remainder of membership with identifiable affiliations. MFIN members are distributed across the state with predictably higher engagement in population centers, primarily in western and southern Michigan, and lowest engagement in the Upper Peninsula.

MFIN works to connect institutional foodservice directors and buyers with local food vendors, to leverage the collective demand from institutions to drive changes in the upstream supply chain, and to help consumers at institutions identify, value, and enjoy local foods. MFIN's local food purchasing campaign, Cultivate Michigan, was launched at the same time as the network to help institutions reach the 20 percent by the 2020 local sourcing goal. As part of its promotion of Michigan agriculture, the campaign features a Michigan food in each of the four seasons. When institutions sign up to participate, they receive free promotional materials for the featured foods for use in their cafeterias or dining areas and purchasing guides to help with sourcing and using the products. In turn, they are asked to submit data on their purchases of Michigan foods. Cultivate Michigan membership is restricted to institutions, defined as ECE programs, K-12 schools, colleges/universities, hospitals, and senior facilities.[1]

Cultivate Michigan members are recruited through the Cultivate Michigan website, and email communications are disseminated through professional networks of advisory committee members and staff of partner organizations, and conference sessions and exhibits. In addition, eight Cultivate Michigan Ambassadors provide regional or sector-specific touch-points, information on how to join, and technical assistance to new and existing members. In 2018, MFIN staff, along with partners from the Michigan Departments of Education and Agriculture and Rural Development, and partners from local communities, will host meet and greet type events in four regions across the state to help current members connect to more local food vendors within their regions and recruit new members.

Financial support from private and public funders has been essential to enabling the activities of these support organizations and, therefore, to facilitating the rise of farm-to-institution activity. Key sources of funding include:

- *Private foundations.* The W.K. Kellogg Foundation, based in Battle Creek, Michigan, has played a major role in advancing farm-to-institution activity by providing funding to develop and promote the Michigan Good Food Charter, to operate the MI Farm to School Grant Program and to support the salaries of non-profit staff at multiple organizations.

---

1. Other types of food buyers, including camps, retailers, and catering operations, have asked to join Cultivate Michigan and track their local food purchases, but were only provided membership with the understanding that Cultivate Michigan's tracking tools are designed for the primary audience of institutions.

- *Federal government.* USDA funding, which increased its support for local food efforts under the Obama administration, has also been crucial in advancing farm-to-institution efforts. Over the past decade, MFIN members and partners received a USDA Agriculture and Food Research Initiative (AFRI) farm-to-institution research grant, a Specialty Crop Block Grant for a culinary training program, and a USDA Farm to School Grant, which was awarded to the Michigan Department of Education in partnership with MFIN personnel.
- *State government.* The state legislature appropriated funds for a farm-to-school pilot program, called 10 Cents a Meal for School Kids and Farms, for the first time in 2016.

Although these funding sources have had crucial impacts, funding alone is not sufficient to drive farm-to-institution activity. Institutional foodservice programs need supportive policy, technical assistance, and changes in supply chain structures in order to make local purchasing more than a token activity. This is largely due to structural challenges and other market failures in agricultural supply chains that make it challenging for local producers to sell competitively to institutional markets and for institutional foodservice to purchase consistent supply from local producers. Ultimately, it is through concerted efforts to develop interconnected areas of support that has both allowed Michigan organizations to be successful in receiving funding and enabled the financial investment in farm-to-institution activities to be effective. The next section outlines the three key areas that make up this interconnected web of support for institutional farm-direct purchasing: state government policy, non-profit and agriculture extension agency facilitation, and supply chain partnerships.

STATE GOVERNMENT POLICY

State government policy—both legislation and state agency practices—has played a pivotal role through farm-to-school legislation, data tracking, and funding for innovative programming. State agency staff from the Michigan Departments of Education, Agriculture and Rural Development, and Health and Human Services began farm-to-school conversations in 2001. The state legislature delivered a clear message of support in 2008 when a farm-to-school bill package was signed into law. As farm-to-school activity was budding, it became clear that the state's low small-purchase threshold for food procurement could limit programs from flourishing. Public demand inspired state legislators to compile a farm-to-school package of three bills, which were signed into law in 2008 by former Governor Jennifer Granholm. Public Acts 343 and 344 of 2008 increased the small-purchase threshold for food purchases up to $100,000 (the federal threshold at the time) for school districts and intermediate school districts, respectively. Public Act 315 of 2008, the farm-to-school procurement act, required the Michigan Departments of Education, Agriculture and Rural

Development, and Health and Human Services[2] to collaborate and cooperate on providing farm-to-school support. Today, staff from these agencies weave farm-to-institution into their program provision and serve on the MFIN leadership team.

With support from Michigan Department of Education staff, CRFS took on the task of tracking farm-to-school interest, activity, challenges, and opportunities by surveying school foodservice directors, in 2004, 2009, 2012, 2013, and 2014 (Table 1). This data proved to be valuable for legitimizing farm-to-school efforts, tracking impacts, contributing to MFIN's development, and even spurring and directing business development to provide local foods. However, conducting surveys, including on an annual basis between 2012 and 2014, required substantial staff time and never yielded a response rate higher than 38 percent. Starting in 2014, this tracking effort was institutionalized at the Michigan Department of Education. At that time, the staff added optional questions about local food procurement into the electronic application that all school foodservice directors must complete in order for their school or district to participate in the USDA National School Lunch Program. A summary of the 2014 results from these questions showed that 54 percent (470 of 878) of applicants responded affirmatively that they purchased local foods for their school food program. This tracking will continue to be useful to understand the rate and extent of farm-to-school adoption in the state, particularly when paired with the more comprehensive results of the USDA Farm to School Census.

State government policy provided further leverage for farm-to-institution activity in 2016 when the state legislature appropriated funds for a procurement incentive pilot program for schools. A "10 Cents a Meal" pilot was initially recommended as a priority in the 2010 Michigan Good Food Charter. Based on the success of a regional pilot in northwest lower Michigan administered by the Groundwork Center for Resilient Communities using a combination of public and private funds, the 10 Cents a Meal for School Kids and Farms pilot gained traction with some state senators. As a result, $250,000 was appropriated in the state K-12 school support budget to implement a new state-supported pilot administered by the Michigan Department of Education, with $40,000 supporting administration. The initial state pilot for the 2016–17 school year provided $210,000 in matching reimbursements for the purchase of Michigan-grown fruits, vegetables, and dry beans; reinvesting a total of $420,000 in the Michigan economy.

For the first year of the pilot (2016–17), sixteen school districts were accepted into this competitive program. They were located in eight counties in northwest and western Lower Michigan, in areas ranging from rural to urban, including Muskegon and Grand Rapids metropolitan areas. Together, these

---

2. When Public Act 315 was enacted, the Michigan Department of Agriculture and Rural Development was referred to as the Michigan Department of Agriculture and the Michigan Department of Health and Human Services was referred to as the Michigan Department of Community Health.

districts serve nearly 48,000 students more than 3.7 million meals. Seven of the sixteen participating districts, representing about one-third of the total students reached through this pilot program, had free and reduced-price meal eligibility rates of 50 percent or greater. Two of the largest participating school districts (more than 10,000 students each) had free and reduced rates below 50 percent.

By December 2016, only mid-way through the pilot, 86 different Michigan farms and sixteen other businesses located in 28 Michigan counties received business as a result of the pilot (Michigan Department of Education et al., 2017). At least 50 different types of Michigan-grown produce were served to children, including 30 foods new to school food programs. Foodservice directors indicated that the variety of produce served, certainty for planning purchases, and purchasing power are all increased by participating in the pilot (Michigan Department of Education et al., 2017).

As a result of a legislative expansion of the program, funding was increased to $375,000 and school districts from an additional six-county region in southeast Michigan (which includes Ann Arbor) were also eligible to apply to participate in the second year of the pilot (2017–18). Thirty-two school districts in 29 counties were accepted into the program. Together these districts enroll 95,000 students, nearly double the number of students served the prior year. Sixteen of the school districts, representing over one-third of the students impacted by the program, have free and reduced rates of 50 percent or greater.

By passing legislation to remove local sourcing barriers, institutionalizing state agency engagement in farm-to-school discussions, incorporating farm-to-school tracking into existing procedures, and allocating funds for pilot farm-to-school programs, Michigan's state government has played a growing role in legitimizing farm-to-school activity, which has spurred interest in farm-to-institution activity more broadly. These policy changes have helped pave the way for more active facilitation support from agriculture extension agencies and non-profits in Michigan.

NON-PROFIT AND MSU EXTENSION FACILITATION

Surveys of Michigan foodservice directors between 2009 and 2014 revealed the main barriers to sourcing local food, particularly in the K-12 school sector. These included budget constraints, federal and state procurement regulations, lack of producers in the area from whom to purchase, limited seasonal availability, and food safety concerns. This research allowed MSU Extension and other non-profit organizations to provide targeted technical assistance to help foodservice directors overcome these challenges and initiate farm-to-institution programs.

Much of this technical assistance has come in the form of educational resources. CRFS staff developed a suite of step-by-step guides for farm-to-school (Izumi and Matts, 2008; Matts, 2010; Colasanti et al., 2010b; Koch and Matts, 2016) and farm-to-ECE (Murphy et al., 2015; Harper and Matts, 2017). To address the seasonal availability challenge, a team of community food systems extension educators, who are also active MFIN members, developed and implemented culinary training focused on seasonal Michigan menus for school foodservice staff. In 2016–17, CRFS provided a twelve-part Farm to Early Child Care Mini-Webinar Series[3] that offered ten-minute "bites" of topics related to local food procurement and supporting activities, including on purchasing from different local food vendors (farmers, food hubs, and distributors), seasonal menu planning, taste tests, lesson plans, gardens for experiential learning, and engaging parents. The series of short webinars, specifically designed for busy program staff, has resulted in more frequent requests for technical assistance.

In response to survey data showing that budget was one of the top challenges Michigan school foodservice directors face in purchasing local foods (Izumi et al., 2006; Colasanti et al., 2012), CRFS, with funding from the W.K. Kellogg Foundation, launched the MI Farm to School Grant Program in 2011. Year-long mini-grants, typically from $500 for smaller ECE sites to $2000 for school districts, are awarded to eligible schools and ECE food providers who apply, meet program requirements, and are accepted based on the application review score. In order to focus its impact on vulnerable children, the grant program requires free and reduced-price meal eligibility rates of 50 percent or greater for schools and the equivalent for ECE programs. Grantees are eligible to participate in the program for up to three years, ideally moving from a planning year into two years of implementation upon successful re-application.

In the first three years of the MI Farm to School Grant Program, 50 mini-grants were awarded—35 to K-12 schools and fifteen to ECE programs—that together served more than 40,000 children. During that time, grant awards went to schools and ECE programs in nineteen of Michigan's 83 counties, including six in Wayne County, which is highly urbanized, and two in the Upper Peninsula, which is primarily rural.[4] By the 2016–17 grant year, more than $200,000 had been disbursed through about 110 grants to 70 different programs through CRFS. Grantees' reporting showed that the majority of grant funds were typically reinvested in communities through local food purchases. Follow up interviews with previous grantees conducted in 2015 showed that those who participated in one year each of planning and implementation grants were most likely to maintain or even grow their farm-to-school programs after the grant program. By providing a package of financial assistance combined with training and technical assistance, this grant program has benefited kids, farms, and communities, and helped link farm-to-school practitioners to one another through participation in MFIN events.

---

3. Available from: http://www.canr.msu.edu/news/farm_to_early_child_care_mini_webinar_series (Accessed 21 December 2017).
4. As the second phase of the MI Farm to School Grant Program is still ongoing, cumulative information after the first three years is not yet available. Data from the first three years of the grant program is available in a summary report is available from: http://foodsystems.msu.edu/resources/mi_farm_to_school_grant_program_report (Accessed 21 December 2017).

Finally, non-profit staff help institutions understand federal and state regulations addressing local food procurement and provide procurement tools to help institutions meet the requirements. An MSU Extension educator works to provide farm-to-institution-specific education around food safety for both growers and buyers, including the impending rules for the federal Food Safety Modernization Act. In 2016, MFIN initiated partnerships with eight ambassadors who, as part of their professional positions, most as extension educators, provide a personal point of engagement with the network for institutions in their region or sector. In addition to recruiting new institutions to the campaign, the ambassadors stay in touch with participating institutions, welcoming them, connecting them to network resources and reminding them to complete quarterly surveys documenting local purchases. MFIN staff have also begun to meet with food hubs and distributors directly to convey the demand for local food on behalf of partner institutions and to help equip institutions to ask for sourcing documentation from their food distributors and suppliers.

As farm-to-institution work in Michigan has broadened to include more institutions and more sectors, non-profit and extension staff have learned that understanding the differences between and within institution types is essential to providing relevant technical assistance. Farm-to-institution activity can look very different based on an institution's mission, structure and size; level and type of agricultural activity in the surrounding community; whether the foodservice program is self-operated or managed through an external contract; and the range of different foodservice operations—for example, dining halls, catering, restaurants, food courts, convenience stores, coffee shops or stands, and concessions. Some of these differences include:

- *ECE programs.* These sites tend to have flexible curricula and an emphasis on parent involvement, making this sector more likely to integrate local food procurement with educational and parent engagement activities. Because the programs are often small, they can be a good fit for small-scale and beginning farmers interested in trying out institutional marketing.
- *K-12 schools.* Schools have particularly tight foodservice budgets and are constrained by the USDA National School Lunch Program and other Child Nutrition Program funding structures. However, schools are the easiest for non-profit staff to communicate with through a single administrative entity because, with the exception of schools that may opt-out, they all report to the Michigan Department of Education (MDE) in order to participate in federal Child Nutrition Programs.
- *Hospitals.* Hospitals are more likely to serve local foods in their retail settings than in patient meals, partially because of the marketing possibilities at the point of sale. Hospitals tend to be particularly concerned about food safety audits and assurances because they serve many immuno-compromised individuals.

- *Higher education.* Colleges and universities often have broader sustainability plans into which they integrate local food procurement, meaning they look for vendors that can also meet environmental standards.

Although targeting farm-to-institution facilitation efforts to respond to identified barriers and the unique contexts of different institutions takes more time and more resources, it has allowed the efforts to be more effective.

SUPPLY CHAIN PARTNERSHIPS

Supply chain partnerships represent the third key element underlying the growth of farm-to-institution activity in Michigan, after state government policy and non-profit and MSU Extension support. MFIN co-coordinators have recognized from the beginning that, given the complex and interconnected nature of food systems, scaling up farm-to-institution requires simultaneously building up other components of the food system. Alongside direct facilitation of farm-to-institution activity, MFIN leaders have pursued synergistic partnerships with networks and initiatives focused on other components of the food system.

For example, in 2014, CRFS helped expand an innovative farmer support program to include farm-to-school. The Hoophouses for Health program, administered by the Michigan Farmers Market Association (MIFMA) with technical assistance provided by the MSU Department of Horticulture and with funding from the W.K. Kellogg Foundation, provides zero-interest loans for farmers to build a new hoophouse.[5] Participating farmers repay their loans by providing food to vulnerable children and families in their communities, through vouchers or electronic benefits redeemed at farmers' markets and, after the program expansion in its fourth year, through "sales" to ECE and K-12 school food programs. The partnership creates new relationships between farmers and foodservice buyers while growing the supply of food available to institutions year-round. In 2015 and 2016, 28 Hoophouses for Health farmers established farm-to-school/ECE relationships as part of their loan repayment, collectively providing $74,221 of food to participating schools and ECE programs, including more than $67,000 in fruits and vegetables, translating to more than 100,000 school meal fruit and vegetable servings.

In 2015, MFIN began an annual joint meeting with the Michigan Food Hub Network, recognizing the natural connections between the respective members. For many of Michigan's food hubs, institutions are potential customers. And for institutions, food hubs create an opportunity to purchase from small-scale farmers who do not individually have the capacity to provide the quantities large foodservice programs need. Over the course of three joint meetings to date,

---

5. Hoophouses are outdoor structures, typically covered in plastic, that offer many benefits to gardeners and farmers. For example, they shield food crops from cold weather.

participants have had opportunities to learn more about the constraints and possibilities for food hub sales to institutions.

MFIN has also begun to cultivate support from some of the more than 80 different commodity groups in Michigan (Michigan Department of Agriculture and Rural Development, 2017), many of which are dedicated to specialty crops. Unlike other Midwestern states, Michigan's agriculture is known for specialty crop production, so these agricultural commodity organizations have the potential to play a strong role in supporting farm-to-institution activity. By highlighting one specific featured food or food category per season, Cultivate Michigan provides a specific platform for MFIN staff to engage commodity organizations' leaders about their particular crop of interest.

Since the launch of Cultivate Michigan, several commodity organizations have provided financial support. Two large Michigan commodity organizations—the Michigan Cherry Committee and the United Dairy Industry of Michigan—provided financial sponsorship to help support the production and distribution of Cultivate Michigan materials (posters, guides, and window decals) as well as educational tours to help institutional buyers learn more about the production of these agricultural products. Others have supported educational tours to farms and food processing facilities. According to Mike Wenkel, the general manager of the Potato Growers of Michigan, Inc., which co-sponsored an educational tour, Cultivate Michigan "has provided a way to tell our story to businesses statewide. It's helping to link up Michigan potato growers with institutions that are able to buy a large number of potatoes, while ensuring that they were grown in Michigan" (Greening of the Great Lakes, 2016). These examples are promising, but MFIN leaders will need to continue to focus on these relationships, including engaging individual commodity group leaders, in order to realize the potential connections between the producers represented by the commodity organizations and institutional foodservice buyers.

CASE STUDIES

In order to understand the evolution of local food sourcing, it is important to look not only at the network of support activity statewide, but also at the experiences of individual institutions. The following three case studies, developed from interviews the authors conducted with foodservice directors in three institutions between March and April 2017, offer stories of institutions that have worked to embed local sourcing into the culture of their organizations. Though their motivations differ, the foodservice personnel in each case have aligned the foodservice program goals with broader organizational goals in order to generate buy-in from administrators and systematize buying Michigan-grown and produced foods. Representing three different types of institutions operating at different scales, the case studies show how motivated foodservice directors can develop farm-to-institution models that can be disseminated and replicated through state government policy, non-profit facilitation, and supply chain partnerships.

Case Study 1: Baxter Child Development Center
Sector: Early Childcare and Education
Average Meals Served Per Day: 90

Baxter Child Development Center (CDC), located in the urban Baxter neighborhood of Grand Rapids, Michigan, offers programs for 60 children each year, ranging from infants to school-aged children up to age twelve. According to its website, "more than 30 percent of children served have families with an annual income less than $15,000."[6] The center has year-round programs for infants and toddlers, as well as a preschool program for students aged two-to-four. Its school-aged programs, which include after-school programming and summer camp, enroll about 20 students. Each age group has opportunities to eat meals cooked from scratch with locally sourced foods and interact with food and gardening in various age-appropriate capacities.

Baxter CDC has been highly engaged with MFIN, helping network staff understand the unique dynamics of ECE foodservice programs, and sharing with network members its experience building local sourcing into their organizational culture. Baxter CDC staff, in turn, value MFIN as an avenue to share their knowledge and experience and to connect with others doing similar work. Staff from Baxter CDC have co-presented at multiple conferences, served as featured speakers on webinars, contributed to a new resource for Farm-to-ECE (Harper and Matts, 2017), and participated in planning conversations on how to move farm-to-ECE forward in the state of Michigan. Beyond Michigan, Baxter CDC has been funded with a W.K. Kellogg foundation grant to create resources that support the development of farm to ECE nationwide.

Baxter CDC's journey toward an organization committed to sourcing and using local food began in 2014, when Baxter CDC's director at the time, Starr Morgan, temporarily took on the role of food program manager. Morgan began planning for a foodservice program that used local, healthy foods. She started with small steps to reduce costs and move toward new purchasing priorities, cutting out individually packaged items, such as muffins, and replacing them with a made-from-scratch version. She also began replacing canned produce with frozen options. Through these replacements and adjustments, the program was able to realize considerable cost-savings, freeing up more funds to purchase fresh produce and food items they prioritized, such as grass-fed beef and free-range eggs. These small steps were helpful to justify the expansion of the farm-to-ECE program to Baxter CDC administrators who expected food costs to remain flat (Morgan and Kwasteniet, 2017, pers. comm., March 21). Working to address chronic diseases, including diet-related diseases, among the residents of the Baxter community is in line with the mission of the organization, and keeping food program costs stable increased the likelihood of support from the administration.

---

6. Source: http://www.wearebaxter.org/cdc/ (Accessed 21 December 2017).

Baxter CDC then hired a food program manager to dedicate full-time attention to local food. That manager, Becky Kwasteniet, made connections to local farms, food businesses, and a local food hub. After two years under Kwasteniet's direction, the food program at Baxter CDC is currently sourcing nearly 80 percent of its food locally, all without an increase to the food budget. Kwasteniet attributes this neutral impact on the budget to eliminating high cost items from the menu, reducing food waste, and cultivating connections to local suppliers to keep its local product prices low.

Kwasteniet believes that one of the first steps in creating a successful and sustainable program is to offer wages that reflect the talent and credentials needed to start or maintain a well-designed farm to ECE program. Baxter CDC's next purchasing goal is diversifying its suppliers to reflect the racial background of the community in which it is located. Baxter CDC is actively seeking out vendors of color and is willing to cut costs in other areas to pay a little extra for these contracts if necessary. This goal is a new iteration of the long-standing racial equity work that the Center was founded on.

At Baxter CDC, education around food goes beyond what is on each child's plate. The center has a garden program that gives all children a chance to explore how food is grown. From toddlers sifting through soil and seeds, to school-aged children planting, harvesting, and tasting food grown in the greenhouse, Baxter CDC tries to bring nutrition education full circle for its students. Moving forward, Kwasteniet is excited to share their work with others who are interested in starting farm-to-ECE programs. While Baxter CDC grew its farm-to-institution program without external technical assistance, MFIN now offers a platform through which Kwasteniet can share their approach and provide technical assistance to other network members.

Case Study 2: Calumet Center

Sector: Juvenile Justice System

Average Meals Served Per Day: 645

The Calumet Center, located in Highland Park, Michigan, is a privately managed therapeutic residential treatment program housing up to 88 young people between the ages of thirteen and 20. The center provides a treatment program for juvenile sex offenders, chronic high-risk juvenile offenders, and juvenile offenders with acute substance abuse behaviors. Treatment is individualized and residents go through seven stages of treatment during their stay, which is typically at least one year. About two years ago, Calumet introduced a culinary program. Residents are eligible to participate in this optional program once they have reached the fourth or fifth stage of treatment.

The staff from the Calumet Center have played an important role in MFIN by bringing the perspective of a residential treatment program, sharing their story and practices, and working to bring other treatment centers into the conversation.

In 2017, Calumet Center entered its third year as a MI Farm to School Grantee, which provided the springboard for their local food procurement planning and implementation efforts. MFIN has helped staff from the center elevate their stories by offering them opportunities to present at farm-to-school training sessions, supporting their travel to a national conference, and inviting them to MFIN networking and educational events.

Executive Chef and Foodservice Manager, Xaviar Jaramillo, is the architect of the culinary program and local food champion for the foodservice program, which participates in the USDA National School Lunch Program as a Residential Child Care Institution (RCCI). He has integrated local and healthy food in all aspects of the treatment program, from resident education and activities, to cafeteria meals. He believes his work is important because it teaches the residents the importance of a healthy diet and equips them with the knowledge to select and prepare nutritious foods. During his eight years as Foodservice Manager, Jaramillo has observed that many of the residents enter the treatment program overweight and leave at a healthy weight. He attributes this to the foodservice meals, which focus heavily on fruits and vegetables grown in Michigan, and whole grains.

Another goal of the culinary program is to build skills in food handling and preparation in order to increase participants' employment opportunities after completing treatment. Residents that complete the program receive a certificate and a food handlers' card. Jaramillo also hopes that graduates of the program will share their new culinary and nutrition knowledge with their families when they leave.

The culinary program consists of two month sessions, cumulatively reaching about 80–100 residents per year. Students learn about the importance of purchasing food fresh from a farm; processing whole foods; cost-saving measures, such as using whole chickens rather than precut pieces; as well as growing food; including planting, harvesting, weeding, and watering. The Calumet Center has a garden with six to eight raised beds, growing mainly lettuce, tomatoes, and herbs. Throughout the program, students have the opportunity to learn about various meat cuts, prepare a meal, and make a salad from produce that they harvest from the garden. According to Jaramillo, the students remark on the brightly colored produce from the garden and the fresh taste of the Halal chicken they prepare. Some of the students express their appreciation for the chance to participate by writing notes to Jaramillo and his staff. Jaramillo makes a point to share these remarks and letters with his foodservice staff. These sentiments connect the staff to the rewards of offering this program.

Gaining support for local sourcing from the Center's administration was a fairly easy step, according to Jaramillo. As he put it, the Executive Director has a great love for Michigan and fresh Michigan produce, so when Jaramillo outlined his desire to increase local purchasing and start a culinary program, she was excited to put it in the budget. Most of the food budget is covered through general funds, though a portion is reimbursed through USDA Child

Nutrition Programs. Because some of the local products used in the foodservice program cost more than items that were previously purchased, the center has partially offset the additional expenses by leveraging grant funds and connecting with a Hoophouses for Health farmer for "free" produce as part of her loan repayment. The administration acknowledges the health benefits that residents receive from consuming fresh foods, and the value of investing in Michigan's food and agriculture industry. Now all of the center's promotional materials feature its farm-to-table approach, school gardens, and the culinary program. As a result, the Calumet Center is receiving recognition from public officials, such as the Mayor of Highland Park, the surrounding community, and other treatment centers interested in implementing similar programs. The Calumet Center demonstrates how an initial, small amount of financial support for local sourcing, and accompanying technical assistance, can pave the way for significant changes in a foodservice program; by the 2016–17 school year, more than 30 different locally grown and -raised food products, ranging from kohlrabi to rutabaga, and chicken to maple syrup, worth more than $21,000 were purchased and served in the center's food program.

Case Study 3: Bronson Methodist Hospital

Sector: Healthcare

Average Meals Served Per Day: 800–900 In-patient Meals and 3000 Retail Transactions

Bronson Methodist Hospital is the flagship of Bronson Healthcare, a non-profit healthcare system serving all of southwest Michigan and northern Indiana. Located in Kalamazoo, Michigan, the facility has 434 beds and a self-operated foodservice program. Based on their tracking, Bronson Methodist currently purchases about 42 percent of its food from Michigan farmers and producers. Another campus, Bronson Battle Creek Hospital, has a contracted foodservice management company that requires negotiations with the contract partner to source local options. Even so, the Bronson Battle Creek foodservice department is innovatively navigating this relationship to reach the current level of about 14 percent of food from local sources.

Bronson Methodist began its efforts to support local producers in 2008 by hosting guest chefs, an on-site farmers' market, and a CSA pick-up. According to Grant Fletcher, System Director of Healthy Living & Sustainability at Bronson Methodist, these efforts allowed the hospital to build relationships with producers in the area and develop a culture of prioritizing local foods. Small steps paved the way for big changes and now Bronson Methodist uses whole, free-range eggs in all of its egg products; locally raised, grass-fed beef in all of its hamburger patties; and is working on transitioning all of its chicken to free-range.

For Bronson Methodist foodservice staff, sourcing locally is a natural extension of the organization's mission to "advance the health of our communities." Fletcher sees their food sourcing efforts serving as a catalyst to bring better nutrition to Bronson patients, employees, and the community, as well as contribute to the local economy. Similarly, Sustainability Coordinator Brendan Molony states, "if our purchasing decisions foster the growth of local suppliers, who, as a result, are able to sustainably support themselves and their families, the community as a whole will benefit."

Tracking local purchases is an integral part of the farm-to-institution efforts at Bronson Methodist. The foodservice team uses an account dedicated to local purchasing to facilitate tracking. Every time a buyer purchases food from a local supplier, the pounds of product and the amount spent is recorded. Only purchases from producers with whom Bronson foodservice staff have direct, face-to-face relationships are included in records of local food purchases. While they receive Michigan products through their broadline (full-service) distributors, they do not count these items because these distributors do not provide information about the specific farms from which they source. Keeping detailed records has allowed the foodservice team to define a specific budget for local sourcing and offer detailed reports to the administration. This detailed accountability helps to ensure ongoing administrative support to continually grow local purchasing. The tracking also helps chart progress toward Bronson's overall goal of sourcing 60 percent of its food locally.

The additional labor needed to process whole foods is the hardest part of local sourcing for Bronson Methodist. However, community partnerships and key equipment upgrades are allowing them to overcome some labor challenges. The new Food Innovation Center on the Bronson Healthy Living campus at neighboring Kalamazoo Valley Community College was developed to provide the processing of local products that institutions like Bronson need. With this new partnership, Bronson Methodist uses cut salad greens grown in the food hub's greenhouse, and diced beets, diced squash, and coin-cut carrots processed by the food hub. It has also switched from instant mashed potatoes to mashing potatoes in-house using potatoes peeled at the hub. After the 2015 Avian flu outbreak created a price spike in liquid eggs, Bronson Methodist began to purchase whole, pasture-raised eggs from nearby farmers. To avoid the labor costs of cracking individual eggs, the hospital purchased a liquid egg extractor. By making this switch, Bronson Methodist helped two egg farmers expand their businesses beyond farmers' market sales.

Ultimately, these innovations have helped Bronson Methodist improve the quality of the food it serves and increase profitability, even while purchasing more Michigan food. As Molony said, "Everyone thinks that a local or sustainable product is going to be more expensive. We've never been more profitable as a foodservice department – ever." The supply chain partnerships Bronson Methodist has developed also created a model for other institutions that MFIN can help to promote. For example, after seeing how Bronson Methodist began sourcing eggs directly from an egg farmer, a foodservice director of two school districts in West Michigan made a similar change—he purchased a liquid egg extractor and began purchasing eggs from a farmer in the area.

CONCLUSION

Reflecting on the progress of farm-to-institution activity in Michigan reveals that while it may be intuitive for institutions to buy food from farmers nearby, the nature of the dominant food system is such that it does not happen easily. Nearly two decades of dedicated funding, technical assistance, advocacy, and strategic partnerships underlie the progress in the state. Farm-to-institution facilitation in Michigan largely began as site-specific support. Increasingly, these efforts are moving toward leveraging partnerships with commodity organizations, state agencies, and distributors in an effort to embed local sourcing as standard practice across all supply chain channels—farm direct sales as well as sales through food hubs, grower cooperatives, and broadline vendors. MFIN staff are also starting to work more with other institution types, such as senior living facilities. Recognizing that this is likely to be a growing sector as baby boomers age, network staff are laying plans to learn more about the needs and opportunities and to develop relationships at the sixteen Area Agencies on Aging, which represent senior living facilities in different regions of the state (Area Agencies on Aging Association of Michigan, 2015).

The three keys to progress and the three case studies together illuminate several key points:

- *More than a trend.* In the different ways the highlighted foodservice directors have prioritized local sourcing based on a larger organizational mission, the case studies show that the interest in buying from local growers stems from a commitment to rethinking the values underlying both food supply chains and foodservice operations.
- *Multisector support networks are necessary.* Despite growing interest, farm-to-institution activity cannot expand beyond individually motivated institutions without supportive state policy, organizations providing technical assistance, and networks focused on reshaping the supply chain.
- *Equity is the next frontier.* Institutional purchasing has the potential to play an important role in high-poverty areas by catalyzing the economies of farm communities and serving healthy foods to food insecure individuals. However, to date, there is very little data quantifying these impacts or distinguishing impacts in low-resource versus high-resource communities. As the farm-to-institution movement grows, parsing out these equity dimensions will be essential to ensure the movement delivers on its promises and directs efforts to where they are most needed.

Given the size and breadth of institutional foodservice programs across the United States, institutions have an opportunity to drive the reshaping of the food system and to be catalysts for health in their communities—both for the people they serve food and the people from whom they source food. Realizing this opportunity, however, requires supportive policy, organizations that can provide technical assistance, and supply chain partnerships.

REFERENCES

Area Agencies on Aging Association of Michigan, 2015. 2015 annual report. Available from: http://www.mi-seniors.net/pdfs/4AM%20Annual%20Report%209-15.pdf. Accessed July 27, 2017.

Beery, M., Joshi, A., 2007. A growing movement: a decade of farm to school in California. Center for Food & Justice, Urban and Environmental Policy Institute. Available from: http://scholar.oxy.edu/cgi/viewcontent.cgi?article=1381&context=uep_faculty.

Colasanti, K., Matts, M., 2013. Farm to Institution in Michigan: a summary of research on local food purchasing by institutions. Michigan State University Center for Regional Food Systems. Available from: http://foodsystems.msu.edu/uploads/file/resources/fti-summary.pdf.

Colasanti, K., Thompson, M., 2016. 2016 cultivate Michigan data brief. Michigan State University Center for Regional Food Systems. Available from: http://foodsystems.msu.edu/resources/cultivate-michigan-2016-data-brief.

Colasanti, K., Cantrell, P., Cocciarelli, S., Collier, A., Edison, T., Doss, J., George, V., Hamm, M., Lewis, R., Matts, C., McClendon, B., Rabaut, C., Schmidt, S., Satchell, I., Scott, A., Smalley, S., 2010a. Michigan good food charter. C.S. Mott Group for Sustainable Food Systems at Michigan State University, Food Bank Council of Michigan, Michigan Food Policy Council. Available from: http://www.michiganfood.org/reports_and_resources.

Colasanti, K., Matts, C., Blackburn, R., Corrin, S., Hausler, J., 2010b. Putting Michigan produce on your menu: how to buy and use Michigan produce in your institution. Michigan State University Center for Regional Food Systems. Available from: http://foodsystems.msu.edu/resources/mi-produce-institution.

Colasanti, K.J.A., Matts, C., Hamm, M.W., 2012. Results from the 2009 Michigan Farm to School Survey: participation grows from 2004. J. Nutr. Educ. Behav. 44 (4), 343–349.

Feenstra, G., Allen, P., Hardesty, S., Ohmart, J., Perez, J., 2011. Using a supply chain analysis to assess the sustainability of farm-to-institution programs. J. Agric. Food Syst. Community Dev. 1 (4), 69–84.

Greening of the Great Lakes, 2016. Cultivate Michigan' partnership expands sourcing opportunities for locally-grown potatoes Ag report, October 28, 2016. Available from: http://agleadersmi.com/inthenews/industrynews/Pages/Ag-Report---%27Cultivate-Michigan%27-partnership-expands-sourcing-opportunities-for-locally-grown-potatoes-(10.29.16).aspx.

Hardesty, S., Allen, P., Feenstra, G., Ohmart, J., Perkins, T., Perez, J., 2010. Institutional food distribution systems: bringing students, farmers, and foodservice to the table. J. Food Distrib. Res. 41 (1), 58–63.

Harper, A., Matts, C., 2017. Local food for little eaters: a purchasing toolbox for the Child & Adult Care Food Program. Michigan State University Center for Regional Food Systems. Available from: http://foodsystems.msu.edu/resources/local-food-procurement-toolbox.

Izumi, B., Matts, C., 2008. Purchasing Michigan products: a step-by-step guide. Michigan State University Center for Regional Food Systems. Available from: http://foodsystems.msu.edu/resources/mi-purchasing-guide.

Izumi, B., Rostant, M., Moss, M., Hamm, M., 2006. Results from the 2004 Michigan farm-to-school survey. J. Sch. Health 76 (5), 169–174.

Izumi, B.T., Wright, D.W., Hamm, M.W., 2010a. Market diversification and social benefits: motivations of farmers participating in farm to school programs. J. Rural Stud. 26 (4), 374–382.

Izumi, B.T., Wright, D.W., Hamm, M.W., 2010b. Farm to school programs: exploring the role of regionally-based food distributors in alternative agrifood networks. Agric. Hum. Values 27 (3), 335–350.

Izumi, B.T., Alaimo, K., Hamm, M.W., 2010c. Farm-to-school programs: perspectives of school foodservice professionals. J. Nutr. Educ. Behav. 42 (2), 83–91.

Koch, K., Matts, C., 2016. Garden to cafeteria: a step-by-step guide. Michigan State University Center for Regional Food Systems. Available from: http://foodsystems.msu.edu/resources/garden_to_cafeteria.

Matts, C., 2010. Marketing Michigan products: a step-by-step guide. Michigan State University Center for Regional Food Systems. Available from: http://foodsystems.msu.edu/resources/marketing-mi-products.

Matts, C., 2013. Farm to School in Early Childcare in Michigan: exploring the opportunity for local food access and awareness. Michigan State University Center for Regional Food Systems. Available from: http://foodsystems.msu.edu/resources/fts-early-childcare.

Matts, C., Colasanti, K., 2013. Local food interest by institutions in Southeast Michigan: a report for Eastern Market Corporation. Michigan State University Center for Regional Food Systems. Available from: http://foodsystems.msu.edu/resources/fti-report-se-mi.

Matts, C., Smalley, S., 2013. Farm to School in Michigan: 2012 survey shows interest in purchasing local foods continues to grow. Michigan State University Center for Regional Food Systems. Available from: http://foodsystems.msu.edu/resources/fts-2012-survey.

Matts, C., Smalley, S., 2014. Farm to School in Michigan: still going strong. Michigan State University Center for Regional Food Systems. Available from: http://foodsystems.msu.edu/resources/mi-fts-going-strong.

Matts, C., Conner, D.S., Fisher, C., Tyler, S., Hamm, M.W., 2015. Farmer perspectives of farm to institution in Michigan: 2012 survey results of vegetable farmers. Renew. Agric. Food Syst. 31 (1), 60–71.

Michigan Department of Agriculture and Rural Development, 2014. Michigan agriculture facts and figures. Available from: http://www.michigan.gov/documents/mdard/MI_Ag_Facts__Figures_474011_7.pdf.

Michigan Department of Agriculture and Rural Development, 2017. Michigan Agricultural Commodity Organizations. Available from: https://www.michigan.gov/documents/mdard/Commodity_Directory_February_2016_-PUBLIC_525946_7.pdf.

Michigan Department of Education, Networks Northwest, Michigan State University Center for Regional Food Systems, Groundwork Center for Resilient Communities, Michigan Department of Agriculture and Rural Development, 2017. 10 Cents a Meal for school kids and farms 2016–2017 legislative report. Available from: http://foodsystems.msu.edu/resources/10-cents-a-meal-for-school-kids-and-farms-2016-2017-legislative-report.

Murphy, J., Smith, J., Matts, C., 2015. Farm to Early Childhood Programs: a step-by-step guide. Michigan State University Center for Regional Food Systems. Available from: http://foodsystems.msu.edu/resources/farm_to_early_childhood_guide.

Pirog, R., Miller, C., Way, L., Hazekamp, C., Kim, E., 2014. The local food movement: setting the stage for good food. Michigan State University Center for Regional Food Systems. Available from: http://foodsystems.msu.edu/resources/local-food-movement-setting-the-stage.

Ralston, K., Beaulieu, E., Hyman, J., Benson, M., Smith, M., 2017. Daily access to local foods for school meals: key drivers. United States Department of Agriculture Economic Research Service. Economic Information Bulletin Number 168. Available from: https://www.ers.usda.gov/publications/pub-details/?pubid=82944.

Thompson, M., Matts, C., 2015. Farm to school in Michigan: statewide response shows widespread activity. Michigan State University Center for Regional Food Systems. Available from: http://foodsystems.msu.edu/resources/2014_survey_brief.

Thompson, M., Colasanti, K., Matts, C., 2016. Understanding how Michigan institutions define local food. Michigan State University Center for Regional Food Systems. Available from: http://foodsystems.msu.edu/resources/understanding-how-michigan-institutions-define-local-food.

United States Department of Agriculture, 2016. 2015 Farm to School census. Food and Nutrition Service. Available from: https://farmtoschoolcensus.fns.usda.gov/.

United States Department of Agriculture, Agricultural Marketing Service, 2017. Farmers Market Promotion Program 2016 Highlights report. Available from: https://www.ams.usda.gov/reports/farmers-market-promotion-program-2016-highlights.

Vallianatos, M., Gottlieb, R., Hasse, M.A., 2004. Farm-to-school: strategies for urban health, combating sprawl, and establishing a community food systems approach. J. Plan. Educ. Res. 23 (4), 414–423.

Vogt, R.A., Kaiser, L.L., 2008. Still a time to act: a review of institutional marketing of regionally-grown food. Agric. Hum. Values 25, 241–255.

FURTHER READING

Barlett, P.F., 2011. Campus sustainable food projects: critique and engagement. Am. Anthropol. 113 (1), 101–115.

Richman, N., Ha-Ngoc, T., Oberholtzer, L., Obadia, J., Haskins, K., Allison, P., 2017. Campus Dining 101: benchmark study of farm to college in New England. Farm to Institution New England. Available from: http://www.farmtoinstitution.org/blog/new-fine-report-reveals-local-food-buying-stats-colleges-universities.

Kathryn Colasanti is a specialist at the Michigan State University Center for Regional Food Systems (CRFS). Since joining CRFS in 2009, she has supported the Michigan Good Food Charter Initiative. She has conducted research on healthy food access, farm-to-institution, urban agriculture, beginning farmers, farmers' markets, and school food supply chains. She leads the Research and Impacts Subcommittee of the Michigan Farm to Institution Network and coordinates the Michigan Good Charter Shared Measurement Project. She holds a BS in Biology and BA in Spanish from Calvin College and a MS in Community, Agriculture, Resources and Recreational Studies from Michigan State University.

Colleen Matts is the Farm to Institution specialist at the Michigan State University Center for Regional Food Systems (CRFS), where she has worked throughout Michigan to support local food purchasing programs at institutions since 2007. She serves as the Michigan Core Partner for the National Farm to School Network, coordinator for the Michigan Farm to Institution Network, and advisory committee member of the CRFS Shared Measurement Initiative. Colleen has an MS in Agriculture, Food, and Environment with a specialization in community development from the Friedman School of Nutrition Science and Policy at Tufts University.

Kaitlin Koch Wojciak serves as a Community Food Systems educator in southeastern Michigan, focusing on institutional purchasing, food policy councils, farmers' markets and food systems education. Wojciak serves on the Michigan Farm to Institution Network's Tech-Ed team, developing marketing and sourcing resources for Michigan products and trainings to address the needs of institutional purchasing stakeholders. Wojciak also works with the Macomb Food Collaborative to provide food systems education and local food business development resources. She holds a BS in Ecology and Evolutionary Biology from University of Michigan and an MS in Community Sustainability from Michigan State University.

Chapter 8

Farm to Institution New England: Mobilizing the Power of a Region's Institutions to Transform a Region's Food System

Nessa J. Richman*, Peter H. Allison†, Hannah R. Leighton†
*Rhode Island Food Policy Council, Kingston, Rhode Island, United States, †Farm to Institution New England, Hartland, VT, United States

Chapter Outline

Introduction	175	Farm-to-College	186
History of Farm to Institution New England	177	Case Study: The "Maine Food for UMaine System" Project	188
The Fine Network Today	178	Fine's Research Into the Supply Chain	188
Core Functions: Convening, Communication, and Metrics	180	Distributors	189
Farm-to-School	182	Producers	189
Regional FTS Efforts	183	Fine's Impact as a Regional Network	191
Case Study: Northeast Farm to School Institute	184	Strategic Objectives of Fine Moving Forward	191
Farm-to-Hospital	185	References	192
Case Study: Sea-to-Campus at Boston Medical Center	185	Further Reading	194

INTRODUCTION

New England's schools, institutions of higher education, and hospitals spend hundreds of millions of dollars on food and beverages every year (Farm to Institution New England, 2015; Health Care Without Harm, 2017; United States Department of Agriculture (USDA), Food and Nutrition Services, 2016). If these institutions increase the amount of locally produced and processed foods they purchase, they could increase access to fresh, healthy, local food for hundreds of thousands of children, adults, and elderly people throughout the region, while supporting local producers and generating tens of millions of dollars in new economic activity for the region. This chapter tells the story of Farm to Institution New England (FINE), a six-state network of non-profit, public, and private entities working to mobilize the purchasing power of institutions to facilitate these outcomes and transform the regional food system.

The initial idea for FINE grew out of a need identified by many institutions, organizations, and individuals across New England who were already deeply involved in farm-to-institution work. They wanted to find an effective and efficient way to work across state lines and institutional types to identify new opportunities for collaboration and inspiration. FINE was created to fill this gap. FINE is careful to act in an additive manner—to support rather than duplicate work within states or institutional sectors. Consistent attention paid to the strength of partners and other stakeholders who make up the network has helped FINE grow its reputation as the trusted backbone network for regional farm-to-institution efforts across the six New England states: Connecticut, Maine, Massachusetts, New Hampshire, Rhode Island, and Vermont. Other factors contributing to FINE's success include a shared New England culture, the small size of the states in the region, and the fact that many producers, distributors, foodservice management companies, and other farm-to-institution supply chain actors work across state lines.

FINE's work is rooted in a desire to create a more local and regional food system.[1] It is estimated that about 5 percent of the land in New England is dedicated to food production today. There are nearly 35,000 farms in the region (USDA, 2014) and thousands of fishermen, processors, aggregators, distributors, and other related businesses in the supply chain. Yet, 90 percent of food consumed comes from outside the region. In addition, an estimated 10–15 percent of New England households are considered food insecure, meaning they are without reliable access to a sufficient quantity of affordable, nutritious food (Donahue et al., 2014).

On the institutional demand side of the equation, New England is home to 4628 K-12 schools, 210 colleges and universities with dining services, and 256 hospitals that feed an estimated 3.8 million dining customers a day (American Hospital Directory, n.d.; Bureau of Labor Statistics, 2014; American Hospital Association, n.d.; U.S. Department of Education, n.d.), not to mention additional institutions that serve food such as correctional facilities, nursing homes, and more.

While FINE is focused on the institutional sector, it operates in the larger context of regional food systems development, which is presented most cohesively by the Food Solutions New England "New England Food Vision." This

---

1. FINE focuses its efforts on local food, but intentionally avoids assigning a set definition of "local." Rather, FINE supports the establishment of clear definitions by stakeholders so that their farm-to-institution efforts can be tracked and measured. Stakeholders across New England define "local" in a variety of ways, and these definitions will be explored throughout the chapter.

vision calls for the region to build the capacity to produce at least 50 percent of its own food by 2060 while supporting healthy food for all, sustainable farming and fishing, and thriving communities.

FINE works to connect supply and demand via three core functions: convening, communications, and metrics. These core functions increase awareness, alignment, and action among stakeholders, building the capacity of the regional farm-to-institution movement. Convenings bring local food sellers, institutional food buyers, and supply chain facilitators together to discuss successes, challenges, and plans for the future. Communication, via FINE's e-newsletter, blog, social media, and virtual calendar are used by individuals, agencies, institutions, other industry members, and non-profit organizations to learn and to let others know about local purchasing efforts. Metrics combine these stories of challenges and successes with data-based findings and recommendations that are shared regularly with key audiences, including funders, elected officials, government staff, and other decision makers on state, regional, and national levels. FINE staff measure the impact of these functions through an evaluative program that assesses factors such as how the organization's efforts lead to an increase in local food procured by New England institutions and progress on key policy and regulatory topics.

Beyond FINE's core functions, the network takes on special projects. These projects address select situations where FINE has identified a gap and made a plan to provide direct services to fill it. Project work (usually focused on a particular type of institution and/or limited to a single state) allows FINE to dive more deeply into a problem and then to use its learnings to develop evidence-based best practices to share across the network. One example is the "Maine Food for UMaine Project," in which FINE collaborated with a number of partners to influence the Request for Proposals (RFP) and contract process for the University of Maine system's foodservice management operations in order to increase local and sustainable food procurement (see the following case study). FINE was founded on the belief that creating regional, cross-sector networks is one of the keys to changing New England's food system to be healthier and more sustainable. As members of the staff of FINE, we also believe that the concepts and tools shared in this chapter are applicable and replicable in other states and regions.

HISTORY OF FARM TO INSTITUTION NEW ENGLAND

FINE launched as a funded entity in 2011. It originated as a partnership among regional farm-to-school (FTS) leaders and the six New England agricultural commissioners, as well as nongovernmental farmland preservation organizations, and public and private funders. At the time, the National Farm to School Network's (NFSN) Northeast Regional Steering Committee (RSC), made up of state leads from the six New England states and New York, served as the voice for the FTS movement (focused primarily on K-12 schools) in the region.

The RSC, then supported by the non-profit, NFSN, met regularly after its inception in 2007 and held regional conferences in 2008 and 2009. Since its inception, the Steering Committee has been coordinated by Vermont Food Education Every Day (VT FEED), a project of Shelburne Farms and the Northeast Organic Farming Association of Vermont (NOFA-VT).

The Northeast Regional Steering Committee has a strong history of successful collaboration and has been a collective seven-state (including New York) voice in conversations with foodservice management companies, policymakers, and regional partner coalitions. It has held national and regional conferences, forums and trainings, and professional development opportunities for schools including the Northeast Farm to School Institute, and has worked to develop large-scale procurement projects to identify and overcome barriers to sourcing local food for schools. The group recognized a need to expand its interventions beyond public schools to other institutional sectors to further capitalize on the buying power of institutions in supporting the local food economy. The members also acknowledged the value of collaboration on regional projects in addition to information sharing and relationship building.

In 2010, the RSC presented on the potential for farm-to-institution activity to provide markets for food producers in the region at a meeting of the New England governors and the Chief Agricultural Commissioners of the six New England states (excluding New York). At the time, the agricultural commissioners had been developing a series of strategies to keep farmland in production and had identified institutional markets as a potential area of growth for regional producers (New England Governor's Conference Inc., 2010). The Commissioners determined a set of priorities for the region based in part on the RSC's presentation, including increased FTS, farm-to-college, and farm-to-hospital partnerships. The RSC argued that a broader farm-to-institution network, which would require substantial funding, should be established to meet these goals. The non-profit, American Farmland Trust, helped pave the way for initial funding proposals, which led to an eventual cooperative agreement from the USDA (Deas, 2015). The John Merck Fund and the Henry P. Kendall Foundation then matched the public funds. FINE officially began funded work in early 2011, when it hired a part-time coordinator, distributed project funds to state partner organizations, and formed a nascent network structure. The initial budget for the project was approximately $500,000, with $250,000 from USDA and the remainder from The John Merck Fund and the Henry P. Kendall Foundation.

THE FINE NETWORK TODAY

Since its inception, FINE has served as a cross-sector and regional connector with a mission to mobilize the power of the region's institutions to transform the food system. Today, FINE is a fiscally sponsored program of TSNE MissionWorks, a charitable 501(c)(3) non-profit. FINE's staff is composed of five dedicated individuals (3.5 full-time equivalent). The FINE network includes non-profit organizations, government agencies, institutions, foundations, farms, food distributors, food processors, foodservice operators, and others. These partners share a common agenda around local food purchasing and participate in a variety of convenings, educational activities and training sessions, and projects. The Network receives regular communication from FINE through a variety of platforms.

FINE's eighteen-member Network Advisory Council (NAC) serves as a strategic and governing body, and is comprised of leaders from across the six New England states who share FINE's vision to change the food system through institutional procurement. The NAC has representatives from the three institutional sectors and organizations along the supply chain including farms, government agencies, and regional non-profits. The FINE Network Diagram (Fig. 1) depicts the full range of staff, advisors, functions, programs, and stakeholders that participate within FINE.

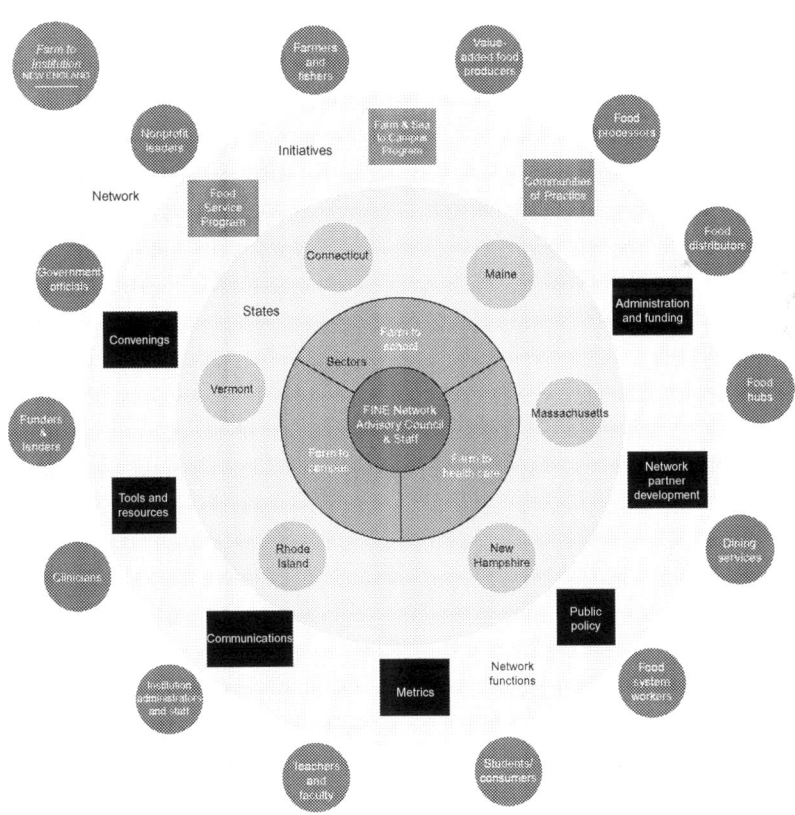

FIG. 1 The FINE Network Diagram.

FINE's values and vision are aligned with other food system entities in the region, including Food Solutions New England's *New England Food Vision*. FINE is dedicated to contributing to a region that is moving toward self-reliance, and has developed a set of core values to ensure that it is consistently working toward that goal. These core values include collaboration, dedication to community and place, diversity, equity, healthy ecosystems, strategic disruption, the right to food, thriving local economies, and transparency. They are the foundation for all of FINE's work.

The three primary institutional sectors upon which FINE focuses are public K-12 schools (referred to as "FTS"), colleges and universities (referred to as "farm-to-campus"), and health care (referred to as "farm-to-hospital"). FINE's role is unique in each of these sectors, demonstrating the different ways the organization helps mobilize the farm-to-institution movement. In the K-12 sector, FINE amplifies the work of the regional network Northeast Farm to School Collaborative through communications, partnerships on events and funding requests, and more (the Collaborative will be discussed in greater detail in "Farm-to-School" section). In the healthcare sector, FINE works with the regional team at the non-profit Health Care Without Harm on metrics efforts and joint events and training sessions. For institutions of higher education and private secondary schools, where there was a gap in regional coordination, FINE has taken a leadership role by developing the Farm and Sea to Campus Network (FSCN) program. These three types of institutions were prioritized due to the market opportunity they present, their economic importance, and the large numbers of people they feed. To a lesser extent, FINE is a resource for other institutional stakeholders working in correctional facilities, early childhood care and education settings, and senior centers. Plans are underway to expand slowly and intentionally to serve other types of institutions, pending funding and capacity.

FINE realizes that true change will not occur on the demand side (institutions) without changes on the supply side (e.g., producers, processors, and distributors). Thus, FINE also supports and advises groups of producer service providers and leaders in regional food processing and distribution when their work intersects with institutional markets.

Core Functions: Convening, Communication, and Metrics

FINE supports the growing New England farm-to-institution network by convening and communicating with stakeholders, creating tools, conducting research, and disseminating resources. The resulting case studies, survey research reports, and toolkits have promoted a better understanding of how institutions are harnessing their purchasing power to support local producers and build a more resilient regional food system. This body of work has also revealed major and minor barriers to increasing local food purchasing across different types of institutions and across the supply chain. Identifying these barriers, sharing recommendations for addressing them with key partners along the food supply chain, and advancing farm institution policy that will support local food procurement is an essential part of FINE's cross-sector role.

FINE's convening and communication functions are integrally interconnected. FINE provides regular opportunities for stakeholders to come together to build relationships and identify opportunities for collaboration around their common interest in growing the farm-to-institution movement via conferences, in-person meetings, webinars, social media, and other communication strategies. Educational training, webinars, toolkits, case studies, websites, newsletters, social media, and workshops allow regional farm-to-institutional stakeholders to share their stories of building supply chains to bring local food to institutions.

FINE's flagship event—the biennial New England Farm to Institution Summit—was held for the second time in April 2017 in Leominster, Massachusetts, and was attended by approximately 450 regional stakeholders from across the food system. The Summit offers sessions, speakers, activities, cooking demonstrations, field trips, exhibitors, entertainment, and open time for networking related to a variety of topics, including supply chain management, food waste, racial equity, and skill building. There is also space for specific interest groups to meet and work together, such as a regional group of food systems funders, lenders, and investors; an emergent food hub network; and a meeting of state agency staff, USDA, and non-profits addressing FTS. Planning for the 2019 summit is underway. FINE also hosts smaller convenings focusing on specific issues in the supply chain, including farm-to-college events that have served different target populations (e.g., small colleges, colleges that operate their own dining services, colleges in individual states, and colleges that operate larger-scale farms).

In addition to hosting its own events, FINE staff present workshops and trainings at state, regional, and national forums hosted by like-minded groups, including state food system and farm-to-institution organizations. These include the biennial Harvest New England conference, the annual Northeast Sustainable Agriculture Working Group "It Takes a Region" Conference, Food Solutions New England events, regional conferences, the Association for the Advancement of Sustainability in Higher Education, the National Association of College and University Food System's national conferences, the NFSN's biennial National Farm-to-cafeteria conferences, and others. While these regional and national forums broadcast FINE's work far and wide, the network's participation in smaller state-based convenings keep the network aligned with the goals and objectives of the practitioners, policymakers, and programmatic leaders from within New England states.

FINE's focus on metrics started in 2014, when network staff convened an Advisory Team and began collecting, analyzing, and sharing farm-to-institution data from across the food chain. By collecting existing national data and working in partnership with other organizations to produce original survey research, FINE has created an overview of the New England farm-to-institution landscape. Using these data, FINE makes recommendations designed to drive farm-to-institution progress forward. These data are available through various deliverables including a series of three detailed research reports with findings from original surveys of distributors, producers, and institutions of higher education; state profiles for the six New England states that highlight key state-level indicators; and an interactive farm-to-institution metrics dashboard. The dashboard also features a resource page, links to original survey instruments, clean datasets with identifiers removed, and more.

Metrics project data for the six New England states shows that institutions who responded to surveys spent a total of $123 million on local food in their most current fiscal year[2]: Of this total, approximately $57 million was spent in the college and university sector (FINE, 2015), $25 million in the K-12 school sector (USDA, Food and Nutrition Services, 2016), and $42 million in the hospital sector (Health Care Without Harm (HCWH), 2017). Approximately 70 percent of school districts said they will buy more locally sourced food in the future, while 98 percent of colleges said they will increase local food buying over the next three years. Adding to these findings, 88 percent of the 56 surveyed distributors reported that they believe their sales of local food to institutions will increase over the next three years (USDA, Food and Nutrition Services, 2016; FINE, 2016a,b).

FARM-TO-SCHOOL

FTS programs enrich the connection communities have with fresh, healthy food, and local food producers by changing food purchasing and education practices at schools and early care and other education sites. These programs provide experiential learning with school gardens, classroom cooking, healthy eating curricula, and farm field trips. FTS is the only farm-to-institution sector that is extensively supported by the federal government (through the USDA Farm to School program, which was established by the Healthy Hunger Free Kids Act of 2010), and has also been supported by a national non-profit, the NFSN, since 2007. According to the NFSN (n.d.), FTS "empowers children and their families to make informed food choices while strengthening the local economy and contributing to vibrant communities." FTS implementation differs by location; the Northeast follows a model of connecting classrooms, cafeterias, and communities with local farms, in an integrated effort to move FTS efforts forward.

In 2013, the USDA conducted the first nationwide Farm to School Census with the goal of establishing baseline data around FTS activity and local food procurement in K-12 schools. In 2015, the USDA conducted a second Farm to School Census to measure progress toward reaching this goal. While the data

---

2. The response rate for colleges and universities was 50 percent (105/209 colleges and universities in New England with Dining Services). The response rate for K-12 schools was 72 percent (727/1015 school districts in New England). The response rate for hospitals was 36 percent (54 of the 150 hospitals in the HCWH network).

resulting from this work has limitations, it has informed FINE in developing key indicators for FTS activity in New England. According to the most recent census data, 2489 schools (and 541 school districts) reported FTS activities in the 2013–14 school year across the six New England states. Schools reported spending $25 million on local food, and more than 1.1 million students in the region had access to FTS activities during this time.[3] This survey also asked respondents to cite the reasons why their district does not source more local products. The top three responses were availability of local foods throughout year, the price of local foods, and local items not being available from primary vendors (USDA Food and Nutrition Services, 2016).

There is a robust network of state FTS leaders and USDA staff in New England and New York that is served by the Northeast Farm to School Collaborative (an evolution of the original Northeast Regional Steering Committee of the NFSN) coordinated by staff at Shelburne Farms in Vermont. Northeast Farm to School Collaborative partners meet monthly by phone or in-person to share learning, coordinate FTS efforts across the region, and drive collective action. Their work includes resource sharing, event planning, developing policy agendas, collective policy advocacy strategy, shared messaging, and developing training opportunities for stakeholders. The Collaborative plays a key role in developing the FTS track of the New England Farm-to-Institution Summit, and organizes events such as the annual Northeast Farm to School Institute, an FTS professional learning program for schools from the seven Northeast states. Within their respective states, some members of the Collaborative also coordinate statewide networks that include state agency partners, practitioners, advocates, educators, and producers, and others who work together to advance state-specific FTS goals. This region is a leader in the nation in FTS work, with many in the New England states serving as true pioneers.

Regional FTS Efforts

While the six New England states share a regional identity, there is variation in state culture, geography, population, and politics that influences their FTS programs. For example, Vermont, which is home to less than 5 percent of the regional population, hosts many statewide, county, and local FTS efforts. Vermont is fortunate to have engagement from multiple state agencies, and more than ten years of government support for FTS programs, including creation of the first FTS grant program in the country. Maine, the largest state in New England in terms of landmass, has limited government support, but strong leadership from community health organizations. Massachusetts, which contains half the population of New England, has a statewide organization called Massachusetts Farm to School Project that receives state, federal, and private funding.

---

3. Respondents were asked to provide their definition of "local," for which the top three responses were "within state," "within region," and "within 50 mile radius" (USDA Food and Nutrition Services, 2016).

In Rhode Island, the smallest state in New England in terms of landmass, the statewide non-profit Farm Fresh Rhode Island runs the Rhode Island Farm to School Project. In Rhode Island, all but one of the school districts are served by foodservice management companies, while in other states, many school food programs are self-operated. In Connecticut, the Department of Agriculture has legislation that allows it to run a statewide FTS program with the Department of Education. In addition, the University of Connecticut Extension Service and Food Corp Connecticut are organizing a nascent FTS Network. And in New Hampshire, the statewide FTS program is housed within the University of New Hampshire, which is implementing a pilot project targeted at specific schools called the New Hampshire Farm to School Beacon Community Project.

Case Study: Northeast Farm to School Institute

In 2010, Vermont FEED convened the first Farm to School Institute to support Vermont schools in developing farm-food-nutrition programs. States in the northeast began to express interest, and in 2015, it became a Northeast FTS Institute, serving the six New England states and New York. The Institute evolved into a year-long professional development program for school teams that are implementing FTS programs. Twelve school teams of four to seven members, including teachers, school nutrition staff, administrators, and community partners participate annually. The Institute model incorporates a three-day residential summer training program, sixteen hours of ongoing support from a coach and mentor, and involvement in local and regional networks—designed around a "3C" model of change (connecting community, classroom, and cafeteria). Participants draft an FTS action plan for implementation with the support of their coach during the school year that is integrated into a comprehensive strategy to help students establish a positive relationship with healthy, local food. Vermont FEED provides technical assistance, free online resources, and support to regional partners. This model builds capacity among the states to mentor a growing number of FTS programs within their borders. Once embedded in the school, the FTS programs become models for other schools, locally and regionally. There are plans underway for FTS institutes in Massachusetts, New York, and New Hampshire.

The following elements are key features of the FTS Institute:

- Development of well-functioning, school-based teams with diverse membership,
- A three-day residential training program,
- Multiple contacts through the school year with trained coaches experienced in FTS,
- Support for developing and implementing school team action plans,
- Opportunities for team members to implement strategies, reflect on progress, revise course, and identify training and support needs, and
- Team documentation of progress toward goals with descriptions of evidence-based promising practices (Vermont FEED, n.d.).

FARM-TO-HOSPITAL

Farm-to-hospital (FTH) programs address food-related chronic disease through increased access to fresh, healthy, local foods in hospitals. These programs also feed the families and friends of patients, hospital staff, and other community members; allow patients to enjoy on-site gardens; establish on-site farmers' markets and Community Supported Agriculture (CSA) subscription programs; and advocate for prevention-based health care as a means to a healthy food system (Clinton et al., 2014).

The non-profit Health Care Without Harm (HCWH) is the primary actor in the FTH sector in New England. The non-profit estimates that their network (150 of the 256 hospitals in the region) spent $42 million on local food in 2017.[4] HCWH's Healthy Food in Health Care program works to address issues such as the "overuse and misuse of antibiotics in animal agriculture, aligning dietary guidelines with health and sustainability principles, harmful chemicals in the food system, the climate-food connection, and opportunities for health care to make upstream investments in public health" (Health Care Without Harm, 2017). In New England, this program has 150 health care facilities. Hospital leadership teams, comprised of staff from participating health care facilities in each of the six New England States, bring together these engaged institutions to leverage collective buying power, share innovative strategies, and learn about key issues.

FINE works with the Healthy Food in Health Care program to catalyze sustainable procurement efforts, train and support clinician advocates, and inspire health care institutions to become leaders in shaping a sustainable food system that supports prevention-based health care. FINE partners with them on research and project concept development, and each organization acts as an advisor to the other. FINE also uses HCWH's annual Healthy Food in Health Care survey as its main source of farm-to-hospital data.

Case Study: Sea-to-Campus at Boston Medical Center

Boston Medical Center (BMC) in Boston, Massachusetts, is a 498-bed facility that serves 360,000 patient meals and 980,000 cafeteria meals per year. In 2010, BMC signed HCWH's Healthy Food in Health Care Pledge and has since developed a sustainable and local seafood program. In partnership with the Gloucester Fishermen's Wives Association and Cape Ann Fresh Catch (who coordinate with local seafood vendor Ocean Crest), the program supports the Gloucester fishing community and brings local, sustainably caught fish to the hospital's patients and cafeterias. BMC prepays for a share of sustainably caught fish fillets and in exchange receives various types of locally caught fish

---

4. HCWH (2017) defines "local" as farms, ranches, and production/processing facilities located within a 250-mile radius of the facility. For processed foods with multiple ingredients, more than 50 percent of the ingredients must fit into the preceding definition.

throughout the year. This community-supported fishery (CSF) allows the producers and vendors to choose the varieties they will provide in order to prevent overfishing of certain species. Cape Ann Fresh Catch works with small boats landed in Gloucester that primarily catch ground fish in the gulf coast of Maine. The product is highly traceable, and orders are identified by type of fish, fishing boat, date, and location of the catch.

Through the CSF, BMC serves about 6000 pounds of fish such as Atlantic pollock, white hake, redfish, and cape shark per year. The program helps BMC further its mission in two ways: First, serving foods with high nutritional content such as wild-caught seafood is beneficial to the health of patients, visitors, and faculty who eat in the hospital every day. Second, the program supports BMC's mission to increase sustainability efforts and support local fisheries. One hundred percent of BMC's seafood spending is currently local (Horwitz, 2016).

FARM-TO-COLLEGE

There are 210 colleges and universities with dining services in New England. They have relatively high buying power, menu flexibility (because there are fewer direct requirements to meet federal nutrition standards and more dollars to allocate per meal), and the greater likelihood of having fully equipped in-house kitchen facilities when compared with other types of institutions. These characteristics increase the potential for colleges and universities to procure food locally (Murray, 2005). Additionally, college and university students on many campuses are increasingly interested in knowing the origin of the food they are served in campus dining halls. This interest is in line with the growing attention that local food is garnering with consumers as a whole, and is amplified by organizing efforts of groups such as the Real Food Challenge, and support from sustainability coordinators, interested faculty, and as part of a growing number of academic food systems programs on campuses.

Unlike the FTS and hospital sectors, there is no national equivalent to USDA or HCWH working to obtain data in the farm-to-college (FTC) sector. As such, FINE plays a larger role in this sector compared with the other two. FINE has long recognized the need to connect the diverse and growing number of FTC organizations and associations across the region. FINE established the Farm to College program in 2013. It started by contracting with state leads in Maine, New Hampshire, Vermont, and the greater Boston area to work with two campuses each, while also conducting background research, developing tools and resources, and holding meetings across the region. In 2015, FINE enhanced the Farm to College program with the creation of the Farm-and-Sea-to-Campus Network, a community of higher education and food systems stakeholders who connect, share, and collaborate to develop transparent regional supply chains and educate campus communities about regional food systems.

FINE designed and fielded a 25-question survey in 2015 to establish baseline data on FTC dining trends in New England. The survey was sent to all 210 colleges and universities in the region with on-site dining services and garnered a 50 percent response rate, with 95 percent of respondents reporting that they purchase local food.[5] Respondents also reported serving 65 million meals in their past fiscal year and spending an estimated 21 percent of their annual food budget on local food. This 21 percent contains two separate categories of institutions: those that operate their own foodservice, and those that use foodservice management companies (FSMCs). The two have statistically different success rates for local procurement. Those that are self-operating spent, on average, 27 percent of their budgets on local food, and those with FSMCs average 17 percent. However, it is noteworthy that there are campuses in New England served by FSMCs that are achieving high percentages of local food.

While colleges are devoting higher budgetary percentages to buying local than respondents in the other two institutional sectors, this survey revealed several barriers that colleges face in increasing local procurement. Top barriers included distributors' availability of locally grown foods throughout the year, the price of local foods, sufficient volume from distributors, distributors' availability of locally processed foods, and distributors' variety of local foods (FINE, 2015).

The Farm and Sea to Campus program acts as the primary unifying entity for FTC work in the region. The program aims to increase the amount of local food served in colleges, universities, and residential high schools by bringing together advocates, dining staff, faculty, students, and businesses to spur change in food purchasing patterns, encourage healthy eating, and empower a new generation of food activists.

The Farm and Sea to Campus program has held numerous meetings for stakeholders in the FTC sector since its inception. These meetings connect food producers with buyers and provide educational resources for producers who are interested in reaching more institutional buyers and other wholesale markets. Some gatherings have been state-focused, such as the Farm and Sea to Campus program statewide gathering in Rhode Island in 2016 where more than 70 people convened at the University of Rhode Island. The event allowed college and university dining operators, sustainability staff, faculty, and students to come together with non-profits, distributors, producers, and others to talk about key issues related to FTC work in Rhode Island. Other meetings have focused on topics specific to certain types of institutions, such as the Local Foods for Small Campuses event in 2016 where 32 individuals from seventeen different institutions with fewer than 2500 students participated in a local food event co-hosted by Unity College and FINE. The event brought together foodservice directors, chefs, cooks, and campus farmers from seven colleges, five residential high schools, and two hospitals.

---

5. Respondents were asked to provide their definition of local, for which the top three responses were "within a 250 mile radius," "within the state," and "within New England."

FSCN has also developed close to 20 case studies, including studies on leaders in the field, campus farm stories, local food subscriptions, and a series on bringing local seafood to institutions (FINE, 2017a).

Case Study: The "Maine Food for UMaine System" Project

In 2015, FINE, Real Food Challenge, Environment Maine, and Maine Farmland Trust, along with hundreds of students, farmers, advocates, and community members, came together to create "the Maine Food for the UMaine System Coalition." This work was timely, as the five-year University of Maine foodservices contract was about to expire, creating both an opportunity and a sense of urgency. The goal of the coalition was to create preferential sourcing for Maine- and New England-produced foods in the University of Maine System's 2015–16 foodservice Request for Proposals (RFP). The University of Maine System (UMS) represents seven universities across the state, all of which were a part of this contract, except for the flagship campus in Orono, which is self-operating in relation to foodservice.

Through research, developing recommendations, coalition-building, stakeholder engagement, and media coverage, the group was able to successfully influence the UMS foodservice contracting process over the course of 2015. The RFP included a commitment to reaching 20 percent local and regional foods by 2020, the establishment of a UMS Food Working Group, transparent tracking and metrics, and a commitment to a supply chain partnership with Maine producers. The inclusion of these recommendations and commitments in the RFP required competing vendors to submit proposals that reflected their methods for reaching these goals. By challenging foodservice vendors—whose contracts can be in place for several years—to meet the demands set forth by the coalition, vendors were forced to think differently about food sourcing and accountability to consumers and to the local food system.

The UMS continues to work with some of the original coalition members to inform their contract governance and metrics development (Neugebauer, 2016). Three bids were submitted for the UMS contract, two from well-established foodservice management companies, the incumbent Aramark and Sodexo, as well as a new entity, the Maine Farm & Sea Cooperative. Sodexo ultimately won the contract and began serving UMS in July 2016. In its first year, Sodexo organized an advisory group, set local food priorities and procurement targets, branded the local foods program "the Maine Course," and hired a local food coordinator.

FINE'S RESEARCH INTO THE SUPPLY CHAIN

While FINE focuses on institutions in most of its work, research extends beyond institutions to study how food producers and distributors intersect with institutional markets. FINE has stepped into this role because, while a great deal of research has been conducted on these issues, there are still important questions the network wants answered about sales to institutions in particular. As such, it has conducted extensive surveys of supply chain entities in the region.

Distributors

FINE has taken a lead in researching the role of distributors in the farm-to-institution landscape. Early efforts gathered information on distributors in the region, which were published along with guidelines for institutions on how to involve distributors in sourcing and reporting local food purchases (FINE, 2016b). In 2015, FINE fielded a 21-question survey, designed by FINE and HCWH, to explore distributor perspectives on institutional demand for local products, and identify the challenges and opportunities they face in serving this segment of the food market. The survey was sent to 87 distributors and 56 responded, a 64 percent response rate.

Survey results show that New England food distributors play a significant role in how people eat both inside and outside of institutional dining facilities. Respondents transported more than a billion dollars of food in 2014 (FINE, 2016a). While the proportion of institutional sales to total sales varies widely among individual distributors, larger food distributors generally see institutions as a more integral part of their businesses. All survey respondents who served institutions offered local products, but varied in their definition of local. Notably, local product sales usually represented a high proportion of gross sales for smaller food distributors. Local sales in relation to gross sales declined as gross sales increased. Overall, survey respondents had an overwhelmingly positive outlook on future sales of local products to institutions, suggesting that institutions will continue to grow and strengthen their farm-to-institution strategies in the coming years.

Producers

Agricultural producers, especially operators of small and midsized farms, often see institutional foodservice operations as promising markets. Efforts to collect data demonstrating the scale of farm-to-institution sales and their related impacts, while limited in the past, are growing. To gain a better understanding of the opportunities and challenges for New England producers in these markets, FINE conducted a survey that received 223 producer respondents from across the six New England states in early 2016.[6] This survey examined the differences in characteristics between producers who sell direct-to-institution and those who do not; it delved deeper into the practices, and the perceived benefits

---

6. The producer survey was sent to as many farmers in the region as possible through a collaborative process undertaken with state departments of agriculture, producer organizations, local food distributors, and others. The invitation went out via a combination of direct emails, newsletters, and social media.

and challenges, of producers who sell direct-to-institution. It also explored sales to institutions through intermediaries such as food distributors, food hubs, and foodservice management companies.

Almost three-quarters (71 percent) of respondents reported that they were either selling or interested in selling their products direct-to-institution, defined as "direct sales to users such as K-12 schools, colleges, hospitals, prisons, and other institutions." Specifically, 26 percent reported that they sold products direct-to-institution, 25 percent reported that they were interested in selling their products direct-to-institutions in the future, and another 20 percent said that they may be interested.

Farms that sell direct-to-institution generally had higher gross sales than those who did not. While a majority of farms in the sample were small, defined as those with gross sales of less than $350,000, fewer of the producers selling direct-to-institutions (69 percent) can be characterized as operating small farms than other respondents (83 percent). On average, more than half of all the respondents' gross sales were composed of either fresh vegetables or fruit. However, fresh fruits made up a greater proportion of sales for respondents selling direct-to-institution (25 percent) than for other producers (13 percent) who responded to the survey. The top five products (by value) that respondents sold to institutions were tomatoes, apples, meat, carrots, and potatoes.

In terms of motivations and barriers for selling direct to institutions, producers who already sell direct-to-institution most frequently listed the reasons for doing so because institutions provided:

- An additional market,
- A stronger relationship with the community,
- A stable and fair price, and
- Large volume orders.

Producers selling direct-to-institution reported that the biggest barriers to selling to institutional markets were:

- Seasonality of their products,
- Low level of customer interest in/awareness of their products, and
- Low purchase price offered by the institutions.

Notably, producers who are interested, but not yet selling into institutional markets, perceived barriers to the market as more problematic than those with some experience selling direct-to-institution. This suggests that there are some inaccurate perceptions about supplying institutions among producers and, therefore, that there is some potential for education and awareness-raising activities to encourage more local sales to institutions.

Data was also collected from farmers who sell to institutions through intermediaries (wholesale buyers such as food distributors, food hubs, or food auctions) that may also be supplying the institutional market. However, producers often do not have knowledge about the end consumers of their products when selling to intermediaries. Thus, many producers may be selling to intermediaries that in turn sell their product to institutions, but producers are unaware of the final buyer. The FINE survey sought to examine whether farmers selling to intermediaries knew whether any of their products were headed to institutions or not. Overall, 18.7 percent reported that they sell to institutions through an intermediary, while 13.1 percent reported they do not know if the intermediary sold their products to institutions. Given that many producers may not know where their products ultimately end up, this analysis likely underestimates the number of producers whose products are being sold to institutional settings through intermediaries (FINE, 2017b).

FINE'S IMPACT AS A REGIONAL NETWORK

FINE works to understand the farm-to-institution sector in order to communicate with key audiences to catalyze positive change. Sharing in the collective experiences of stakeholders has allowed FINE to establish key leverage points that form the foundation for specific recommendations. Establishing a regional farm-to-institution movement is not the sole work of one organization or even one network. In order to contribute to a truly regional and sustainable food system, it is critical to understand, support, and connect the existing organizations and networks that work up and down the supply chain. It is when stakeholders come together and communicate with each other that transformational ideas form. Many of FINE's recommendations revolve around creating a space for this type of collaboration. For example, FINE recognizes a need in the region for more wholesale readiness training for farmers. FINE does not work directly with the farmers, but rather, works to better understand the need, and then partners with producer service providers who can develop and implement appropriate training.

FINE realizes that working with supply and demand is not enough on its own to change the food system. Good public policy in support of farm-to-institution activity is critical to making systemic change. FINE uses research findings and experience from the field to make the case to government officials and policymakers. State policy allowing or requiring preferential purchasing of local food at the institutional level will make a major impact, as will state support for producer wholesale readiness training. Other state support in the form of targeting grant funds and technical assistance to institutional local food programs can serve to shift the dialogue and make sustainable farm-to-institution supply chains the new norm.

STRATEGIC OBJECTIVES OF FINE MOVING FORWARD

FINE recently completed a three-year strategic plan, which will inform its work going forward. The goals outlined in the 2017–19 strategic plan are grounded in the mission, vision, and core values developed to reflect the aspirations of the network stakeholders and to guide FINE in carrying out its mission to transform the food system. FINE has defined its primary role for the next three years as to: (1) serve as the backbone for the farm-to-institution network in New England, (2) catalyze collaborative projects that address key barriers in the New England institutional supply chain, and (3) advance a policy and programmatic agenda.

FINE's overarching goal is to get more local food served at institutions. In order to meet this goal, FINE recognizes the need to advance other goals as well. As such, the strategic plan outlines the following five goals:

- More local food served at institutions,
- A more developed regional network of individuals and entities across all parts of the food system that are mobilizing the power of institutions to transform the food system,
- Broader stakeholder understanding of the role of farm-to-institution,
- Stronger state and federal policies in support of farm-to-institution, and
- A stronger organizational foundation to support FINE's mission-driven work.

In the coming years, FINE intends to significantly advance the impact of the farm-to-institution network in New England by continuing to assist institutions in overcoming key barriers to local and sustainable food procurement. FINE also recognizes the importance of sharing materials and best practices with a larger audience, and will continue to prioritize its role as a platform for convening and communications around farm-to-institution activity in the region.

More recently, FINE has started to reach out beyond New England to collaborate with farm-to-institution organizations outside of the region. Since 2016, FINE has been acting as the convener of a growing network of national farm-to-institution stakeholders, coordinating a national farm-to-institution metrics collaborative that brings together leaders across the country to share experiences, materials, and insight. This group is of interest to the USDA and has hosted staff from various USDA departments, including Rural Development, the Agricultural Marketing Service, the Economic Research Service, and the Food and Nutrition Service Farm to School program, in order to ensure they are informed about this work and the potential impact it has for our nation's small- and medium-sized farms. The collaborative currently focuses on metrics, but as the farm-to-institution conversation grows nationally, this group has the potential to serve as a foundation for a broader national network.

It is FINE's vision that, by 2030, institutions such as K-12 schools, hospitals, and colleges and universities will be leading the region toward a sustainable regional food system. This vision honors the power of a diverse network, connected through core network functions, to support thriving local communities, healthy ecosystems, and the right to food for all people.

REFERENCES

American Hospital Association n.d. Annual survey of hospitals. Hospital Statistics, 1976, 1981, 1991–2011 editions. Chicago, IL. (Copyright 1976, 1981, 1991–2011: used with the permission of Health Forum LLC, an affiliate of the AHA.) See Appendix I, American Hospital Association (AHA) Annual Survey of Hospitals. http://www.cdc.gov/nchs/data/hus/2011/116.pdf.

American Hospital Directory n.d. "Numbers of staffed beds are taken from a hospital's most recent Medicare cost report (W/S S-3, Part I, col.1). https://www.ahd.com/state_statistics.html.

Bureau of Labor Statistics, 2014. State Occupational Employment Statistics Survey. May, http://kff.org/other/state-indicator/total-health-care-employment/.

Clinton, S., Stoddard, J., Perkins, K., Peats, B., Collins, A., 2014. New england healthy food in health care: leading the charge to a healthy, sustainable food system. Available from: https://noharm-uscanada.org/sites/default/files/documents-files/2723/NE%20HFHC%20 Report%202014.pdf.

Deas, M., 2015. Farm-to-institution New England: building a network. Available from: https://www.farmtoinstitution.org/sites/default/files/imce/uploads/FINE%20Case%20Study.pdf.

Donahue, B., Burke, J., Anderson, M., Beal, A., Kelly, T., Lapping, M., Ramer, H., Libby, R., Berlin, L., 2014. A New England Food Vision. In: Food Solutions New England. University of New Hampshire, Durham, New Hampshire. Available at: http://www.foodsolutionsne.org/sites/default/files/LowResNEFV_0.pdf.

FINE, 2015. Campus Dining 101: Benchmark Study of Farm-to-college in New England. FINE. Available from: http://www.farmtoinstitution.org/sites/default/files/imce/uploads/FINE%20 Farm%20to%20College%20Report_1.pdf.

FINE, 2016a. Getting there: understanding the role of New England food distributors in providing local food to institutions. FINE. Available from: http://www.farmtoinstitution.org/sites/default/files/imce/uploads/FINE%20Distributor%20Report_3.pdf.

FINE, 2016b. A toolkit for institutional purchasers sourcing local food from distributors. FINE. Available from: https://www.farmtoinstitution.org/sites/default/files/imce/uploads/1-Toolkit%20for%20Institutional%20Purchasers%20Seeking%20Local%20Produce%20thru%20a%20Distributor%202013.pdf.

FINE, 2017a. FINE Team. About us/Farm-to-institution New England. Available at: http://www.farmtoinstitution.org/about#team.

FINE, 2017b. Producer perspectives: the New England farm-to-institution market. FINE. Available from: http://www.farmtoinstitution.org/sites/default/files/imce/uploads/FINE%20Producer%20Report.pdf.

Health Care Without Harm (HCWH), 2017. Healthy food in health care survey. Unpublished. Available from: healthyfoodinhealthcare.org.

Horwitz, S., 2016. Sea to campus case study. Available from: https://www.farmtoinstitution.org/sites/default/files/imce/uploads/Seafood%20Case%20Study_BMC.pdf.

Murray, S.C., 2005. A survey of farm-to-college programs: history, characteristics and student involvement. Thesis for College of Forest Resources University of Washington.

National Farm to School Network (NFSN) n.d. About National Farm to School Network. Available from: http://www.farmtoschool.org/about

Neugebauer, R., 2016. Maine Food for the Umaine System. Available from: http://www.farmtoinstitution.org/blog/case-study-maine-umaine-system.

New England Governor's Conference Inc., 2010. Blue Ribbon Commission on Land Conservation 2010 Report to the Governors. 17–20. Available from: http://www.ct.gov/deep/lib/deep/forestry/2010__clc_rpt-final.pdf.

U.S. Department of Education n.d. National Center for Education Statistics, Integrated Postsecondary Education Data System (IPEDS)—SY 2009-2010 Available from: http://nces.ed.gov/programs/stateprofiles/index.asp.

USDA, 2014. 2012 Census of Agriculture, United State Summary and State Data, Volume 1, Geographic Area Series, Part 51, Table 53 Selected Practices 2012. USDA, Washington, DC. Available from: https://www.agcensus.usda.gov/Publications/2012/Full_Report/Volume_1,_Chapter_2_US_State_Level/usv1.pdf.

USDA, Food and Nutrition Services, 2016. The Farm to School Census. U.S. Dept of Agriculture, Washington, DC. Available from: https://farmtoschoolcensus.fns.usda.gov/home.

Vermont FEED n.d. Vermont Feed: a farm to school project of NOFA-VT and Shelburne farms. Available from: https://vtfeed.org/.

FURTHER READING

FINE, 2017c. Research reports. New England farm-to-institution metrics dashboard. Available from: http://dashboar.farmtoinstitution.org/research-reports/.

Healthy Food in Health Care/Health Care Without Harm n.d. Healthy food in health care/health care without harm. Available at: http://noharmuscanada.org/issue/us-canada/healthy-food-health-care.

Nessa J. Richman served as Director of Research and Evaluation at Farm to Institution New England from 2014 to 2018. She initiated the New England Farm to Institution Metrics Project, which measures the impact of the farm-to-institution market across the supply chain and three institutional sectors: schools, colleges, and hospitals. Nessa coordinated the National Farm to Institution Metrics Collaborative. Nessa is now Network Director of the Rhode Island Food Policy Council. She serves as an advisor to the Rhode Island Local Agriculture and Seafood Act Grants Program and the Social Enterprise Greenhouse Local Food Accelerator Program. Nessa holds a Master of Public Policy from the Harvard University John F. Kennedy School of Government.

Peter Allison is the Executive Director for Farm to Institution New England, a six-state cross-sector regional network that is transforming the food system by mobilizing the power of New England institutions. Peter is a frequent speaker at state, regional, and national food and farm gatherings, and serves on advisory boards for Food Solutions New England, and several state farm and food oriented organizations. Three decades of prior work includes leadership roles on innovative environmental and sustainability initiatives in the non-profit, business, and public sectors. Peter holds an M.A. in Urban and Environmental Policy from Tufts University.

Hannah Leighton is the Research and Evaluation Manager at Farm to Institution New England (FINE). She holds a Masters degree in Sustainability Science with a concentration in Sustainable Agriculture and Food Systems from the University of Massachusetts, Amherst and a Bachelors degree in Creative Writing from the New School University. Prior to her work with FINE, she spent a decade working across the food supply chain in the food service industry, on diversified vegetable farms, and as a food writer.

Chapter 9

Montana's Beef-to-School Project: Making Connections to Enhance Local Agriculture

Carmen Byker Shanks[*], Thomas M. Bass[†], Joel B. Schumacher[‡]

[*]*Food and Health Lab, Department of Health and Human Development, Montana State University, Bozeman, MT, United States,* [†]*Animal & Range Sciences Extension Service, Montana State University Extension, Bozeman, MT, United States,* [‡]*Department of Agricultural Economics & Economics, Montana State University Extension, Bozeman, MT, United States*

Chapter Outline

Introduction	195
Background	197
Cattle Production Basics	197
General Description of Processing and Yield	198
Local Beef in Montana's School Food Programs	199
Governance of Beef-to-School Activities	199
Federal and State Nutrition Standards	199
Federal Food Safety Standards	201
Beef-to-School Decisions: School and District Levels	201
Case Studies	201
Hinsdale, Montana: Vertically Integrated Model	205
Livingston, Montana: Producer Contract Model	206
Flathead Valley Region: Processor Contract Model	207
Somers, Montana: Processor Contract Model	207
Whitefish, Montana: Processor Contract Model	208
Kalispell, Montana: Processor Contract Model	209
Dillon, Montana: Donation and Community Member Involvement Models	210
Case Study Analysis: Overarching Themes	**211**
Motivations for Beef-to-School	211
Challenges in Implementing Beef-to-School Models	211
Conclusion	**215**
References	**216**
Further Reading	**217**

INTRODUCTION

At the heart of the local food movement, including farm-to-school, are communities with unique and often place-based motivations. The authors of this paper, researchers based in Montana, investigated what motivated the state's local beef-to-school (B2S) programs through case study interviews, secondary data, and field observations.[1] Overall, we found that a community-based sense of identity; specifically, a strong cattle-ranching culture throughout the state, was a significant motivation to engage in B2S. We also found that the perception that B2S would lead to particular social, economic, educational, and product benefits increased willingness to engage in B2S. Many times, schools and producers discussed the benefits of supporting B2S in light of national and global issues in beef production, including lack of transparency in the beef supply chain, concerns with sustainability, and a perception of lower-quality products. The purpose of this chapter is to highlight the grassroots efforts in Montana to serve local beef in schools and to distill barriers and opportunities in the hopes that other communities seeking local beef can benefit from lessons learned in this state.

The definition of "local" is debatable among food scholars, because strict geographic boundaries or radii from production to consumption do not always reflect the extent of embedded social relationships in the system. Though physical geography and proximity are important to defining what is considered "local," the social relationships between stakeholders and supply chain actors from production to consumption also play a large role (Hand and Martinez, 2010; Selfa and Quazi, 2005). In their definition of "local," the United States Department of Agriculture (USDA) encourages consumers (schools included) to "buy farm products that originate from known, local farms that preserve the identity of the farm for each item (USDA National Agricultural Library, n.d.)." This notion of a place's "identity" proved to be a significant motivator for schools to pursue local beef in Montana. The majority of B2S respondents represented herein interpreted "local" to mean "within Montana" and physically proximate to their specific community. B2S supply chains in Montana were characterized by multiple community, family, and social relationships between producers, processors, schools, and consumers.[2]

---

1. We are grateful to many individuals that contributed to this work. The authors would like to give distinctive acknowledgement to the following individuals for their contributions to the case study research: Janet Gamble, Aubree Roth, Demetrius Fassas, Katie Halloran, Mallory Stefan, Jeremy Plummer, Jay Stagg, Robin Vogler, Karla Buck, John Polacik, and Jennifer Montague. Furthermore, the authors would like to thank the case study participants, Montana Beef to School Coalition, and other experts locally and nationally that contributed greatly to the content and review of materials.
2. Southwest Montana is physically closer to beef production in neighboring Idaho than beef production in Eastern Montana; however, the case study supply chains in this chapter were completely located in Montana, with 250 miles (including live cattle shipping to processing, and meat returning to a school district) being the greatest distance traveled. This suggests that definitions of local are not always defined consistently and objectively in terms of distance, but also by the nature of cultural and economic ties in a specific place.

BACKGROUND

Montana is the fourth-largest U.S. state in land area, and the eighth least-populated state with an estimated one million residents (U.S. Census Bureau, 2017). The human population is contrasted with an inventory of 2.6 million head of cattle and calves (USDA NASS, 2016) spread throughout the state's mountain valleys, river plains, and prairies. Generally, seven cities are recognized as the major population centers, with only one of these exceeding 100,000 residents; they also include the largest school districts. The seven counties containing these cities represent more than 60 percent of the state's population (US Census Bureau, 2010). Therefore, compared with most other states, Montana is sparsely populated, and while most of the population is geographically concentrated in a few small cities, the rest of the residents are dispersed throughout a large rural area.

Though cliché, there is no doubt that common associations with Montana include ranchers and cattle, farmers and wheat fields, and mountains and a big sky. Such associations are reinforced by the presence of many highly visible equipment dealers, grain elevators, and cattle auctions throughout the state, even within Montana's population centers. Agriculture dominates land use in Montana: more than 94 percent of Montana's land is utilized for pasture or cropland. Cattle and wheat are among the top agricultural products. There are more than 12,000 farms and ranches in Montana, with gross sales of more than $25,000. A large number of these operations are cow-calf operations, and beef cattle production represents more than a third of the state's gross agriculture sales annually (USDA NASS, 2016). In short, agriculture is prominent in shaping Montana's culture, economy, and way of life.

Cattle Production Basics

Only a small number of Montana-raised cattle are finished and slaughtered in the state. However, producers in the United States raise and move cattle around during their lifecycles in a number of ways. Beef cattle begin on a cow-calf operation, where the producer has a herd of cows bred to produce calves that are then sold at six to ten months of age. In this cow-calf stage, cattle receive their nutrition by grazing pastures and rangeland; cow-calf operations are found across the country.

Sold calves are typically shipped for the finishing stage, generally somewhere in the Midwest or lower Great Plains, where there is widespread availability of feed grain crops, such as corn and soybeans, leading to conventional cattle feeding. In this dominant beef system, the finishing phase occurs in a feedlot where cattle grow to slaughter weight on grain or other concentrate feeds. Cattle generally reach slaughter weight between twelve and 22 months of age; most conventionally or grain-finished cattle are at the lower end of this range (Lowe and Gereffi, 2009). Cattle are then slaughtered and processed near these feedlots. Refrigerated or frozen meat is then distributed in national and international supply chains.

An alternative model is for the calves to forage on grass or hay (grass-finished) until they reach slaughter weight (USDA-SARE, 2012; UC Davis, 2004). Cattle finished in this manner tend to be older at slaughter. Grass-finished beef is generally associated with local or regional supply chains, and cattle may stay on their ranch of origin until they reach slaughter weight. Both cow-calf operations and grass-fed beef operations also supply cull cows to the market. Cull cows have either aged out of the breeding herd or are being removed for other performance reasons. They may range from two to twelve years of age.

Montana, with its vast grass and range resources, is known as a cow-calf state. Nearly all steer calves produced in Montana are sold and moved to the finishing phase, and eventually slaughter, out of state. The majority of heifer calves are sold as replacement breeding stock, and some are also sold and finished in the Midwest. A small share of Montana-raised cattle are actually finished and slaughtered in Montana. Approximately 20,000 head are processed in the state (Montana 2016 Agricultural Statistics, 2016). These may be grain-finished, grass/forage-finished, or cull cattle not in a finishing program. Of these 20,000 head, most are custom-exempt slaughtered and processed for the personal consumption of ranch families or shareholders, rather than retail. These cattle—raised, finished, and slaughtered in Montana—are the ones that are important to our B2S story.

General Description of Processing and Yield

"Processor" is a blanket term, which generally describes a facility capable of slaughtering animals and further processing and packaging meat products for consumers. Beef yield varies based on weight at slaughter, breed, fatness, feeding regime, age, and muscling. The yield of beef from cattle can be estimated at 40 percent of live-weight. The average 1200 pound live-weight cattle yields 400–500 pounds of beef. The producer or processor markets various cuts of boneless and trimmed beef to consumers, including ground beef, roasts, and steaks.

Even though the typical Montana cattle operation does not focus on producing slaughter-ready cattle that are finished, the supply of slaughter-ready cattle within the state has been sufficient to meet the needs of local markets. The success of any local food system, including farm-to-school, is dependent on not only producers and markets, but also intermediaries such as processors, aggregators, and distribution channels. Montana has approximately 20 small processing plants that are inspected for producing retail products suitable for direct marketing, grocers, restaurants, and institutions. Three of these processors are USDA inspected, and the remainder are under state inspection by the Montana Department of Livestock (USDA-FSIS, 2017; MT-DoL, 2017). The proportionally large number of small processing facilities in the state is unique, and a key component to successful B2S procurement programs.

Local Beef in Montana's School Food Programs

Schools in Montana tend to be small, but they provide many meals to their students. In the 2015–16 academic year, there were more than 145,000 students educated in 821 schools within 410 school districts in Montana (Facts about Montana Education, 2016). Fifty-four percent of districts had fewer than 100 students enrolled, 39 percent had 100–499 students enrolled, and 6 percent of school districts had 500 or more students enrolled. Altogether, Montana schools served almost 13.5 million school lunches in 2015 (Montana School Nutrition Programs, 2015). That means that, on average, Montana schools are serving about 93 lunches per student, per year.

Farm-to-school programs, including B2S, are on the rise in Montana. This trend is intertwined with a robust local food movement, in general. Montana is ranked fourth on the Locavore Index (2016), published by Strolling for Heifers, a Vermont based non-profit that tracks such data. The USDA's Farm to School Census Report (2015) notes that 40 percent of the 115 Montana schools surveyed (out of 821 total schools in Montana) participated in farm-to-school programming, 14 percent plan to begin farm-to-school programming in the future, and 41 percent plan to increase local food purchases. Examples of local food education programs and resources in Montana include the Harvest of the Month local beef curriculum, Montana Farm to School/Beef to School Programs, and FoodCorps.

Governance of Beef-to-School Activities

The process of getting local beef onto school children's plates at school is governed at the federal- and state-levels, as well as at a local- or district-levels. Although supporting local producers is a significant part of a farm-to-school platform, each school, producer, and processor in the B2S partnership must follow specific protocols and meet certain requirements in order to continue providing beef for school lunches. This section describes how farm-to-school beef is governed, first from the federal- and state-levels, and then at the school- and district-levels. It is important to keep the following safety and nutritional standards in mind when interpreting proceeding case studies, because they help explain the high level of coordination necessary in the supply chain in order to meet these regulatory standards—and why several distinct models of coordination have emerged.

Federal and State Nutrition Standards

School food is an especially highly regulated segment of the food system due to the need to ensure that children have access to healthy and safe food. In the 1940s, school food programs were formalized through the National School Lunch Act (Gunderson, 2017) "as a measure of national security, to safeguard the health and well-being of the Nation's children and to encourage the domestic consumption of nutritious agricultural commodities and other food." Since then, school food programs have grown beyond just lunch to include the School Breakfast Program, Afterschool Snacks Program, Seamless Summer, Farm to School, and others. The USDA regulates the meal patterns and nutrition standards of school foods, including specific regulations governing the amount of protein that schools must offer in school lunches. For school lunch, a school must offer five "components:" meat/meat alternate, vegetable, fruit, grain/bread, and milk. Students must choose three of these meal components, with at least one component being a half cup of a fruit or vegetable. Meal components are designed to meet the nutrition standards set forth by the Dietary Guidelines for Americans.

Meat and other protein foods, also called "meat/meat alternates," are a central meal component of school meal requirements. Under the Healthy Hunger Free Kids Act of 2010, the National School Lunch Program requires schools to offer at least two ounces of meat/meat alternates per day for grades nine through twelve, and at least one ounce of meat/meat alternates per day for kindergarten through eighth grade (USDA School Meals, 2017).[3] To provide perspective about the size of beef servings in school meals, three ounces of meat is about the size of a deck of cards. Many recipes use two to three ounces of beef per serving. Therefore, approximately 12.5–18.8 pounds of beef is required for a school lunchroom serving 100 students per meal.

Schools participating in the National School Lunch Program through the USDA are eligible for federal reimbursements per meal served that meets nutrition requirements (Federal Registrar, 2016). The amount of reimbursement per meal is based on a child's eligibility for free, reduced, or paid meals, which is determined by household income.[4] Schools also receive a certain amount of food at no cost through USDA Foods, which are items that the USDA procures from agricultural producers and food processors; these foods are referred to as entitlement foods. Beef products from the commodity market are often included in entitlement foods (USDA Foods Available List, 2017). The combination of low federal reimbursement rates and availability of entitlement beef products through the USDA poses a challenge to justifying the added expense of purchasing local beef.

---

3. Schools are also required to provide at least eight to nine ounces per week for grades K-5, at least eight to ten ounces per week for grades six through eight, and at least ten to twelve ounces per week for grades nine through twelve. There are no maximum requirements for how much protein is served on the lunch tray, but school foodservices must adhere to maximum calorie requirements for designated grade levels.

4. In the 2016 through 2017 academic year, for a meal served to a child approved for free benefits, a school receives $3.22 per student in federal reimbursement to cover the cost of labor and meal production. Schools receive $2.82 per student in federal reimbursement for reduced benefits and $0.36 per student in federal reimbursement for fully paid lunches.

Federal Food Safety Standards

The beef supply chain is also structured by food safety regulations, which impacts B2S in Montana. The USDA and the U.S. Food and Drug Administration (FDA) have oversight of certain important production practices, such as animal traceability, approved therapeutic and subtherapeutic pharmaceuticals, and feed sources. Additionally, many producers participate in Beef Quality Assurance (BQA) programs; these are supplementary voluntary programs that help ensure the quality of eventual meat products, and preserve the economic value of cattle sold into the food system. Most of the significant food safety protocols and rules that allow beef to enter retail and institutional markets reside or begin in the slaughter and processing phase.

Retail beef must be slaughtered and processed at a state- or federally inspected processing facility. State-inspected facilities are inspected by a state agency and can sell products within the state; federally inspected facilities are inspected by the USDA and can sell products inside and outside of the state. An actual inspector is present at the time of slaughter and has authority over all processes in the plant.

Other processing facilities are "custom exempt," meaning that the facility processes animals only for the animal's owner and the meat is not sold for retail—these facilities cannot market their beef to schools (USDA FSIS, 2014). There is a minor caveat to this system where a custom exempt plant, or other entity, with a retail inspection by a local authority (such as a health department) could further process meat purchased from an inspected slaughter facility for schools or other retail (however, it cannot use their own uninspected slaughtered animals for any retail market).

Beef-to-School Decisions: School and District Levels

In Montana, the school foodservice director in each school district usually makes decisions about where to source foods; depending upon the administrative structure, some school boards and school administrators also heavily weigh in on decisions about food sourcing. Decisions about where to purchase foods are most heavily influenced by foodservice's budget and the cost of the food, in addition to other factors, including input from the school board, administration, parents, other stakeholders, delivery schedules, kitchen capacity, and relationships with vendors. Based upon these factors, schools have the opportunity to work with a state- or federally inspected beef processor and/or producer to buy beef locally, purchase beef from larger food distributing companies, or acquire beef from USDA Foods at a minimal cost, if any.

CASE STUDIES

The data and food stories presented here are the combined results from six Montana B2S case studies conducted in 2015 through 2017 by the Montana

Beef to School Project. The Montana Beef to School Project is a three-year collaborative effort that includes:

- Beef producers and processors,
- Schools,
- Researchers at Montana State University,
- National Center for Appropriate Technology,
- Montana Department of Agriculture,
- Montana Department of Livestock,
- Montana Team Nutrition Program,
- Office of Public Instruction, and
- Community partners in the Montana Beef to School Coalition.

The Montana Beef to School Project is funded by the USDA Western Sustainable Agriculture Research and Education program from 2015 to 2018.

The following combined case studies are designed to highlight the needs of all B2S supply chain partners, including consumers, and inform strategies to make local beef available in Montana schools. Case study research involved 20 in-depth interviews with individuals who represented entities involved in the Montana B2S supply chain. We analyzed the case study data by conducting and recording interviews with producers, processors, and school foodservice directors; transcribing them, and identifying themes and subthemes. We collected additional data where gaps in knowledge existed from other B2S stakeholders (e.g., administrators, community members, other B2S programs, and state and federal agencies) and field observations (e.g., internet research of valid resources and site visits).[5] The partnerships represented in this case study span six school districts that include 28 schools (range of two to twelve schools per district) representing 11,149 students, two producers, two processors, and one vertically integrated producer and processor (Fig. 1).

From the case study analysis, we found that several procurement models existed to make B2S partnerships successful. These procurement models included what we are calling: (1) vertical integration, (2) producer contract, (3) producer donation, (4) processor contract, and (5) community member involvement (Fig. 2). These models are expanded upon as follows. The schools represented in our case studies differ in size, location, staffing, budget constraints, community involvement, and other attributes. These differences lead to the unique and contextual model each case represents. However, practices and opportunities identified in these models may fit other scenarios in similar, or alternate combinations. These models should not necessarily be considered set packages, but modular collections of solutions.

---

5. In some cases, schools, producers, or processors did not provide financial information about the sale or purchase prices of local beef because they viewed it as proprietary.

Schools

Dillon School District	Hinsdale School District	Kalispell School District	Livingston School District	Somers Lakeside School District	Whitefish School District
• Parkview School, 477 • Dillon Middle School, 235 • Beaverhead County High School, 332 • Total 1044	• Hinsdale School, 32 • Hinsdale 7–8, 13 • Hinsdale High School, 20 • Total 65	• Cayuse Prairie School, 193 • Cornelius Hedges School, 373 • Creston School, 83 • East Evergreen School, 418 • Edgerton School, 659 • Elrod School, 290 • Fair-Mont-Egan School, 138 • Helena Flats School, 174 • Evergreen Junior High, 166 • Kalispell Middle School, 1016 • Flathead High School, 1474 • Glacier High School, 1343 • Total 6327	• Winans School, 374 • East Side School, 294 • Pine Creek School, 30 • Sleeping Giant Middle School, 327 • Park High School, 500 • Total 1525	• Lakeside Elementary School, 350 • Somers Middle School, 172 • Total 522	• Muldown School, 651 • Whitefish Middle 5–8, 535 • Whitefish High School, 480 • Total 1666

Total = number of students in attendance school year 2015–16

Processors

Ranchland Packing (Butte) L and S Meat Processing (Dell)	Bear Paw Meats (Chinook)	Lower Valley Processing (Kalispell)	Ranchland Packing (Butte) Stillwater Processing (Columbus)	Lower Valley Processing (Kalispell)	Lower Valley Processing (Kalispell)

Producers

4-H and local ranchers around Dillon	Bear Paw Meats (Chinook) and local ranchers around Chinook	Local ranchers around Flathead Valley	Lazy SR (Wilsall) Muddy Creek (Wilsall)	Lazy SR (Wilsall) Muddy Creek (Wilsall) Local ranchers around Flathead Valley	Local ranchers around Flathead Valley

FIG. 1 Schools, processors, producers participating in Montana Beef to School case study research.

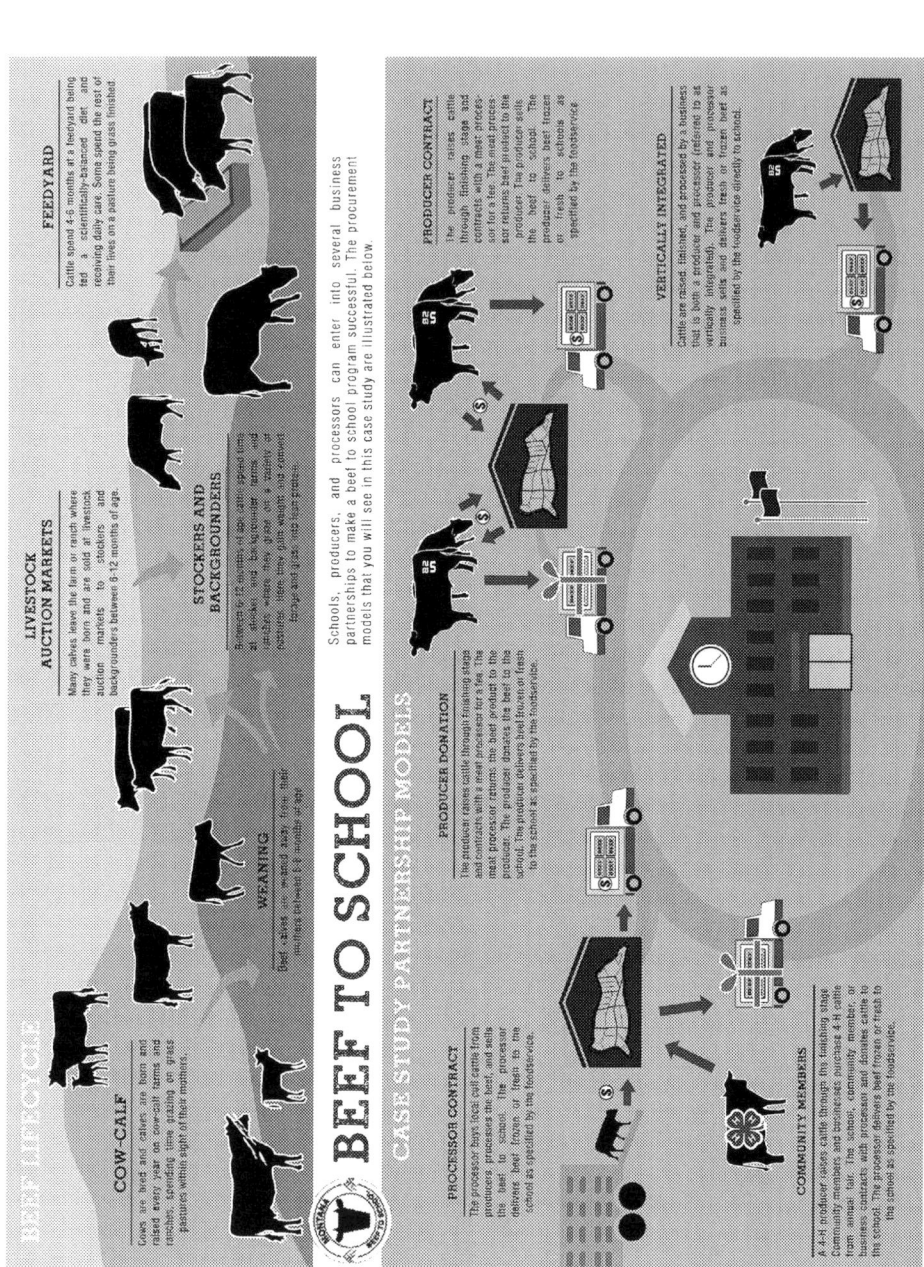

FIG. 2 Common case study partnership models in Montana Beef to School.

Hinsdale, Montana: Vertically Integrated Model

Hinsdale School District is a small district located in Hinsdale, Valley County, Montana. The district currently sources beef from Bear Paw Meats. Hinsdale School District began purchasing local beef in 2011 following an educational lesson on "food miles."[6] The lesson focused on how far food traveled before reaching student lunch trays. Foodservice staff were involved in helping with the lesson and surprised to learn that the beef they were serving was traveling to Hinsdale from Houston, Texas—more than 1700 miles away from Hinsdale. This number did not count the trucking of cattle before they become meat and the miles from the feedlot to the packer then to the distributor, Houston Sysco. The whole meal (including beef and other food items) added up to more than 7000 miles. With more cattle than humans residing in the state, the foodservice staff thought that Montana must have the capacity to offer high quality locally produced beef to schools such as Hinsdale. Moreover, many Hinsdale students live on cattle ranches or have family connections to the ranching industry. Even though the economics of the current system were not in question, they felt, for social reasons, it was time to try serving local beef.

For a short time, Hinsdale struggled to understand the "how to" of serving local beef, including finding a producer or processor, cost, and logistics. The B2S process became much easier after making a connection through a phone call with Bear Paw Meats. Today, 100 percent of the beef that is served at Hinsdale School is from Bear Paw Meats, located in Chinook, Montana. Bear Paw Meats is a family-owned and vertically integrated cattle feeding, auction, processing, and retail meat enterprise that serves the central Hi-Line region of Montana.[7] Bear Paw Meats operates as a cow-calf operation in which the calves are often raised to market weight in a small feedlot also owned by Bear Paw. It also raises some of the feed that is utilized by the feedlot. These finished cattle are then processed in the processing plant also owned by Bear Paw Meats. The operation is a state inspected facility, allowing their products to be sold to any buyer (restaurants, institutions, grocers, and individual consumers) within the state of Montana. Bear Paw Meats sells finished beef cuts to individuals, restaurants, and schools.

Since 2008, Bear Paw Meats has been selling products to K-12 schools in its region for use in the school lunch program. Bear Paw Meats began approaching schools after the owner's observations led the company to believe that Bear Paw's beef was higher quality (e.g., less moisture and fat run off when cooked) than what was being served at the schools, and the meat was coming from fewer cattle that were fully traceable to local origins. At the time of the case study, the business processed about 200 animals, or 100,000 pounds of beef per year for

---

6. "Food miles" refers to the distance that food has traveled before ending up on a consumer's plate.
7. "Vertical integration" means that the company owns and operates the supply chain for the final product (e.g., produces feed, feeds cattle, and processes meat).

all retails sales, which includes schools. During the 2015–16 school year, one fourth of all retail beef that Bear Paw Meats processed went to area schools. Schools primarily utilize Bear Paw Meats' ground beef in their menu offerings. Items such as tacos, lasagna, sloppy joes, hamburgers, spaghetti, and a variety of casseroles are regulars on school menus. Some schools that work with Bear Paw Meats also utilize roast beef in addition to ground hamburger.

Livingston, Montana: Producer Contract Model

Livingston School District is located in Livingston, Park County, Montana. The District has worked with Muddy Creek Ranch and Lazy SR Ranch, which are both producers and distributors of beef (that contract out for processing) in Park County, to source local beef since 2012. These ranches work directly with federally inspected processors, Ranchland Packing and Stillwater Processing, to provide the types of beef products needed for schools. The foodservice first purchased local beef sporadically for special events and celebrations only, and used USDA beef for regular meals. After receiving competitive prices and using all of the USDA beef in storage, Livingston School District began to source approximately 90 percent of its beef from a local ranch.

Muddy Creek Ranch raises and produces seed stock, semen, and embryos for shipment across the country, maintains a commercial cow operation, and also operates a small grass-finished beef business. The commercial cow operation produces around 90 market ready animals each year. It contracts with a USDA inspected processor to slaughter and package its beef before taking possession of the meat for distribution. Muddy Creek Ranch sold and donated beef to local schools in Livingston and Shields Valley in the past, before other components of the ranch's business grew to beef supply capacity.

Lazy SR Ranch evolved as an additional B2S partner at the same time that Muddy Creek Ranch was unable to meet the quantity needed for the foodservice. The school foodservice director at the time connected with Lazy SR by asking local food system advocates about producers that could provide a consistent product to the school. Lazy SR Ranch offers customers beef products and other meats. It also raises heritage poultry, Berkshire pigs, and Leicester lambs, in addition to the cattle. The animals are not given hormones or subtherapeutic antibiotics, according to the owners. The beef, lamb, and pork are processed by outside processors. The turkeys and chickens are processed in a facility owned and operated by Lazy SR.

Schools have been an important client for Lazy SR for the past few years, most notably supplying Livingston School District with its beef products. One of the Lazy SR co-owners believes that supporting a local school is worthwhile as students should know and learn about where their food comes from. Lazy SR invites students and school groups to their ranch and learn about meat production. Schools are also important to the Lazy SR business model because the company often markets its higher value cuts of beef (steaks) to restaurants in the area. Schools typically purchase ground burgers, which contains cuts of beef of lower value, which helps Lazy SR market its beef cuts.

Flathead Valley Region: Processor Contract Model

Lower Valley Processing, Inc. is a family owned meat processing facility near Kalispell, Flathead County, Montana. Lower Valley Processing is state inspected, allowing the business to market its products anywhere in the state of Montana. Lower Valley typically processes around 20 cattle each week in addition to hogs, lamb, and wild game (seasonally). Lower Valley has been working with multiple school districts in the Flathead Valley for the past ten years. Featured school districts include Somers-Lakeside, Whitefish, and Kalispell.

Lower Valley's core business is custom processing animals for its clients. Additionally, Lower Valley Processing purchases and processes animals to sell directly to local restaurants, bars, individuals, several local K-12 schools, and occasionally daycares. These cattle processed for restaurant, bar, and retail sales are usually between eighteen and 24 months old. However, older cull cattle are the primary animals processed for ground beef sales to schools. Cull cattle tend to sell at a 15–30 percent discount to prime age animals which makes the beef more affordable for schools.

Lower Valley Processing management and employees have stated they are proud of the positive feedback that they have received regarding their B2S program from multiple stakeholders. Lower Valley Processing opens its doors to the public through field trips and tours. After a school field trip, students often bring their families back to show where the school's meat comes from. Lower Valley's owner and stakeholders see value in their B2S model. Healthy cattle are selected for processing just days before the beef is delivered to the school. The beef usually comes from a single animal, is delivered fresh, and is usually cooked that week; only occasionally are small amounts of extra beef frozen for later use. The extra business generated from selling to schools has also allowed Lower Valley to add an employee and upgrade its hamburger patty machine.

Somers, Montana: Processor Contract Model

Somers-Lakeside School District is located in Somers, Flathead County, Montana. In 2006, the USDA issued a beef recall for a large portion of the beef that had been distributed through the USDA Foods program. This event motivated the foodservice director to find local beef products by connecting with a local rancher. A culled cow was processed at a local processor, White's Meats, and purchased as ground beef to serve in the cafeteria as hamburgers. At first, the school board scrutinized the increased expense of local beef purchases as compared with other nonlocal beef available through USDA Foods or national foodservice distributors. However, generating support from community members, parents, and students helped to mitigate the board's concerns.

Today, the District's foodservice uses both local beef and nonlocal beef available through USDA Foods or national foodservice distributors. It observes that local beef cooks better, without as much fat and water content[8] as other nonlocal beef products. In Montana, there is no requirement or best practice statement about the percent of fat content when sourcing beef in school meals. USDA Foods beef products range reported fat content of less than 20 percent for all its beef products (USDA FSIS, 2017). Case study participants also reported less than 20 percent and, in most cases, less than 15 percent, for their beef products.

The foodservice staff additionally noted that the quality of nonlocal beef has improved over time. The school's foodservice now purchases local hamburger patties or other beef products processed at Lower Valley Processing in Kalispell, Montana when it fits within the budget, usually once per month. Students provide positive feedback about the local burgers and other beef dishes. According to foodservice staff, local beef tends to be a favorite menu item for these Montana students. The Sommers-Lakeside foodservice director shared, "When kids turn on to fresh food, they don't want to go back. So, when you start feeding them well, they know the difference (pers. comm.)."

Whitefish, Montana: Processor Contract Model

Whitefish School District is located in Whitefish, Flathead County, Montana. Local beef use in Whitefish Schools became a consideration due to the foodservice director's curiosity about the conventional beef supply chain, including potential concerns about where and how animals were being produced and processed. During 2012, the District's foodservice purchased one cow and had it processed at Lower Valley Processing as an experiment in using local beef. Four years later, in 2016, Whitefish planned to purchase five head of cattle for processing, or an estimated 3000 pounds—resulting in almost exclusive use of local beef for beef menu items.

Whitefish School District's foodservice director finds that purchasing an entire cow is usually more cost effective than buying specific beef products. In 2016, the local beef typically cost $3.00–$3.50 per pound, or 43 cents per two-ounce portion of beef. Local beef is a large portion of the meal cost because a lunch tray costs the school foodservice a maximum of $1.20. Occasionally, the foodservice has purchased a steer and sold other cuts such as New York Strip steaks to staff at the school or to local restaurants to offset the price of local beef cuts served in the cafeteria. This was a convenient and

---

8. Many processors in Montana dry-age their beef, which reduces the amount of moisture in the final product (Montana Department of Agriculture, 2012). In contrast, most national food service distributors sell schools wet-aged beef, which is processed to retain moisture. Thus, the nonlocal beef product is more likely to yield high amounts of water content in the cooking process.

popular method for balancing the budget. The current foodservice director, Jay Stagg, suggests,

> Start small... I wouldn't jump in thinking that you're going to replace every single one of your beef products on your menu, like this month. Start with doing your marinara, ya know, and figure that out. And if it works okay, then go to taco meat. Just kind of evolve into it. I've had a few things where I've jumped too far into them and it ended up being more work and hassle and more expensive than I thought in the long run. So start small, and I mean even if you're a really small school... just find someone who has ten pounds of ground... Start there. Try one thing (pers. comm.).

Kalispell, Montana: Processor Contract Model

Kalispell School District is located in Kalispell, Flathead County, Montana. In 2011, the former school foodservice director wanted to build a strong farm-to-school program, and serving local beef was a top priority. With the help of a FoodCorps service member, foodservice staff contacted local beef producers and processors to determine if these vendors could provide the products and quantity that the school district needed, at the right price. Through market research, foodservice staff determined some of their specifications had to change to accommodate what was available locally, such as the ability to accept fresh hamburger patties rather than the individually quick frozen (IQF) patties that the foodservice staff had accepted previously.

Foodservice staff started purchasing beef from Lower Valley Processing in spring 2012 on a limited basis to create smooth processes for ordering, delivery, and preparation; in fall 2012, the Lower Valley beef was placed in regular rotation on the school menu. Today, the foodservice staff at Kalispell Public Schools are committed to purchasing beef locally for many reasons. They include perceptions that: the beef is a higher quality product than what they used in the past; the cattle have been raised in more humane conditions and are higher quality by being grass-finished rather than finished in a feedlot; and the beef is more visually appealing, and tastes better. Additionally, they report local support of students, staff, and community members for purchasing from local vendors, reducing food miles and simplifying the processes between producer and consumer, and keeping money within their community's economy.

Initially, foodservice staff were not necessarily supportive of the switch to the local products because of the increased labor involved to learn new recipes, prepare products, and create new inventory systems. However, the staff have become the strongest advocates for the B2S program despite the additional logistics, including the time it takes to order and prepare. Foodservice staff observed that when the district started purchasing local beef, the overall acceptance of the meals increased, especially when coupled with the education and marketing efforts within the district in cooperation with FoodCorps service members. Teachers and other school staff have taken notice as well, and comment on the appealing aromas of the local beef product cooking for lunch. They are more likely now to eat school meals. While students do not comment on the taste difference of local beef, they do note the appearance of the local product is more appealing because they can identify the beef in meals.

Dillon, Montana: Donation and Community Member Involvement Models

Dillon Elementary District is located in Dillon, Beaverhead County, Montana. Dillon Elementary District partners with their local 4-H program and ranchers to obtain donated beef and works with Ranchland Packing in Butte, Montana. The Dillon School District began to use local beef for hamburgers on their school menus years ago when businesses in the community began bidding on and purchasing show animals from 4-H students during the fall fair.[9] Once purchased, some businesses did not know how to use the animal and decided to donate them to the local school district. Beef and pork are donated to the school district every fall through 4-H. Beef has also been donated toward the end of the school year from local ranchers when stocks run low. Some donors also pay for the processing, while other times the school pays for the processing. Regardless, this program would not exist without the cattle being donated.

Ranchland Packing has worked with schools in Montana since 2011. Ranchland Packing is a full-service processor offering slaughter, fabrication (breaking the carcass or primals into cuts), and further processing. For Montana, Ranchland Packing is a larger processor (processing approximately 50 cattle each week); its USDA federal inspection status allows for both intrastate and interstate commerce. It serves a range of clients that include restaurants, grocers, institutions, and community members in the greater Montana, Wyoming, and Idaho regions.

Ranchland provides processing services on a weight-based, set fee schedule. When Ranchland Packing sells meat to schools, it looks at the livestock market for current prices and then sets its margin at a bare-minimum profit. Ranchland Packing stands behind the quality of its beef and it sees value in the fact that the origin of the cattle is also traceable. Generally, the owners of Ranchland Packing do not report any internal challenges working with schools, and believe that farm-to-school programs, including beef, are positive overall for producers, processors, and schools.

---

9. 4-H is a youth leadership program with an arm that is dedicated to teaching youth how to raise and market animals.

CASE STUDY ANALYSIS: OVERARCHING THEMES

The next section will discuss themes that emerged from the case studies. These include motivations for participating in B2S and challenges that schools encountered when trying to implement it.

Motivations for Beef-to-School

The case study analysis revealed a variety of motivations, perceptions, and beliefs about starting the B2S movement in Montana. The most common motivations for B2S reported were founded on social, economic, educational, and product reasons:

- *High quality product:* School foodservice staff perceived local beef as higher in quality when compared with other, nonlocal and conventional beef sources due to the taste, freshness, and noticeably less water and fat, which cooks off during preparation. Case study subjects perceived local beef as more nutritious because they knew where and how the beef product was produced and processed, with traceable and identifiable ingredients. This is a topic for future research.
- *Community:* Schools, producers, and processors were proud of their community's local agriculture. The B2S supply chains highlighted were represented by individuals and businesses known to the community and through familial and social relationships. Supporting a B2S program was one way that these partnerships felt that they could engage with the community's local agriculture.
- *Food literacy:* Schools, producers, and processors believed that learning about Montana beef increased food and agriculture literacy among students, school staff, and communities. Menu labeling of local beef and explicit educational tie-ins were practiced to make connections between cattle on the range and beef on the plate.
- *Local identity:* This theme was apparent throughout all factors motivating B2S programs in Montana. Among schools, producers, and processors in the case studies, local was defined as close to the school as possible, although many perceived anywhere in Montana as local. In our case studies, stakeholders commonly had awareness of each other because of being local, if not through direct familial or community relationships. For example, in the Dillon case study, student consumers, a family member, or a neighbor raised the beef served in the B2S program.

Challenges in Implementing Beef-to-School Models

Multiple models or contract types in the case-studies are presented as place and scenario-based examples that may contain concepts and lessons exportable to other communities. Several common challenges emerged from the case studies, as well as strategies that schools used to overcome them. These are as follows:

Higher Cost

By far, respondents reported that the cost of procuring local beef posed the biggest challenge to establishing and sustaining a B2S program. Local beef tends to be priced higher than other sources of beef that schools have access to, including USDA Foods Program beef. Some districts require their school lunch program to operate as a separate budget unit that must break even or operate at a profit each year. Other districts will provide additional revenue to the lunch program from other parts of the school budget. School districts have limited funds with which to address many resource needs (teaching staff, tutoring, books, supplies, etc.), so an extra investment in the lunch program must produce justifiable benefits.

The fee charged to students for a school lunch is often less than three dollars per meal in Montana. To put this in perspective, if hamburger is selling for four dollars per pound then a 1/5 pound hamburger patty would cost $0.80; the patty alone is over 25 percent of the meal cost. At the time of this research, case study participants shared that the average cost for beef from USDA foods and nonlocal markets for schools ranges greatly from free (through USDA Foods) to three dollars per pound. Depending upon the local B2S model, the cost for local beef ranged from free (donation models) to four dollars per pound at the time of the case study. Current prices for local beef range 15–20 percent higher than conventional beef, but this percentage changes regularly depending on market factors. Several strategies to bring local beef into schools at a feasible cost for the school, producer, and processor are being utilized successfully in Montana.

First, producers and processors can utilize schools as a market for lower value cuts, such as those used for hamburger, while selling the higher value cuts to other markets. A single beef animal yields a range of steaks, stew meat, hamburger, roasts, and other lower value cuts (tongue, ox tail, heart, ribs, etc.). Schools provide a large market for lower value cuts as they utilize hamburger almost exclusively. Restaurants and specialty grocers may provide markets for the high value steaks, but not necessarily all of the hamburger an animal may yield. Schools may become a valuable partner in the local beef supply chain and purchase hamburger at a price that fits within budget constraints.

In the processor contract model, with Lower Valley Processing, cull animals are utilized to overcome cost. Cull are animals eliminated from the breeding herd, typically over 36 months of age. These animals tend to sell for lower prices than prime age (eighteen to 30 months of age; range for grain to forage finish) animals. A recent weekly auction in Montana resulted in cull animals selling at a 15–30 percent discount to prime age steers (Montana Livestock Auction, Market Reports, n.d.). When purchasing beef product from a cull animal, the purchase price tends to be lower than purchasing a prime age steer.

Donation models are another option; though used by fewer communities, this model is gaining in popularity. In this model, donation can include the animal itself, the cost of processing, or financial support to source local beef. Donations can occur occasionally, or on an ongoing basis, and help the school significantly in balancing the budget. Donation models rely on relationship building and the generosity of one or more parties.

Like most proteins, local beef is a costly portion of the school meal. In many schools, the foodservice staff balance the cost of local beef by pairing more expensive items such as the local hamburger patties with less expensive items such as USDA Foods sliced carrots within a meal; some attempt to balance the cost of local beef across multiple meals and longer timeframes. They also cut costs by using their staff's creativity to add less expensive ingredients to meals (e.g., lentils to beef chili) or reducing the number of preprocessed items the foodservice uses in general.

Technology and Kitchen Convenience

The technology limitations of Montana processors limit the production of certain products that may be required at schools that have inadequate kitchens. For example, a school may not have kitchen or staff capacity to brown large amounts of ground beef, and would require precooked crumble. In such a scenario, a food-hub with inspected kitchen facilities may be a source for local precooked beef crumble. Montana schools have also had success with frozen or fresh hamburger patties separated by wax paper; however, some schools may prefer individual quick freeze (IQF) patties (without wax-paper separators). There is currently no IQF line in the state of Montana.

In many cases, kitchen staff will not need additional training or new recipes to utilize local beef. Switching from purchasing ten pound frozen hamburger packages from a large distributor to ten pound frozen packages from a local beef supplier would not change or impact kitchen processes; but, switching from premade seasoned meatballs to making meatballs from scratch is not as simple. Recognizing which beef products are available in similar packaging is helpful in identifying opportunities for a kitchen staff to switch to a local product.

School Size and Meal Participation

In Montana, student enrollments range from fewer than 100 students to more than 5000 students per district. Nationally, even Montana's large districts would still be considered small. School lunch programs also vary by the number of schools served by a single kitchen, which changes the model for producing and distributing fresh beef products and the scale at which beef can be purchased and stored. In some school districts, a central kitchen may provide lunch for several district schools. In others, each school within the district may have a standalone kitchen and program. Some schools may also have hybrid models, in which some items are prepared in a central kitchen while others are prepared at each school. These differences can substantially impact the amount of beef required for a typical order. School lunch program attributes such as size and district kitchen networks will have to be considered on case-by-case basis when investigating local beef procurement and use.

Another factor that impacts the number of lunches a particular lunch program will serve is the historic participation rate of students. Rarely do enrollment numbers equal lunches served, as some students choose to bring their own lunch and some schools have an "open campus" policy that allows students to leave school property for lunch. Both of these options reduce the number of students participating in the school lunch program, and may also influence purchasing decisions for local foods, including beef. In the case studies, Montana schools reported better lunch participation on local beef days, which presents an opportunity for local beef advocates. This was not an initial motivation.

Supply Chain Logistics and Product Availability

In all case studies, at least one school, producer, processor, or stakeholder demonstrated a significant amount of grit and commitment to overcoming the challenges of logistics and availability. Most foodservice directors are able to easily order a wide variety of products, in large or small quantities, from a national food distributor that will deliver to the school on a regular basis. Ordering an extra case of an item or trying a new item is as simple as adding to their order. Switching to a local supplier for some or all meat products involves placing an order with an additional supplier and arranging for an additional delivery to take place. Even if this process is smooth and efficient, it is still an extra step in the process.

Some schools will also find that a local meat supplier is not readily available. Definitions of "local" vary, but it is probable that some areas of the United States will not have a processor that meets the geographic definitions of local. In this case, a regional product may be the most preferable option; however, the credence value of local and the inherent relationships of local systems could be diminished. In larger and more complex markets, the roles of alternative distributors and food hubs may be elevated in the pursuit of local beef purchasing.

Larger school lunch programs may also have to consider the volume of beef they will be purchasing. A school serving 3500 lunches per day may require several thousand pounds of beef each month while a district serving 300 lunches per day will require only a small fraction of that amount. Local processors tend to be smaller than regional processing facilities, and thus may not have large quantities of beef available at all times. In Montana, many of the local meat processors process fewer than 25 cattle per week.

B2S research also found that Montana school lunches provide students with nearly 1.5 million pounds of protein over the course of the school year. We calculated that if one third of the protein were to be fulfilled by beef, then about 500,000 pounds of beef would be required by Montana school districts each year. Given that an average cow will yield between 500 and 600 pounds of edible beef products, if all Montana school lunch programs utilized local beef, then approximately 850–950 animals would be required annually. Another way to represent this is that during each week of the school year, approximately 25 head of cattle would need to be processed to supply the beef needs for all Montana schools, an achievable number to serve 100 percent local beef in Montana schools, given the number of cattle produced and processed in the state.

CONCLUSION

B2S in Montana is a growing movement that will continue to encounter successes and challenges, with success predicated on hard work and relationship building throughout local supply chains. An emphasis on relationship building and case-by-case problem solving appeared to be a key driver of B2S program success in Montana. Overall, Montana B2S partnerships exist because of the commitment of schools, producers, processors, and stakeholders to preserve a community's agricultural identity through local beef products. Based on case study research in the context of Montana's beef supply chains, local identity was formed around characteristics such as a ranching and farming heritage, pride in local solutions, and positive associations with community, business, familial, and social relationships.

Motivations for B2S are based upon the social, economic, educational, and product benefits embedded in these partnerships among stakeholders. For example, respondents shared positive cooking experiences with local beef because of lower fat and less cooked-off water; satisfaction with the option for some producers and processors in these local supply chains to market whole animals (i.e., premium cuts going to higher end markets, and more abundant ground beef going to schools); the perception that B2S can lead to growth within a food production-related business (such as the Lower Valley Processing case); improved student participation in school lunch programs; and opportunities for philanthropy as evidenced by the donation model in the Dillon School District.

Distinct challenges to B2S include the cost of local beef, technology required, foodservice structure, school size, logistics, and availability. Key recommendations exist for starting and sustaining a local B2S program, which help to overcome challenges.

- School foodservice directors overwhelmingly recommend starting small by choosing just one day during the school year or one dish where local beef is always featured, and then expanding to weekly or monthly local beef meals once a sustainable B2S partnership is established.
- Schools and processors must discuss and determine equipment needs to process the raw beef product into more efficient forms for the school foodservice (e.g., hamburger patties or beef crumbles).
- Producers and processors must adhere to food safety protocols established by the school and local sanitarian. Schools must train employees to handle and cook a safe and desirable beef product and have enough storage to source quantities of beef for high volume prices.

- Montana producers and processors have the capacity to increase B2S partnerships. Schools have the desire, but cost remains the main barrier to B2S. Communities, schools, producers, and processors must be creative to fund B2S programs so that they are sustainable for all partners involved.
- Maintaining and increasing the number of state and federally inspected processing facilities in states is paramount to continuing B2S efforts.

With such strong emphasis placed on social networks as a variable to motivation and success, the findings from this chapter may be most applicable to school systems and communities in states with strong social connections and proximity to cattle production, and agriculture in general. While the B2S movement in Montana has shown promise for institutional purchasing to connect local producers, processors, and schools, there are several future directions that purchasers, policymakers, and foodservice researchers could explore for expanding it further and maximizing benefits for the local and regional economy. It is also important to pinpoint state and national policy levers to encourage B2S, such as using the USDA Foods allotment toward local beef purchases or developing a statewide reimbursement program for local purchases.

Further research is needed to understand the economic, equity, and environmental impacts of B2S for various scales of producers and processors. For example, accounting data from producers, processors, and schools engaged in B2S programs would be useful to understand specific monetary gains and losses. Minimum profitability needs should be identified and also balanced with social gains for producers, processors, workers, and schools engaged in B2S programs.

Since the case study research was completed, several new B2S programs have emerged in rural and semiurban communities, and the B2S project has provided technical assistance to many. The knowledge gained by researchers will be translated into tools and materials to help maintain and add stability to existing programs while hopefully providing schools easier entry into new B2S procurement programs.

Building B2S programs that support Montana beef producers' and meat processors' business viability and sustainability, while increasing the availability and consumption of local beef in schools and communities requires a commitment to finding solutions in each local food system. The work presented in this chapter explores the feasibility, opportunities, and challenges of establishing and sustaining B2S programs across Montana by featuring school, producer, and processor stories from the field. This information will assist other communities interested in supporting their agriculture identity, helping them find successful strategies to work together with producers and processors to use local beef in area schools.

REFERENCES

Facts about Montana Education, 2016. Available from: http://opi.mt.gov/pdf/Measurement/EdFacts2016.pdf (Accessed May 10, 2017).

Federal Registrar, 2016. Administrative reviews in the school nutrition programs. Available from: https://www.federalregister.gov/documents/2016/07/29/2016-17231/administrative-reviews-in-the-school-nutrition-programs.

Gunderson, G., 2017. National School Lunch Act (history). Available from: https://www.fns.usda.gov/nslp/history_5.

Hand, M.S., Martinez, S., 2010. Just what does local mean. Choices 25 (1), 13–18.

Lowe, M., Gereffi, G., 2009. Duke-CGGC, value chain analysis of the U.S. beef and dairy industries. Available from: https://ai2-s2-pdfs.s3.amazonaws.com/f95e/e9adb-444757b5704a240ce465da859f19f1e.pdf.

Montana 2016 Agricultural Statistics, 2016. Available from: https://www.nass.usda.gov/Statistics_by_State/Montana/Publications/Annual_Statistical_Bulletin/2016/Montana_.

Montana Department of Agriculture, 2012. Available from: https://farmtocafeteria.ncat.org/wp-content/uploads/2012/11/MT-Beef-in-Schools_Processor-List.pdf.

Montana Livestock Auction, Market Reports. n.d. Available from: http://www.montanalivestock-auction.com/market-reports (Accessed 31 July 2017).

Montana School Nutrition Programs, 2015. Annual report. Available from: http://opi.mt.gov/pdf/SchoolFood/15SNPAnnualReport.pdf (Accessed May 10, 2017).

MT-DoL State Plant List, 2017. Available from: http://liv.mt.gov/content/MI/2017_05_State_Plant_List.pdf.

Selfa, T., Qazi, J., 2005. Place, taste, or face-to-face? Understanding producer–consumer networks in "local" food systems in Washington state. Agric. Hum. Values 22 (4), 451–464.

Strolling for Heifers, Locavore Index, 2016. Available from: http://www.strollingoftheheifers.com/locavoreindex/.

UC Davis, 2004. Sample costs for cow-calf/grass fed beef operation. Available from: https://coststudyfiles.ucdavis.edu/uploads/cs_public/83/84/838417e7-bdad-40e6-bcaa-c3d80ccdcd71/beefgfnc2004.pdf.

US Census Bureau, 2010. American Fact Finder. Available from: https://factfinder.census.gov/faces/tableservices/jsf/pages/productview.xhtml?src=CF.

US Census Bureau, 2017. QuickFacts Montana. Available from: https://www.census.gov/quickfacts/fact/table/MT/PST120217.

USAD NASS, 2016. State agricultural overview Montana. Available from: https://www.nass.usda.gov/Quick_Stats/Ag_Overview/stateOverview.php?state=MONTANA.

USDA Farm to School Census, Montana, 2015. Available from: https://farmtoschoolcensus.fns.usda.gov/find-your-school-district/montana.

USDA Foods Available List, 2017. Available from: https://www.fns.usda.gov/sites/default/files/fdd/schools-institutions-foods-available18.pdf (Accessed May 10, 2017).

USDA FSIS, 2014. Federal Meat Inspection Act. Available from: http://www.fsis.usda.gov/fmia.

USDA FSIS, 2017. Inspection directory. Available at: https://www.fsis.usda.gov/wps/portal/fsis/topics/inspection/mpi-directory.

USDA National Agricultural Library, Local Foods n.d. Available from: https://www.nal.usda.gov/aglaw/local-foods#quicktabs-aglaw_pathfinder=0 (Accessed 31 July 2017).

USDA SARE, Annual Report for FNV12-860, 2012. Available from: 2012 https://projects.sare.org/sare_project/fnc12-860/?ar=2012 (Accessed 31 July 2017).

USDA School Meals, 2017. Available at: fns.usda.gov/school-meals/child-nutrition-programs (Accessed May 10, 2017).

FURTHER READING

Anon n.d. Annual_Bulletin_2016.pdf (Accessed 10 May 2017).

Montana Ag Facts, 2015. Available from: http://agr.mt.gov/AgFacts (Accessed May 10, 2017).

Montana Ag Stats/NASS, 2017. Available from: https://www.nass.usda.gov/Statistics_by_State/Montana/.

Montana Farm to School, 2017. Available from: http://www.montana.edu/mtfarmtoschool/beeftoschool.html (Accessed May 10, 2017).

Montana FoodCorps, 2017. Available from: http://mtfoodcorps.ncat.org (Accessed May 10, 2017).

Montana Harvest of the Month, 2017. Available from: montana.edu/mtharvestofthemonth/ (Accessed May 10, 2017).

Richard, B., 1946. Russell National School Lunch Act, 1946. P L. 396—79th Congress, June 4. . 60 Stat. 231.

School Lunch Participation Pattern, Montana, 2012. Available from: http://opi.mt.gov/pdf/SchoolFood/forms/FS/MealPattern_Lunch.pdf.

USDA FNS, NSLP USDA Food Products Information Sheet, 2017. Available from: https://www.fns.usda.gov/nslp/nslp-material-fact-sheets-meatmeat-alternates.

USDA Nutrition Standards School Meals, 2017. Available from: https://www.fns.usda.gov/school-meals/nutrition-standards-school-meals (Accessed May 10, 2017).

Carmen Byker Shanks is an associate professor in the Department of Health and Human Development where she leads the Food and Health Lab at Montana State University. Her research program aims to increase dietary quality and decrease nutrition-related health disparities, while contributing to healthier food environments and systems. She partners with communities, researchers, students, and stakeholders to measure and implement contextually specific strategies in food systems that facilitate dietary change and positive health outcomes. She holds a BS in Dietetics and a PhD in Behavioral and Community Science from Virginia Tech and is a Registered Dietitian.

Thomas Bass is an associate extension specialist and doctoral student at Montana State University. His areas of outreach and research include conservation and environmental management for livestock and poultry production, agricultural and rural community emergency management, agriculture and food waste management, and local food system supply chain structure and performance. Thomas has degrees in animal science and agricultural education from the University of Georgia and Montana State University, respectively. He hopes to complete his doctoral dissertation on spatial and temporal performance of Montana's local beef supply chain in 2020.

Joel Schumacher is an associate extension specialist with the Department of Agricultural Economics and Economics at Montana State University. He holds a Master's degree in applied economics and a Bachelor's degree in business from Montana State University. Joel also is an Accredited Financial Counselor. His research interests focus on agricultural finance and consumer economics.

Chapter 10

Institutional Markets Supporting Midsized Farms: A Case Study of Iowa

Kranti Mulik
Fairfax, Virginia, United States

Chapter Outline

Introduction	219	Estimating the Impacts if Midsized Local Farms Supplied Iowa's Food-Buying Institutions	225
Midsized Farming Operations in U.S. Food Systems	222		
A Case Study From Iowa: Local Food Consumption by Institutions as a Niche for Midsized Farmers	224	Conclusions and Policy Recommendations	229
		References	230
		Further Reading	234

INTRODUCTION

Midsized farms have long formed the backbone of rural economies in the United States, but these farms have been disappearing for almost two decades. The "Agriculture of the Middle" initiative—convened and administered by researchers at Iowa State University (ISU) and the University of Wisconsin-Madison—first called attention to the "disappearing middle" of American agriculture. These researchers determined that midsized farms were disappearing because they were too small to compete in globalized commodity markets, but too large and specialized to sell directly to consumers (Kirschenmann et al., 2004).[1] Based on data from the United States Department of Agriculture

---

1. The definition of midsized farms is based on three factors: ownership, principal occupation, and gross farm cash income (GFCI). The USDA defines midsized farms as having gross annual sales of $350,000–$1,000,000. There are differences in the definition of midsized farms across studies. For example, Kirschenmann et al. (2004) and Ahearn et al. (2009) define midsized farms as having gross annual sales of $100,000–$250,000. The USDA also considers farms operating on 50–999 acres to be midsized farms (Key and Roberts, 2007).

(USDA), the estimated number of midsized farms decreased by 56,000 nationally between 2007 and 2012, while the number of large farms (more than 1000 acres) increased (USDA-NASS, 2014a). Some agricultural states were hit particularly hard. Iowa, for example, lost roughly 6000 midsized farms, accounting for nearly 10 percent of all those lost nationally (USDA-NASS, 2014b).

At the same time, consumer attitudes about food are changing. More consumers are interested in healthy eating and want to know how their food is produced and by whom. This has led to the rapid growth of farmers' markets, which more than doubled from 3706 in 2004 to 8284 in 2014 (ERS, 2014), community supported agriculture (CSA)[2] programs, local grocery stores, and "farm-to-table" programs at restaurants, schools, and other institutions. But, midsized farms are not yet supplying this demand. The decline of midsized farms, despite increasing demand for local and healthy food, suggests that there are market failures in the food system that are preventing midsized farms from connecting effectively with consumer markets.

The purpose of this chapter is to assess institutional foodservice as a market opportunity for midsized farms and the capabilities of these farms to meet growing demand from institutions. It argues that connecting more midsized farms to institutional markets can help address market failures and contribute positive effects to the local economy. Furthermore, there are specialty markets such as those served by farm-to-institution initiatives that have a number of distinctive features that make institutional foodservice an appropriate market for midsized farms (Ray and Schaffer, 2008). For example, as Mooney and Tanaka (2015) summarized:

- These markets are too small for large farms and too big for direct marketing niches.
- The farm-to-institution (school, hospital, restaurant, etc.) movement often finds inadequate supplies of local and regional products from small farms.
- Institutional foodservice markets can be served by regional "food hubs" for aggregating production from small and midsized farmers within a region.
- In some cases, institutional foodservice markets also require the development of the technical infrastructure for transporting and processing farm products.
- Regional-level meat processing facilities are beginning to emerge in some areas to meet the needs of these specialty markets.
- Some institutional foodservice markets might best be addressed by co-operative enterprises of midsized farms addressing regional needs and providing "identity-preservation" through branding associated with a particular locale.

---

2. The majority of the literature on the motivation for joining CSAs cites consumers' desire for environmentally sustainable food production as the major reason (Brehm and Eisenhauer, 2008; Goland, 2002; Cone and Myhre, 2000).

This chapter focuses on Iowa as a case study because it is one of the most prominent agricultural states in the United States and has experienced a significant decline in midsized farms. Given that data on the sales to institutional markets and the potential of midsized farms to meet institutional demand are limited, Iowa was also chosen because of the availability of survey data from Iowa State University (ISU), which conducted a survey in 2012 and 2013 to assess the potential economic influence on Iowa's economy if retail and institutional markets in the state increased their purchases of local foods and midsized farms in Iowa were to supply the increased demand. ISU also collected data on the value of sales linked to institutions. The Union of Concerned Scientists (UCS) then analyzed and extrapolated from the survey data to argue that at the state-level, institutional and retail food purchases could have the potential to generate more than three billion dollars in revenue to Iowa's economy, support up to 17,000 midsized farms, and over 355,000 jobs in Iowa (Mulik, 2016).[3] Hence, increased purchases by institutional and retail markets can contribute to the economic viability of midsized farmers and economic development of the state.

The Regional Food Systems Working Group (RFSWG) promotes local and food system development. As a part of this, it has contributed to the development of a few recent statewide surveys—a farmer survey, a buyer survey, and a regional food coordinator survey. ISU's Leopold Center worked with the RFSWG to gather data in Iowa on total local food sales by farmers, retail, and institutional food purchases,[4] associated job creation, and financial resources available to regional food coordinators. The same three surveys were distributed by ISU to farmers, buyers, and regional food coordinators each year.[5] A total of 103 farmers and 74 retail buyers participated in the survey in 2012; in 2013, 120 farmers—small, midsized, and large—and 73 retail buyers participated.[6] The survey respondents were identified based on the social networks of the RFSWG coordinators, and represented a small proportion of the actual local food sales, purchases, and jobs created as a result of local foods commerce in Iowa. Because this subset of respondents are likely to be more motivated and committed to local food commerce than average, any attempts to generalize the results for Iowa as a whole should be interpreted with caution.

---

3. The author is a senior economist at the Union of Concerned Scientists.
4. The 2007 Census of Agriculture (USDA, 2009b) data reported only direct-to-consumer sales. A 2008 Agricultural Resource Management Survey (ARMS) survey (reported by Low and Vogel, 2011) was the only survey to track local food sales to individual and retail markets. However, it did not track sales to institutions. A 2015 Local Food Marketing Practices Survey by USDA did track sales to institutions at the national level, but did not include state-level data (USDA-NASS, 2016b).
5. Bregendahl and Enderton (2013) do not define "regional food coordinators." However, these individuals can be broadly interpreted as those who support local food activities within RFSWG's network. Local food coordinators in Iowa are building the food system through value chain support work that includes: support for farm-to-school programs, improving healthy food access, food rescue, providing technical assistance to farmers' markets to help them grow, developing local food networks and collaborations, etc.
6. Bregendahl and Enderton (2013) do not report the overall survey response rate.

The remainder of the chapter is organized as follows: The first section provides an overview of the characteristics and trends of midsized farming operations in the United States. The second section summarizes the barriers that midsized farms commonly face in meeting institutional and retail demand, as well as barriers that institutional and retail buyers face. This is followed by a detailed case study of Iowa in the context of growing institutional and retail demand and the potential of midsized farms in Iowa to meet this demand. Finally, the chapter concludes by listing several recommendations on how public policies can facilitate better and growing connections between diversified midsized farms and large institutional and retail food buyers.

MIDSIZED FARMING OPERATIONS IN U.S. FOOD SYSTEMS

In 2012, midsized farms were more prevalent in North Dakota, South Dakota, Nebraska, Iowa, Illinois, and Indiana than in other states, based on the percentage of all farms in the respective state. Most midsized farms produce food crops or livestock. Midsized farmers are aging, consistent with broader trends in farming in the United States, with an average age of 56 in 2012 (Burns and Kuhns, 2016; USDA-NASS, 2014a).

Midsized farms tend to generate greater economic benefits to their local community compared with large farms. Small and midsized farms are more likely than large farms to purchase inputs locally—in particular, livestock feed and equipment—keeping more money in their local economies, particularly if they are predominately producing crops (Chism and Levins, 1994).[7] Large farms, on the other hand, seldom buy locally, as the inputs needed by them cannot be supplied locally in desired quantities, consistent quality, time frame, and price required by these farms (Chism and Levins, 1994; Longworth, 2007). Small and midsized farms also employ more people per acre than large, industrialized farms. Therefore, when they disappear, farming jobs evaporate with them, along with indirect employment in the community. In addition, research has shown that areas having more midsized farms and a stronger middle class have lower poverty and unemployment rates, higher average household incomes, and

---

7. Crop farms regardless of size tend to purchase more locally because the basic inputs required for crop production are available locally and do not change as size of farms increase. However, it is different for livestock farms because basic inputs (such as feeder animals and breeding stock) are often not available in the quantities and consistent quality required by livestock producers (Chism and Levins, 1994). According to a 1990 study of Michigan hog operations, larger hog farms spent less locally than smaller farms. The average amount spent locally by a 500-head farm was $67 per head, while a 5000-head farm spent only $47. Thus, the large farm would spend $235,000 locally, compared with $33,550 spent by the smaller farm. But ten 500-head farms would spend more than $100,000 more locally than one 5000-head farm (Abeles-Allison and Connor, 1990). Comparative results were by Gomez and Zhang (2000) for livestock farms.

greater socioeconomic stability (Labao, 1990; Lyson et al., 2001). In contrast, areas with larger farms are associated with lower household incomes, more poverty and economic inequality, downtowns with fewer retail businesses, and less money spent in the community (Pew Commission, 2008).

Between 2002 and 2012, the number of midsized farms declined by more than 7 percent, and midsized farmers experienced high income volatility between 2007 and 2012 (Burns and Kuhns, 2016; USDA-NASS, 2014a). During this time, the number of large farms also declined, but only by 2 percent. The demise of midsized farms has had serious consequences in rural communities. Already, many rural areas across the country have been steadily losing population—the population across 1300 rural counties dropped by half a million people between 2010 and 2014 (ERS, 2015). Once-vibrant rural communities are at continued risk as the loss of job and business opportunities represented by midsized farms continues. While the number of small farms increased by more than 9 percent from 2002 to 2012, the growth in these farms has not been able to compensate for the losses of midsized farms. Midsized farms with beginning farmers, retired farmers, and renters were especially likely to exit farming between 2007 and 2012 (Burns and Kuhns, 2016).

There are several possible explanations for the decline of midsized farms, including technological change (e.g., mechanization), growing market concentration at several stages of production, and shifting financial incentives due to volatility in prices and other financial incentives. Technological change, increased labor productivity, and growing economies of scale have resulted in increased specialization (producing one or two crops) in production (Hallam, 1993; MacDonald and McBride, 2009). Most technological innovations have favored larger farms, because expensive mechanical equipment needs to be used for a minimum number of hours or acres to achieve efficiency (Halloran and Archer, 2008). Mechanization has also reduced labor requirements and increased labor productivity (Sassenrath et al., 2008), which has led to significant decline in the number of workers per acre of production in the past century (Schjonning et al., 2004).

Consolidation in agricultural supply chains plays an important role in determining the position of midsized farms. Market concentration in the processing and distribution stages of agriculture value chains has benefited a small number of highly integrated, national and global firms (Hendrickson and Heffernan, 2007). Thus, the role of midsized farmers has become largely limited to serving as contract growers for larger corporations (Ikerd, n.d.; Longworth, 2007). For example, in the grain sector, the top four firms—Cargill, Archer Daniels Midland Corporation (ADM), Bunge Limited, and Louis Dreyfus Company—control more than 70 percent of the global grain market, and have significant influence over global food systems, farmers' livelihoods, and consumers throughout the world (Clapp, 2015; Murphy et al., 2012). Similarly, in the beef industry, the three largest producers (Tyson Foods, Cargill Meat Solutions, and JBS-Swift) have a great deal of market power, from determining the traits that the producers breed, to marketing specific products in supermarket display cases (Lowe and Gereffi, 2009). The U.S. swine production industry has also become increasingly concentrated, with roughly the same number of hogs produced on fewer, but larger, farms that specialize in a single phase of production. Such operations now account for about 77 percent of the nation's hog production (Key and Roberts, 2007). A single player, Smithfield Foods, controls 27 percent of U.S. pork production (Lowe and Gereffi, 2008). The strong consolidation and influence of the largest supply chain actors leaves growers who are selling to chains or processing companies increasingly vulnerable to price fluctuations, increased competition, and risk—even in niche market segments such as organics (Guthman, 2004).

Volatility in prices of farmland and financial incentives also affects midsized farmers by increasing their risk burden and costs of production. In the United States, federal subsidies have promoted farm concentration with the largest commercial farms benefiting at the expense of small and midsized farms (Imhoff, 2007). Also, in the past fifteen years, farmland inflation rates have increased by more than 150 percent with prices for farmland as high as $10,000 per acre in some states (USDA-NASS, 2016a). As the price of farmland has increased, large farms often have the resources to buy out smaller farms, whereas the escalating land values and high rents have made it difficult for small and beginning farmers in purchasing and leasing land. As farms get bigger, corporations also grow to meet the large-scale input needs of these farms. Consequently, as the number of midsized farms has declined, so, too, has the basic infrastructure that serves that sector of agriculture. Midsized farmers today have fewer small grain elevators, auction barns, and implement dealers to meet the scalar needs of their businesses (Burns and Kuhns, 2016).

A CASE STUDY FROM IOWA: LOCAL FOOD CONSUMPTION BY INSTITUTIONS AS A NICHE FOR MIDSIZED FARMERS

Iowa is an important agricultural state and has the most productive soils in the world. Iowa is the largest producer of corn, soybean, hogs, and eggs in the United States (USDA, 2009a). The majority of Iowa's land area (86 percent, or 30.8 million acres) is allocated to agriculture (Iowa Department of Agriculture and Land Stewardship, 2009). While local food markets have increased in many places throughout Iowa, it is prominent in Northeast Iowa and around metropolitan areas such as Des Moines, Waterloo, Iowa City, and Sioux city. A number of resources are available to farmers for local food business development. These include three Regent institutions, multiple private colleges, fifteen community college districts, and cooperative extension services covering the state of Iowa. It also includes several farm groups and non-profit organizations, such as the Practical Farmers of Iowa, Iowa's Resource Conservation and Development districts, Iowa farmers'

Union, Iowa Farm Bureau Federation, the Women Food and Agricultural Network, Iowa Buy Fresh Buy Local, and the Iowa Network for Community Agriculture (Leopold Center for Sustainable Agriculture, 2011).

Recent research in Iowa suggests that consumers there are often willing to pay more for local food; that is, food produced in Iowa (DeCarlo et al., 2005; Krouse and Galluzzo, 2007).[8] Between 2007 and 2012, direct-to-consumer farm sales in Iowa increased by 6 percent (from $16.5 million to $17.5 million), even though the number of farms reporting direct sales during that period remained essentially flat (United States Department of Agriculture, 2014b). Further, while Iowans spend more than eight billion dollars for food each year, only 10 percent of that food is produced locally (Tagtow, 2008).[9] As such, there is great potential for further growth in the local food market. One study revealed that farmers' markets in the state generated total sales of $3.4 million and supported 374 direct jobs and 200 indirect jobs in 2009 (Otto, 2010). Another study found that if Iowans each ate five daily servings of fruits and vegetables produced in the state, this change in habit would add $331 million in economic output to the local economy, $123.3 million in labor income, and 4484 jobs (Swenson, 2006).

Local food purchases by Iowa institutions are growing. A 2008 study of eight counties in Iowa found that local purchases had increased from just three institutions purchasing $111,000 worth of local food in 1998 to 25 institutions purchasing $1.8 million worth a decade later (Sharma et al., 2012). Other studies in Iowa have also shown increased demand for local food from retail markets (Gregoire et al., 2005; Strohbehn and Gregoire, 2003). There is tremendous room for further growth, as the total number of institutional and retail food purchasers in the state has been estimated at more than 22,000 (Enderton and Bregendahl, 2014). These potential local food buyers could provide a much-needed opportunity to keep midsized farms in, or return them to rural communities in Iowa and elsewhere.

Estimating the Impacts if Midsized Local Farms Supplied Iowa's Food-Buying Institutions

To understand the potential economic influence on Iowa's economy if midsized farms in Iowa met local food demand from more of Iowa's retail and institutional markets, UCS conducted further analysis of the 2012 and 2013 surveys coordinated by ISU's Leopold Center of Sustainable Agriculture in conjunction with the RFSWG.

---

8. "Local food" is defined by the USDA as "the direct or intermediated marketing of food to consumers that is produced and distributed in a limited geographic area. There is no predetermined distance to define what consumers consider 'local,' but a set number of miles from a center point or state/local boundaries is often used" (USDA NAL, n.d.).

9. Enderton and Bregendahl (2014) estimate Iowa's local food sales to be $322 million.

The RFSWG was established in 2003 to promote local and regional food system development and is organized into fifteen regional groups. Regional food coordinators in each group were asked to invite local food producers and food buyers within their respective regions to complete the survey. Two criteria had to be met: (1) Farmers and buyers surveyed were associated with the work of the regional food coordinators and (2) Farmers (produce, livestock, and dairy farmers) and buyers either produced or purchased locally grown food. The focus of the survey was to gather data on four categories: local food sales by farmers; local food purchases by grocery stores, restaurants, institutions, and others; job creation due to local food production, processing or utilization; and funds leveraged to support the development of regional food systems. In 2012, of the 74 (73 in 2013) buyers participating in the survey, 30 percent (31 percent in 2013) were grocery stores, 26 percent (19 percent in 2013) restaurants, 25 percent (22 percent in 2013) were K-12 schools, 12 percent residential foodservice, (11 percent in 2013), 3 percent non-profits (4 percent in 2013), 10 percent distributors (3 percent in 2013), and others were 4 percent (10 percent in 2013). Over the two-year period, buyers reported purchases of more than eleven million dollars, and farmers reported sales of $11.8 million.[10] The surveys also found that these markets created 171 new jobs over two years, including 63 full-time, year-round jobs.

The UCS analysis used the 2012 and 2013 survey data to determine the potential influence of increased local food purchases by institutions and retail markets in Iowa. Data used included the number of institutional and retail markets in Iowa, average local food purchases by retail and institutional markets, and average jobs supported in Iowa due to increased purchases from buyers as well as increased sales from farmers. The analysis estimated the influence under three different scenarios: when midsized farms supply (1) 25 percent, (2) 50 percent, and (3) 100 percent to institutions and other buyers in Iowa (22,189), making the assumption that these buyers made purchases at the same level as the respondents in the survey (Table 1).[11] Our analysis estimated the potential economic benefits to Iowa's economy in terms of revenue generated, jobs supported, and midsized farms supported as a result of increased purchases of local food by institutional and retail markets as follows:

- Total purchases by retail and institutional markets = total potential markets × percentage of total markets purchasing local food × $150,316.
- Revenue to economy from midsized farms = total purchases by retail and institutional markets/2.

---

10. The 2012 Census of Agriculture (USDA, 2014b) reported that there were 2964 farms in Iowa, with $17.5 million in direct sales, while the ISU 2012 survey reported only $10.5 million in local sales with only 103 farmers reporting. As pointed out by Enderton and Bregendahl (2014), this discrepancy is most likely due to the lack of measurement of sales by farmers to retail and institutional markets.

11. These estimates should be interpreted with caution, because the survey sample included those that were connected to RFSWG networks already, which is likely to mean the sample was biased in their increased willingness to purchase locally.

- Total number of farm jobs supported = total purchases by retail and institutional markets × 106.5/$1000,000.
- Total number of full-time farm jobs supported = total purchases by retail and institutional markets × 29.55/$1000,000.
- Total number of farm jobs supported by midsized farms = total number of farm jobs supported/2.
- Total number of full-time farm jobs supported by midsized farms = total number of full time farm jobs supported/2.
- Total buyer based jobs created = total purchases by retail and institutional markets/$303,555.
- Number of midsized farms supported = revenue to economy from midsized farms/$98,117.

The analysis found the following benefits to Iowa's economy:

- If just 25 percent of retail and institutional markets in Iowa purchased local food at the same level as survey respondents, this would generate more than $800 million annually for the state's economy. At 50 percent, the influence would be $1.67 billion. And if all of Iowa's retail and institutional markets purchased food locally, the influence would increase to more than three billion dollars.
- If, in the 25-percent, 50-percent, and 100-percent cases, Iowa's retail and institutional markets sourced half of their local food demand from midsized farms, from 4249 to 16,997 such farms could be supported.[12] These figures represent 5–19 percent of all farms currently operating in Iowa. Even the low figure would translate to the return of nearly 75 percent of all midsized farms lost by the state between 2007 and 2012.
- Such sales by midsized farms could support between 44,000 and 178,000 total jobs in Iowa.[13]
- The number of full-time farm jobs supported by midsized farms would range from 12,000 to nearly 49,000.
- Full-time buyer-based jobs, such as those associated with procuring, preparing, marketing, and educating about local foods, could range from 2800 to 11,000.

As shown in Table 1, the results indicate that if institutions and retail outlets increased their purchases of local food, that would have the potential to increase the number of midsized farms and benefit Iowa's economy in terms of jobs and livelihoods.[14]

---

12. The Leopold Center's survey defined midsized farms as farms with sales of $50,000–$249,000.
13. Alternatively, Low and Vogel (2011) estimate that for every one million dollars in sales, thirteen full-time jobs are supported. Using the average of the Low and Vogel (2011) estimate and the ISU estimate, full-time jobs supported range from 17,740 to 70,960.
14. Note that these results are subject to variability. The data from the survey represents only a small percentage of the actual total food sales and respondents were selected based on the social networks of the RFSWG coordinators, and because they are incorporating local foods into their operations as a conscious business decision. Therefore, average purchases, jobs created estimates may be higher or lower if a larger representative sample was available. A detailed analysis using an input-output model such as the Impact Analysis for Planning (IMPLAN) was outside the scope of this analysis.

TABLE 1 Estimated Economic Impacts if More Large-Scale Iowa Buyers Purchased Local Food[a]

Estimated Impacts	Scenarios: Percent of Institutional Markets Purchasing Local Food		
	25 Percent	50 Percent	100 Percent
Number of institutions and other buyers purchasing local food[b]	5547	11,095	22,189
Total purchases by institutional markets and other buyers (in billions)[c]	$0.834	$1.67	$3.34
Revenue to economy from midsized farms (in billions)[d]	$0.417	$0.834	$1.67
Total number of farm jobs supported[e]	88,804	177,608	355,215
Total number of full-time farm jobs supported[f]	24,640	49,280	98,560
Total number of farm jobs supported by midsized farms[g]	44,402	88,804	177,608
Total number of full-time farm jobs supported by midsized farms[h]	12,320	24,640	49,280
Total buyer-based jobs created[i]	2747	5494	10,988
Number of midsized farms supported[j]	4249	8498	16,997

[a]This table contains analysis for both local food and local food from midsized farms.
[b]About 56 percent of the total market (22,189) is represented by institutions while the remainder is represented by other buyers (retailers).
[c]Assumes local food purchases of $150,316, the estimated 2012 and 2013 average local food purchases by buyers in the ISU survey.
[d]Assumes that midsized farms can meet half the demand (purchases) institutions and other buyers.
[e]Based on ISU survey data indicating that every one million dollars in farm sales supported an average of 106.5 jobs (average for 2012 and 2013).
[f]Based on ISU survey data indicating that every one million dollars in farm sales supported an average of 29.55 full time farm jobs (average for 2012 and 2013).
[g]See notes d and e.
[h]See notes d and f.
[i]These are jobs created as a result of the purchases made by institutions and other buyers. Calculations based on the 2012 ISU survey, which indicated that spending of $303,555 on local food purchases by institutions and other buyers creates at least one new full-time local food-related job.
[j]Assumes average sales of local food by midsized farms were $98,117 (average of 2012 and 2013 ISU surveys). Assumes that midsized farms can meet half the demand (purchases) by institutions and other buyers.
Based on Bregendahl, C., Enderton, A., 2013. 2012 economic impacts of Iowa's regional food systems working group. Leopold Center for Sustainable Agriculture, Iowa State University. Available from: http://lib.dr.iastate.edu/cgi/viewcontent.cgi?article=1040&context=leopold_pubspapers (Accessed 30 November 2017); Enderton, A., Bregendahl, C., 2014. 2013 economic impacts of Iowa's regional food systems working group. Leopold Center for Sustainable Agriculture, Iowa State University. Available from: http://lib.dr.iastate.edu/cgi/viewcontent.cgi?article=1027&context=leopold_pubspapers (Accessed 30 November 2017).

CONCLUSIONS AND POLICY RECOMMENDATIONS

In 2009, the Northern Iowa Food and Farm Coalition conducted a survey of 180 institutional buyers in Howard, Winneshiek, Allamakee, Clayton, and Fayette counties in Iowa to assess current food purchasing patterns by foodservice marketing, including institutions such as public and private schools, hospitals, and care centers. Sixty-one respondents indicated that they are not purchasing or have stopped purchasing locally grown products because of lack of accessibility or availability, not being approached by local farmers or processors selling these products, or not knowing who to contact for local products (Chase, 2009). This suggests a need for vendor lists and better platforms for information exchange. Other barriers in purchasing local food included state codes and regulations[15] (35 respondents) and the higher cost of locally grown food. Eighty-five percent of the respondents said they would purchase local food if cost barriers were eliminated. Buyers also indicated that they would be willing to pay, on average, 12 percent more for locally grown products (Chase, 2009).

Clearly, barriers exist for institutional sourcing of local food.[16] Despite these challenges, institutional buyers require both large volumes and a diversity of foods, and neither small farms (because of their lack of capacity[17]) nor large farms (because of their lack of flexibility)[18] in places such as Iowa can meet this demand. This leaves midsized farms as untapped resources in the local food movement, with a "comparative advantage in producing unique, highly differentiated products" (Kirschenmann et al., 2004).

---

15. It is not clear if navigating state codes and regulations is an actual or perceived difficulty. Further education or policy implementation can be helpful if this an actual difficulty or a perceived barrier (Chase, 2009).
16. Other barriers, many of which are discussed in this volume, include food safety and liability, access to quality, reliable and adequate supply, seasonality of supply, regulatory barriers and contradictory procurement policies, managing multiple vendors, communicating with different people along the supply chain, low skills of foodservices workers, infrastructure barriers for aggregating distributing and processing and loss of original champion (Barlett, 2011; Conner et al., 2011, 2014; Feenstra and Ohmart, 2012; Harris et al., 2012; Marshall et al., 2012).
17. Small farms are not in a position to meet the growing demand from these markets for a number of reasons. Some small farmers are not interested in marketing beyond the farmers' markets or CSA due to the personal interaction they enjoy with their customers, possibility of reduced profits or quality of life as they move to a higher volume market. In some cases, access to land and capital need for expansion presents a barrier.
18. In the past, farmers who grow federally subsidized crops such as field corn, soybeans, wheat, and cotton were restricted from converting their land to fruit or vegetable production, even for just one year, unless they were permanently willing to give up their right to collect federal payments on that acreage. Thus, reduction in payment began on the first acre of fruit and vegetable planted on base acres. The new farm bill now allows farms with commodity base acres to plant fruits and vegetables on 15–35 percent (depending on the commodity program chosen by farmers, either Price Loss Coverage or Agriculture Risk Coverage, to provide income support in response to adverse price or yield conditions) of base acres without any penalty. Planting fruits and vegetables on base acres above those limits requires a one-to-one reduction in payment acres. For details, see Shields (2014).

Recognizing the benefits that could be achieved with a major shift to local food production and sale, as well as the potential economic and resource challenges farmers must overcome in order to realize that shift, UCS recommends that future federal food and agriculture policies incorporate and emphasize future and consistent investments in infrastructure and coordination to get local food from midsized farms to institutional markets. Midsized farms are often of a size and scale that make it difficult to serve institutional markets. Intermediaries, such as food aggregators or regional hubs, can help farmers overcome this problem. These intermediaries play an important role in delivering the goods produced by midsized farms to local and regional food networks. Local food coordinators may support food aggregation by helping to establish food hubs, and research has shown that a modest public investment in the work of local food coordinators has contributed to job creation in Iowa (Bregendahl and Enderton, 2013; Enderton and Bregendahl, 2014). Successful food aggregators serving the Midwest, Northeast, and Western regions of the United States received initial funding through state and federal grants and are now maintained by state and federal resources. Additionally, our analysis was limited due to the lack of available data on institutional food procurement. While the USDA has recently started tracking sales to institutions, these data are not available at the state level. Annual data that tracks institutional food purchasing by farm size and state would be extremely useful in determining the influence of institutional food procurement on the local economy. UCS also recommends financial incentives to help beginning and transitioning farmers grow the foods institutional buyers want, and research and technical assistance to help farmers adopt midsized systems. We ultimately argue that a comprehensive national food and farm policy can support farmers not only in Iowa, but nationally, and ensure that the food and farm sectors create new jobs and pay fair wages to the millions of people they employ, and enable Americans at all income levels to access affordable, healthful food (Mulik, 2016).

REFERENCES

Abeles-Allison, M., Connor, L., 1990. An Analysis of Local Benefits and Costs of Michigan Hog Operations Experiencing Environmental Conflicts. Department of Agricultural Economics, Michigan State University, East Lansing, MI.

Ahearn, M., Yee, J., Korb, 2009. Producer dynamics in agriculture: empirical evidence. In: Dunne, T., Jensen, K., Roberts, M. (Eds.), Producer Dynamics: New Evidence From Micro Data. University of Chicago Press, Chicago.

Barlett, P., 2011. Campus sustainable food projects: critique and engagement. Am. Anthropol. 113 (1), 101–115.

Bregendahl, C., Enderton, A., 2013. 2012 economic impacts of Iowa's regional food systems working group. Leopold Center for Sustainable Agriculture, Iowa State University. Available from: http://lib.dr.iastate.edu/cgi/viewcontent.cgi?article=1040&context=leopold_pubspapers. Accessed November 30, 2017.

Brehm, J.M., Eisenhauer, B.W., 2008. Motivations for participating in community supported agriculture and their relationship with community attachment and social capital. South. Rural Sociol. 23 (1), 94–115.

Burns, C., Kuhns, R., 2016. The changing organization and well-being of midsize U.S. farms, 1992–2014. United States Department of Agriculture, Economic Research Service. Available from: https://www.ers.usda.gov/webdocs/publications/80692/err-219.pdf?v=42671. Accessed November 30, 2017.

Chase, C., 2009. Northeast Iowa local food survey summary report. Leopold Center Completed Grant Reports, 324. Available from: http://lib.dr.iastate.edu/leopold_grantreports/324. Accessed November 30, 2017.

Chism, J.W., Levins, R.L., 1994. Farm spending and local selling: How do they match up? Minnesota Agricultural Economist. Available from: http://mysare.sare.org/mySARE/assocfiles/990LNC92-048.006.pdf. Accessed November 30, 2017.

Clapp, J., 2015. ABCD and beyond: from grain merchants to agricultural value chain managers. Can. Food Stud. 2 (2), 126–135. Available from: http://canadianfoodstudies.uwaterloo.ca/index.php/cfs/article/download/84/105. Accessed November 30, 2017.

Cone, C.A., Myhre, A., 2000. Community-supported agriculture: a sustainable alternative to industrial agriculture? Hum. Organ. 59 (2), 187–197.

Conner, D., Nowak, A., Berkenkamp, J., Feenstra, G., Van Soelen Kim, J., Liquori, T., Hamm, M., 2011. Value chains for sustainable procurement in large school districts: Fostering Partnerships. J. Agric. Food Syst. Community Dev. 1 (4), 55–68. Available from: https://doi.org/10.5304/jafscd.2011.014.005. Accessed November 30, 2017.

Conner, D., Estrin, H., Becot, F., 2014. High school harvest: combining food service training and institutional procurement. J. Ext. 52 (1). Article 11AW7. Available from: https://joe.org/joe/2014february/iw7.php. Accessed November 30, 2017.

DeCarlo, T.E., Franck, V.L., Pirog, R., 2005. Consumer perceptions of place based foods, food chain profit distribution and family farms. Leopold Center for Sustainable Agriculture, Iowa State University. Available from: http://lib.dr.iastate.edu/cgi/viewcontent.cgi?article=1154&context=leopold_pubspapers. Accessed November 30, 2017.

Economic Research Service (ERS), 2014. Number of farmers' markets continues to rise. United States Department of Agriculture, Washington, DC. Available from: https://www.ers.usda.gov/data-products/chart-gallery/gallery/chart-detail/?chartId=77600.

Economic Research Service (ERS), 2015. Population and migration. United States Department of Agriculture, Washington, DC. Available from: http://www.ers.usda.gov/topics/rural-economy-population/population-migration.aspx.

Enderton, A., Bregendahl, C., 2014. 2013 economic impacts of Iowa's regional food systems working group. Leopold Center for Sustainable Agriculture, Iowa State University. Available from: http://lib.dr.iastate.edu/cgi/viewcontent.cgi?article=1027&context=leopold_pubspapers. Accessed November 30, 2017.

Feenstra, G., Ohmart, J., 2012. The evolution of the school food and farm to school movement in the United States: connecting childhood health, farms, and communities. Child. Obes. 8 (4), 280–289.

Goland, C., 2002. Community supported agriculture, food consumption patterns and member commitment. J. Cult. Agric. 24 (1), 14–25.

Gomez, M.I., Zhang, L., 2000. In: Impacts of concentration in hog production on economic growth in rural Illinois: an econometric analysis. Paper Presented at the American Agricultural Association Annual Meeting, Tampa, FL.

Gregoire, M.B., Arendt, S., Strohbehn, C.H., 2005. Iowa producers perceived benefits and obstacles in marketing to local restaurants and institutional foodservice operations. J. Ext. 43 (1), 1–10. Available from: http://www.joe.org/joe/2005february/rb1.php. Accessed November 30, 2017.

Guthman, J., 2004. Agrarian Dreams: The Paradox of Organic Farming in California. University of California Press, California.

Hallam, A., 1993. Size, Structure, and the Changing Face of American Agriculture. Westview Press, Boulder.

Halloran, J.M., Archer, D.W., 2008. External economic drivers and US agricultural production systems. Renew. Agric. Food Syst. 23 (4), 296–303.

Harris, D., Lott, M., Lakins, V., Bowden, B., Kimmons, J., 2012. Farm to institution: creating access to healthy local and regional foods. Adv. Nutr. 3 (3), 343–349.

Hendrickson, M., Heffernan, W., 2007. Concentration of agricultural markets. Available from: https://www.iatp.org/files/258_2_60439.pdf Accessed November 30, 2017.

Ikerd J n.d., Do mid-sized farms have a future? Available from: http://web.missouri.edu/ikerdj/papers/SFT-Farms%20of%20Middle.htm. (Accessed 30 November 2017).

Imhoff, D., 2007. Food Fight: The citizen's Guide to a Food and Farm Bill. Watershed Media, Healdsburg, CA.

Iowa Department of Agriculture and Land Stewardship, 2009. Iowa agriculture quick facts. Available from: http://www.iowaagriculture.gov/quickFacts.asp. Accessed November 30, 2017.

Key, N., Roberts, M.J., 2007. Commodity payments, farm business survival and farm size growth. Available from: https://www.ers.usda.gov/webdocs/publications/45923/12240_err51_1_.pdf?v=41056. Accessed November 30, 2017.

Kirschenmann, F., Stevenson, S., Buttel, F., Lyson, T., Duffy, M., 2004. Why worry about the agriculture of the middle? Available from: http://agofthemiddle.org/wp-content/uploads/2014/09/whitepaper2.pdf. Accessed November 30, 2017.

Krouse, L., Galluzzo, T., 2007. Iowa's local food systems: a place to grow, the Iowa Policy Project. Available from: https://www.iowapolicyproject.org/2007docs/070206-LocalFood.pdf.

Labao, L., 1990. Locality and Inequality: Farm and Industry Structure and Socioeconomic Condition. State University of New York Press, Albany.

Leopold Center for Sustainable Agriculture, 2011. Iowa local food and farm plan. Available from: http://lib.dr.iastate.edu/cgi/viewcontent.cgi?article=1093&context=leopold_pubspapers.

Longworth, R.C., 2007. Caught in the Middle: America's Heartland in the Age of Globalism. Bloomsbury, New York.

Low, S.A., Vogel, S., 2011. Direct and intermediated marketing of local foods in the United States. United States Department of Agriculture Economic Research Service. Available from: https://www.ers.usda.gov/webdocs/publications/err128/8276_err128_2_.pdf. Accessed November 30, 2017.

Lowe, M., Gereffi, G., 2008. A value chain analysis of the U.S pork industry. Available from: Center on Globalization, Governance & Competitiveness, Duke Universityhttps://gvcc.duke.edu/wp-content/uploads/CGGC_PorkIndustryReport_10-3-08.pdf. Accessed November 30, 2017.

Lowe, M., Gereffi, G., 2009. A value chain analysis of the U.S beef and dairy industries. Available from: Center on Globalization, Governance & Competitiveness, Duke University. https://gvcc.duke.edu/wp-content/uploads/CGGC_BeefDairyReport_2-16-09.pdf. Accessed November 30, 2017.

Lyson, T.A., Torres, R.J., Welsh, R., 2001. Scale of agricultural production, civic engagement and community welfare. Soc. Forces 80 (1), 311–327.

MacDonald, J.M., McBride, W.D., 2009. The transformation of US livestock agriculture scale, efficiency, and risks. USDA/ERS, Economic Information Bulletin No. 43, 36 p.

Marshall, C., Feenstra, G., Zajfen, V., 2012. Increasing access to fresh, local produce_ building values-based supply chains in San Diego Unified School District. Child. Obes. 8 (4), 388–391.
Mooney, P.H., Tanaka, K., 2015. The Family Farms in the United States: Social Relations, Scale and Region. Available from: Village and Agriculture 1.1 (166.1) 45–57. http://ageconsearch.umn.edu/record/230448/files/45_art_WiR%201_1%20calosc.pdf.
Mulik, K., 2016. Growing economies connecting local farmers and large scale food buyers to create jobs and revitalize America's Heartland. Union of Concerned Scientists, Washington, DC. Available from: http://www.ucsusa.org/sites/default/files/attach/2016/01/ucs-growing-economies-2016.pdf. Accessed November 30, 2017.
Murphy, S., Burch, D., Clapp, J., 2012. Cereal secrets: the world's largest grain traders and global agriculture. Oxfam research reports. Available from: https://www.oxfam.org/sites/www.oxfam.org/files/rr-cereal-secrets-grain-traders-agriculture-30082012-en.pdf. Accessed November 30, 2017.
Otto, D., 2010. Consumers, vendors, and the economic importance of Iowa farmers markets: an economic impact survey analysis. Prepared for the Iowa Department of Agriculture and Land Stewardship and the Iowa Farmers Market Association, Des Moines, IA.
Pew Commission on Industrial Farm Animal Production, 2008. Impact of industrial farm animal production on rural communities. Available from: http://www.pcifapia.org/_images/212-8_PCIFAP_RuralCom_Finaltc.pdf. Accessed November 30, 2017.
Ray, D.E., Schaffer, H.D., 2008. In: Lyson, T.A., Stevenson, G.W., Welsh, R. (Eds.), Food and the Mid-Level Farm: Renewing an Agriculture of the Middle. MIT Press, Cambridge.
Sassenrath, G.F., Heilman, P., Lusche, E., Bennett, G.L., Fitzgerald, G., Klesius, P., Tracy, W., Williford, J.R., Zimba, P.V., 2008. Technology, complexity and change in agricultural production systems. Renew. Agric. Food Syst. 23 (4), 285–295.
Schjonning, P., Elmholt, S., Christensen, B.T., 2004. Managing Soil Quality: Challenges in Modern Agriculture. CABI Publishing, Cambridge.
Sharma, A., Strohbehn, C., Radhakrishna, R.B., Ortiz, A., 2012. Economic viability of selling locally grown produce to local restaurants. J. Agric. Food Syst. Commun. Dev. 3 (1), 181–198. Available from: https://doi.org/10.5304/jafscd.2012.031.014. Accessed November 30, 2017.
Shields, D.A., 2014. Farm commodity provisions in the 2014 farm bill (P.L. 113-79). R43448. Congressional Research Service, Washington, DC. Available from: http://nationalaglawcenter.org/wp-content/uploads/assets/crs/R43448.pdf. Accessed November 30, 2017.
Strohbehn, C.H., Gregoire, M.B., 2003. Case studies of local food purchasing by central Iowa restaurants and institutions. Food Res. Int. 14 (1), 53–64.
Swenson, D., 2006. The economic impacts of increased fruit and vegetable production and consumption in Iowa: phase II. Regional Food Systems Working Group, Leopold Center for Sustainable Agriculture, Iowa State University. Available from: http://lib.dr.iastate.edu/cgi/viewcontent.cgi?article=1159&context=leopold_pubspapers. Accessed November 30, 2017.
Tagtow, A., 2008. A vision for 'good food' for Iowa: linking community-based food systems to healthy Iowans and healthy communities. Environmental Nutrition Solutions. Available from: http://www.farmlandinfo.org/sites/default/files/Valuechains_VisionforGoodFood_April2008_1.pdf. Accessed November 30, 2017.
United States Department of Agriculture, National Agricultural Statistics Service (USDA-NASS), 2009a. Iowa's rank in United States Agriculture, Washington, DC. Available from: http://www.nass.usda.gov/Statistics_by_State/Iowa/Publications/Rankings/09Ranking.pdf. Accessed November 30, 2017.
United States Department of Agriculture, National Agricultural Statistics Service (USDA-NASS), 2009b. 2007 census of agriculture: Iowa state data, Washington, DC. Available from: www.agcensus.usda.gov/Publications/2007/Full_Report/Volume_1,_Chapter_1_State_Level/Iowa/st19_1_064_064.pdf. Accessed November 30, 2017.

United States Department of Agriculture, National Agriculture Library (USDA NAL) n.d. Local foods. Available from: https://www.nal.usda.gov/aglaw/local-foods (Accessed 30 November 2017).

United States Department of Agriculture, National Agriculture Statistics Service (USDA-NASS), 2014a. 2012 census of agriculture: U.S. summary and state data. AC-12-A-51, Washington, DC. Available from: http://www.agcensus.usda.gov/Publications/2012/Full_Report/Volume_1,_Chapter_1_US/usv1.pdf. Accessed November 30, 2017.

United States Department of Agriculture, National Agriculture Statistics Service (USDA-NASS), 2014b. 2012 census of agriculture: Iowa state and county data, Washington, DC. Available from: http://www.agcensus.usda.gov/Publications/2012/Full_Report/Volume_1,_Chapter_1_State_Level/Iowa/iav1.pdf. Accessed November 30, 2017.

United States Department of Agriculture, National Agriculture Statistics Service (USDA-NASS), 2016a. Land values 2016 summary, Washington, D.C Available from: https://www.nass.usda.gov/Publications/Todays_Reports/reports/land0816.pdf. Accessed November 30, 2017.

United States Department of Agriculture, National Agriculture Statistics Service (USDA-NASS), 2016b. 2015 local food marketing practices survey, Washington, DC. Available from: https://www.agcensus.usda.gov/Publications/2012/Online_Resources/Local_Food/index.php.

FURTHER READING

Becot, F., Conner, D., Ettman, K., 2016. How to develop a local and regional institutional food buying program? Available from: https://nofavt.org/sites/default/files/files/resources/becot-conner-ettman-developing.pdf.

Bond, C.A., Thilmany, D., Keeling Bond, J., 2008. Understanding consumer interest in product and process-based attributes for fresh produce. Agribusiness 24, 231–252.

Current Research Information System (CRIS), 2011. Table C. National summary USDA SAES and other institutions. In: Fiscal year 2010 funds and scientist years. US Department of Agriculture, Washington, DC. Available from: http://cris.nifa.usda.gov/crisfin/2010/10tables.pdf. Accessed November 30, 2017.

Darby, K., Batte, M.T., Ernst, S., Roe, B., 2008. Decomposing local: a conjoint analysis of locally produced foods. Am. J. Agric. Econ. 90 (2), 476–486.

DeLonge, M.S., Miles, A., Carlisle, L., 2016. Investing in the transition to sustainable agriculture. Environ. Sci. Policy 55 (1), 266–273. Available from: www.sciencedirect.com/science/article/pii/S1462901115300812. Accessed November 30, 2017.

Mailfert, K., 2006. New farmers and networks: how beginning farmers build social connections in France. Tijdschr. Econ. Soc. Geogr. 98 (1), 21–31.

National Center for Appropriate Technology, 2015. Primer on whole-farm revenue protection crop insurance: new for Iowa producers in 2015. Available from: http://practicalfarmers.org/app/uploads/2015/01/Primer-on-Whole-Farm-Revenue_NCAT_12-2014-2.pdf Accessed November 30, 2017.

Ostrom, M., 2006. Everyday meanings of 'local food': views from home and field. Community Dev. 37 (1), 65–78.

Schneider, M.L., Francis, C., 2007. Marketing locally produced foods: consumer and farmer opinions in Washington County, Nebraska. Renew. Agric. Food Syst. 20 (4), 252–260.

Tiernan, S.J., 2013. Can a diversified locally grown food aggregation (hub) facility be economically sustainable in Iowa? Crossroads Resource Center. Available from: http://www.crcworks.org/tiernan.pdf. Accessed November 30, 2017.

United States Department of Agriculture, 2015. USDA announces nearly $18 million to train, educate the next generation of farmers and ranchers. Available from: http://nifa.usda.gov/press-release/usda-announces-nearly-18-million-train-educate-next-generation-farmers-and-ranchers. Accessed November 30, 2017.

Wang, S.L., 2014. Cooperative extension system: trends and economic impacts on US agriculture. Choices. Available from: www.choices-magazine.org/choices-magazine/submitted-articles/cooperative-extension-system-trends-and-economic-impacts-on-us-agriculture Accessed November 30, 2017.

Kranti Mulik is a senior economist at the Union of Concerned Scientists where the emphasis of her work is on transforming the U.S. agricultural system so that it benefits farmers, consumers, society, and the environment at large and policies that can accomplish this goal. Her research has focused on a wide range of topics which include the economic and environmental impacts of diversified agriculture systems, the impact of shifting consumption patterns on land use, and the role of increasing retail and institutional demand for local foods in providing opportunities to declining midsized farms and promoting rural economic development.

Part IV

New Directions for Institutional Foodservice

Chapter 11

Sustainable Food Purchasing in the Health Care Sector: From Ideals to Institutionalization

Kendra Klein[*], Jenna Newbrey[†], Emma Sirois[†]
[*]Friends of the Earth, Berkeley, CA, United States, [†]Health Care Without Harm, Reston, VA, United States

Chapter Outline

Introduction	239
Redefining Healthy Food: An Environmental Nutrition Approach	241
History and Uptake of Environmental Nutrition	243
History of Health Care Without Harm's Healthy Food in Health Care Movement	244
Early Adopters	246
Putting Environmental Nutrition Into Practice in Hospital Foodservice	247
Defining Local and Sustainable Food	247
Purchasing Practices in Hospitals	249
Modeling Food Choices	249
Supply Chain Challenges and Opportunities for Farm-to-Hospital	250
Distributors	251
Group Purchasing Organizations	252
Supply Chain Innovation	253
Case Study: University of Washington Medical Center Purchasing Local Food	253
Purchasing Pathway 1: Northwest Agricultural Business Center	254
Purchasing Pathway 2: The University of Washington Student Farm	255
Challenges and Lessons Learned	255
Conclusion	256
References	257

INTRODUCTION

A growing coalition of health professionals and non-profit organizations are taking on new food initiatives in the health care sector that align with alternative agro-food ideals. These actions are largely inspired and legitimized by rethinking "healthy food" from a systems perspective. Actors are shifting from a traditional nutrition model focused on eating the right balance of nutrients and food groups to an *environmental nutrition* model—examining the public health impacts of social, economic, and environmental factors related to the entire food system. Through this lens, scientific data reveal a host of food system practices that directly and indirectly impact public health, from pesticide contamination on food and synthetic nitrogen pollution in drinking water, to soil erosion and climate change, which undermine the ecosystems that support sustainable food production.

In putting environmental nutrition ideals into action, hospitals are seeking out food that is local, organic, Fair Trade, whole rather than processed, produced by family farmers, and free of a host of agricultural technologies such as antibiotics, growth hormones, and genetic modification (Harvie, 2006; Sachs and Feenstra, 2008; Sirois et al., 2013). Even small shifts in hospital purchasing may have meaningful impacts within the food system. A single hospital can have an annual food budget of one to seven million dollars or more (FSD, 2011), while the health care sector as a whole spends approximately twelve billion dollars annually on food and beverages (Harvie, 2006).

Hospitals offer some unique opportunities and challenges for scaling up alternative food initiatives in relation to other institutional food purchasers. Ostensibly, hospitals have a moral mandate to provide the healthiest food to their patients and visitors. The majority of hospitals in the United States are non-profit, mission-driven organizations (AHA, n.d.), and there is an underlying assumption within the sector that hospitals can and should act as advocates of positive change in the communities in which they are located. The American Medical Association's "Sustainable Food Resolution" demonstrates this belief, stating, "Hospitals should become both models and advocates of healthy, sustainable food systems that promote wellness and that 'first do no harm,'" (AMA, 2008). Like the first hospitals to ban smoking on their grounds, leading hospitals are putting their moral weight behind societal change through new food initiatives.

In contrast to schools, hospitals have the ability to pass on some increases in costs to customers in their cafeterias, they need products year-round, and they need to meet a higher standard of food safety because of the high share of immunocompromised and contagious patients that they serve. Another distinctive characteristic of healthcare food commodity networks is the dominance of Group Purchasing Organizations (GPOs) that act as procurement gatekeepers between hospitals and other food supply chain actors.

The chapter starts by defining the environmental nutrition approach, and then provides a history of the environmental nutrition movement in hospital foodservice. We profile a key non-profit organization that is driving new food initiatives in the health care sector—Health Care Without Harm (HCWH).[1] The next two sections go into more detail about how the approach

---

1. The authors of this chapter have been professionally involved with HCWH. Kendra Klein worked for the California HFHC program from 2012 to 2016, Jenna Newbrey is Northwest Regional Lead for the HFHC program of HCWH, and Emma Sirois is the National Associate Program Director.

is implemented in hospital foodservice and what the main barriers to implementation are. Then we present a case study of the University of Washington Medical Center to demonstrate how implementing an environmental nutrition program works in practice, reflecting on the motivations, achievements, and lessons learned.

REDEFINING HEALTHY FOOD: AN ENVIRONMENTAL NUTRITION APPROACH

A growing body of scientific data on the health, social justice, and environmental impacts of the dominant industrial food system lead to both moral and economic claims for the involvement of the health care sector in food system change. As just one example, the health care sector treats the downstream health burden of agricultural pesticide use in the form of rising rates of cancers, neurodevelopmental and reproductive disorders, asthma, and Parkinson's (Sutton et al., 2011).

Data reveal a boomerang effect of agricultural practices that directly harm human health while undermining the ecosystem functions on which agricultural production depends (Schettler, 2004; Harvie, 2006). Along with health impacts, pesticides also harm soil biota, pollinators, and wildlife. Synthetic fertilizers are associated with blue baby syndrome (Majumdar, 2003), while also contributing to compromised water resources such as an 8500 mile^2 hypoxic "dead zone" in the Gulf of Mexico (Diaz and Rosenberg, 2008). Manure lagoons at Confined Animal Feeding Operations leach pollutants including phosphorus, heavy metals, and ammonia into waterways while creating noxious, asthma-inducing odors (Thorne, 2007). Routine nontherapeutic use of antibiotics in livestock production allows for overcrowded living conditions, while leading to antibiotic-resistant bacteria such as methicillin-resistant *Staphylococcus aureus* (Gilchrist et al., 2007). Production and shipping generate high greenhouse gas emissions, accelerating feedback loops with resultant negative impacts on food production, human health, and ecosystem resilience (Pfeiffer, 2006; Horrigan et al., 2002). These impacts disproportionately affect the health of workers, rural communities, and low-income communities of color (Das et al., 2001; Calvert et al., 2008; Eskenazi et al., 2004; Silbergeld et al., 2008; Gibbs et al., 2006; Heederik et al., 2007; Harrison, 2011; Thu, 2002).

Meanwhile, widely held consensus among medical and public health professionals finds that today's typical U.S. diet contributes to a range of health problems, including obesity, diabetes, cardiovascular disease, cognitive decline and dementia, other neurodegenerative disorders, and various kinds of cancer (Stein et al., 2011; Salas-Salvadó et al., 2011; Divisi et al., 2006).

From this perspective, the health care sector is bearing the economic burden of treating the downstream effects of a "broken" food system. Hospital food purchasing practices often perpetuate this same system because they are deeply rooted in industrial supply chains that provide efficiency and affordability at the cost of the environment and public health.

Advocates within what can be called the Healthy Food in Health Care movement discussed in detail as follows are leveraging this data to redefine healthy food and to inspire and legitimize new health care food procurement initiatives aligned with alternative agro-food goals. They are shifting the definition of healthy food from a traditional biomedical model to an environmental nutrition model.

From a biomedical perspective, nutrition is defined simply as "the intake of food, considered in relation to the body's dietary needs" (WHO, n.d.). In this frame, "healthy food" is defined by measurable food components such as calories, vitamins, and fats, and health interventions are typically aimed at individuals (Mudry, 2009; Scrinis, 2008). Environmental nutrition, on the other hand, examines the public health impacts of social, economic, and environmental factors related to the entire food system. While a traditional nutrition approach asks how much vitamin C and other nutrients are in an apple, environmental nutrition stimulates a broader set of questions, such as whether the apple was grown with toxic pesticides, whether the workers who grew it were treated justly, and which communities had access to purchasing it. From this perspective, not all apples are created equal: A given apple's path from farm to plate can result in greater or lesser health, social, and environmental benefits (Fig. 1).

Rather than continuing to treat the downstream symptoms of health problems associated with the dominant industrial agro-food system, leading health care institutions are using an environmental nutrition approach to look upstream and put food at the center of prevention-based care.

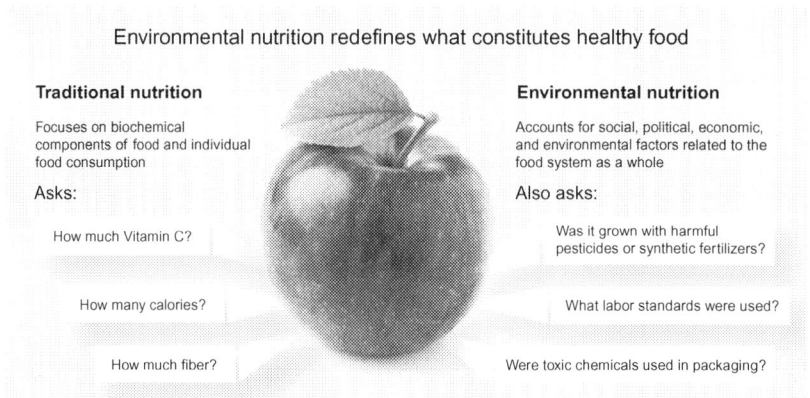

FIG. 1 Not all apples are created equal. *(Courtesy Klein, K., Thottathil, S., Clinton, S., 2014. Environmental nutrition: redefining healthy food in the health care sector. Health Care Without Harm.)*

History and Uptake of Environmental Nutrition

One of the earliest articulations of an environmental nutrition approach emerged in the 1980s from dieticians Joan Dye Gussow and Kate Clancy. In their landmark piece, *Dietary Guidelines for Sustainability*, they make the case that nutrition education should include the effects of food choices on food supply, on agricultural, economic and natural resources, and on the long-term stability of the food system (1986). They argue that without the comprehensive lens of environmental nutrition, crucial aspects of the interconnections between food and health fall outside of our range of vision. For example, a traditional nutrition approach extols the benefits of eating more fish for Omega-3 fatty acids even as fisheries across the globe are collapsing due to unsustainable fishing rates and practices (WWF, n.d.). As our ability to continue producing food is threatened by the common practices and policies of our global, industrial food system, recommendations for a balanced diet become a moot point.

In 1994, an environmental nutrition approach was formalized within a small cohort of dieticians who founded the Hunger and Environmental Nutrition Dietetics Practice Group (HEN) of the Academy of Nutrition and Dietetics. HEN states that "healthy food is not only defined by the quantity and quality of the food we eat, but must come from a food system that conserves and renews natural resources, advances social justice and animal welfare, builds community wealth, and fulfills the food and nutrition needs of all eaters now and into the future" (HEN, n.d.).[2] HEN championed a systems-based approach to healthy food at a time when it was a profound departure from the traditional concept of nutrition. Since then, the Academy of Nutrition and Dietetics, a professional association for more than 70,000 nutrition professionals, has been increasingly receptive to environmental nutrition due to the efforts of HEN members.

Over the past two decades, environmental nutrition has also percolated into national and global dialogues about healthy food. The United Nations first articulated an environmental nutrition approach following the World Food Conference of 1997, and in 2009, released the report, *Agriculture at a Crossroads* (McIntyre et al., 2009). The report was the result of an international three-year collaborative effort involving 900 experts from 110 countries, and it envisioned a new global model of agriculture that addresses the needs of small-scale farms in diverse ecosystems, accounts for the ecological complexity of agricultural systems, and recognizes farmers as producers and managers of ecosystems. In the United States, the U.S. Department of Agriculture (USDA) considered sustainability for the first time as part of the 2015 Dietary Guidelines for Americans. Although it did not make it into the final guidelines,

---

2. The environmental nutrition framework of the HFHC program of HCWH is aligned with this definition developed by the Hunger and Environmental Nutrition Dietetic Practice Group of the Academy of Nutrition and Dietetics.

the Dietary Guidelines Advisory Committee's 2015 Scientific Report declared that long-term food security for Americans should not be separated from the need to promote sustainable dietary patterns that protect our nation's natural resources (DGAC, 2015).

HISTORY OF HEALTH CARE WITHOUT HARM'S HEALTHY FOOD IN HEALTH CARE MOVEMENT

The primary organization responsible for translating an environmental nutrition approach into foodservice changes within the U.S. health care sector is Health Care Without Harm. Since 2005, hundreds of hospitals and health systems have signed onto the HCWH "Healthy Food in Health Care Pledge," which states:

For the consumers who eat it, the workers who produce it, and the ecosystems that sustain us, healthy food must be defined not only by nutritional quality, but equally by a food system that is economically viable, environmentally sustainable, and supportive of human dignity and justice.

(HCWH, n.d.-b)

The rapid uptake of this approach taps into a growing recognition within the health care sector that nutrition plays a significant role in disease prevention and that the health care should not only treat sickness, but should prevent disease and preserve wellness (APHA, 2012).

HCWH was founded in 1996 with the goal of addressing medical waste incineration as the leading source of dioxin, a potent carcinogen, in the environment. It is now an international organization with campaigns in the United States and Canada, Latin America, Asia, and Europe that seek to hold hospitals and health care systems to the Hippocratic oath to "first, do no harm" by changing health care practices that pollute the environment and contribute to disease. Its Global Green and Healthy Hospitals initiative currently includes members from 49 countries from all over the world representing the interests of more than 28,000 hospitals and health centers.

HCWH brings an environmental health perspective to its food work, drawing connections between hospital practices and the broader environment. HCWH argues that this approach is "important for health care food decision-making because the production and distribution of food has a multitude of health-related impacts often removed from the immediate hospital environment." In this framing, healthy food comes from a food system that maintains ecosystem health; has minimal negative impact on the environment; encourages local production and distribution infrastructures; makes nutritious food available, accessible, and affordable to all; and protects farmers and farmworkers (HCWH, n.d.-a). In 2014, the HFHC program formalized this perspective in a white paper, *Environmental Nutrition: Redefining Food in the Health Care Sector*, which addresses the concept in relation to health care engagement and operations (Klein et al., 2014).

In the early 2000s, a set of non-profit organizations began discussing the connections between environment, food, and health care as an expansion of their core interests. In 2002, the non-profit organization Science and Environmental Health Network detailed a new vision for ecological medicine and soon after called on health care institutions to take a leadership role in promoting a more sustainable food system (Schettler, 2002, 2004). In 2005, HCWH expanded the scope of its work to include food with the launch of its Healthy Food in Health Care (HFHC) program. The movement to enact environmental nutrition ideals within the health care sector has since grown from a handful of leading hospitals to a broad-based national movement.

To launch the HFHC program, HCWH and one of its constituent organizations, San Francisco Bay Area Physicians for Social Responsibility, hosted a FoodMed Conference in Oakland, California. The conference brought together individuals from sustainable agriculture, environmental health, public health, and the health care sector to begin talking about how to bring "healthier" food into health care in a way that would mutually benefit everyone along the food chain and contribute to healthier communities and ecosystems (Harvie et al., 2008, 2009). At the conference, HCWH also released the Healthy Food in Health Care Pledge.

Since then, an environmental nutrition framing has spread dramatically within the health care sector, and 580 hospitals have now signed on to HCWH's Pledge and a related Healthy Food Challenge. Additionally, HFHC's regional programs in California, the Pacific Northwest, the Southwest, the Mid-Atlantic, the Upper Midwest, and New England engage a wider set of 750 facilities (including the Pledge-signed hospitals) in innovative and model building activities. Together, these hospitals represent 15 percent of roughly 5000 hospitals in the nation (AHA, n.d.).

HFHC advocates have targeted health professional associations as important locations to establish an environmental nutrition discourse within the health care community. Through their efforts, statements on food and sustainability that echo the front-runner health systems and the Healthy Food in Health Care Pledge have been issued by the American Medical Association, American Public Health Association, American Nurses Association, American Academy of Nutrition and Dietetics, California Medical Association, and Minnesota Academy of Family Physicians, among others. State hospital associations have also endorsed environmental nutrition initiatives. For example, the Michigan Hospital Association launched a Healthy Food Hospitals campaign in 2010, encouraging Michigan hospitals to become models of "exemplary foodservice choices and nutritional selections" (MichiganGreen, 2009).

In the latest iteration of environmental nutrition ideals, the HFHC program is working with health care institutions to become Anchors for Resilient Communities (ARC). Hospitals are "anchored" in the communities they serve; they hold significant investments in real estate and social capital, and are often among the largest employers in their communities. The HFHC program is encouraging hospitals, as anchor institutions, to embrace a commitment to apply their social and economic influence and intellectual resources to better the long-term public and environmental health of their communities. For example, fostering healthy regional food systems to help improve community health outcomes by increasing access to fresh, healthy food and building community wealth by creating "green" jobs (HCWH, 2016).

Early Adopters

Some of the first expressions of environmental nutrition within the health care sector can be attributed to two major health systems within the HCWH network—Kaiser Permanente and Dignity Health. In 2006, shortly after the launch of the Healthy Food in Health Care Pledge, these health systems began their journey to actualizing its vision.

Kaiser Permanente, the largest HMO in the nation serving more than 10.6 million individuals in eight states and the District of Columbia, articulated its own "Comprehensive Food Policy," which committed the health system to providing healthier food in a manner that, "Promotes agricultural practices that are ecologically sound, economically viable, and socially responsible" (2006). In 2007, Dignity Health asserted the following "Food and Nutrition Vision Statement:"

Dignity Health recognizes that food production and distribution systems have wide ranging impacts on the quality of ecosystems and their communities, and so; Dignity Health recognizes that healthy food is defined not only by nutritional quality, but equally by a food system which is economically viable, environmentally sustainable and which supports human dignity and justice.

(Dignity Health, n.d.)

These commitments call on health care institutions to protect and promote individual and community health. They include provisions to procure local food in order to support family farmers, to support farmworker justice, and to purchase food produced without pesticides, synthetic hormones, or the unnecessary use of antibiotics.

The visionary leadership of specific staff was critical in driving the early adopters to pursue sustainable food programs. For Kaiser Permanente, this was Preston Maring, an obstetrician and gynecologist who launched the first hospital farmers' market at Oakland East Bay Medical Center. Kaiser now hosts 52 farmers' markets at facilities in multiple states. At Dignity Health, Mary Ellen Leciejewski championed the connection between sustainable food and the organization's faith-based mission of healing. Her leadership has inspired the adoption of an environmental health framework that states, "As health care providers, we recognize the interdependence between human health and our environment and believe in the caring stewardship of a renewable Earth for the enhancement of all life" (Dignity Health, n.d.). The health system puts this vision into action not only through food initiatives, but through efforts to conserve water and energy, remove toxics in the hospital environment, promote green building, and measure, and report greenhouse gas emissions.

Many of the programs and progress of these early adopters inspired the development of other leaders in the sector to initiate a wide range of activities motivated by environmental nutrition ideals. This included procurement changes for patient meals and cafeterias, hosting farmers' markets and Community Supported Agriculture (CSA) programs, creating hospital gardens and farms, educating supply chain intermediaries about their new preferences, and building momentum in the procurement of local and sustainable food in hospitals.

PUTTING ENVIRONMENTAL NUTRITION INTO PRACTICE IN HOSPITAL FOODSERVICE

The HFHC program played a critical role in facilitating the expansion of environmental nutrition efforts in the health care sector. Despite the variety of interests that may motivate a given hospital foodservice director, administrator, or clinician to participate in environmental nutrition initiatives, the HFHC program has helped to align their actions and shift the norms nationwide by establishing common definitions and materials. The program has generated a significant body of reports, webinars, and other educational materials on issues at the nexus of public health and the food system such as use of antibiotics, arsenic and growth hormones in livestock production, toxics in food and food packaging, agricultural pesticides, and agricultural subsidies. The HFHC program also provides how-to guides for achieving healthier and sustainable hospital foodservice and data collection tools that form the basis of a national awards program for hospitals and foodservice professionals. In addition, the HFHC program provides technical assistance and networking opportunities through roundtables, conference calls, mailing lists, webinars, and national conferences. More recently, these efforts have been disseminated through HCWH's membership arm, Practice Greenhealth, to more than 1200 U.S. hospitals that are pursuing a range of sustainability initiatives at their facilities, thus scaling best practice through an even broader network of hospitals.

Defining Local and Sustainable Food

The HFHC program created guidelines with definitions of "local" and "sustainable" food to guide foodservice changes, track progress, and reward leaders in the field.

To define "sustainable," the program relies on third-party certifications like USDA Organic and federally regulated label claims such as "raised without the use of antibiotics."[3] The program defines "local" as food that is grown or produced on farms or ranches within a 250-mile radius of a facility. An item cannot be counted as local based on the distance from a supplier's distribution center. For processed foods and foods with multiple ingredients, greater than 50 percent of the ingredients, by weight, must be grown and processed within a 250-mile radius of the health care facility.[4]

There are tensions and trade-offs involved in using certifications and labeling to signal that products are sustainable. Certifications and standards typically do not meet the full set of social, environmental, community, and health ideals that hospitals are aiming for in their sustainable food efforts. However, certifications and standardized definitions help to facilitate national efforts and can provide verification that certain production standards are being met. At the regional and local level, HCWH encourages hospitals to source according to a more complex set of criteria.

As one example, the national Healthy Food in Health Care guidelines cite the Marine Stewardship Council as a sustainable seafood criteria, a certification that has been critiqued in terms of transparency and accountability (e.g., Iles, 2007). However, on the regional level, HFHC hospitals and advocates are often involved in more robust sustainability efforts. For example, New England actors have partnered with the Northwest Atlantic Marine Alliance, which has an evolving definition of "ecologically appropriate seafood" that takes into account a range of conditions that a set certification can't capture, such as the scale of demand and the state of the ecosystem. As one non-profit advocate remarks, "Obviously these criteria are not easy to put into a Request for Proposals [from vendors]" (HCWH, 2013).

Likewise, regional HFHC efforts often go beyond the nationally set definition of "food miles" for local food. Spatially based definitions are the flattest reading of what constitutes local foods, as they do not take into consideration ownership structure of farms or agricultural production methods. HFHC hospital efforts tend to align more with underlying agro-food movement goals, either through direct sourcing from independent farmers, working with food hubs or other values-based supply chain intermediaries, or partnering with family-farm-based organizations.

---

3. A 2013 survey conducted by HCWH staff of 85 California hospitals participating in the Healthy Food in Health Care program provides insight on the types of certifications and label claims hospitals are seeking to fulfill their sustainable purchasing goals. Third-party certifications purchased by survey respondents included: USDA Organic (62 percent), Certified Humane Raised and Handled (52 percent), Fair Trade (35 percent), Food Alliance (34 percent), Rainforest Alliance (10.5 percent), Non-GMO Project Verified (8 percent), Marine Stewardship Council (7 percent), Salmon Safe (6 percent), and American Gras-fed (2 percent). Federally regulated label claims included: rBGHfree (73 percent), cage-free eggs (48 percent), raised without antibiotics/no antibiotics administered (48 percent), raised without added hormones/no hormones added (41 percent), USDA grass-fed (7 percent), and no genetically engineered ingredients (3.5 percent).

4. For specific information about the sustainability criteria, see https://www.healthyfoodinhealthcare.org.

Purchasing Practices in Hospitals

Hospitals are undertaking a wide range of activities motivated by environmental nutrition ideals, including procurement changes for patient meals and cafeterias; hosting farmers' markets and CSA programs; creating hospital gardens and farms; educating supply chain intermediaries about their new preferences; and educating patients, visitors, and communities about their alternative food goals. HFHC efforts also often occur in tandem with broader sustainability and wellness goals, and may be driven not only by foodservice leaders, but by hospital Green Teams and Wellness Committees.

A 2017 survey of hospitals in the HCWH network revealed that 20 facilities directed a total of $30,392,923 to foods that met sustainable food goals as defined by HCWH.[5] A 2012 trade journal survey of hospital foodservice directors demonstrates that these trends are sector-wide; of 50 hospitals surveyed, 30 percent reported purchasing some organic products in the previous year, and all reported purchasing local foods (FSD, 2012).

Modeling Food Choices

Within hospital walls, foodservice staff and clinical dietitians are modeling sustainable food choices for patients, employees, and visitors. Like the first hospitals to ban smoking on their grounds, some HFHC hospitals understand that they are putting their moral weight behind societal change through their new food initiatives. "It's not just about the food we serve, it's about the message, the symbolism of it," stated the CEO of one participating hospital (Interview #61, 2011).

In hospitals participating in the HFHC movement, cafeteria displays and patient tray fliers extol the benefits of organic, local, and other targeted purchasing choices. Of the California HFHC hospitals surveyed, many report using a variety of strategies to educate patients, employees, and the community about their alternative food efforts. These include:

- Posting cafeteria signage, such as table tents and posters, and/or using patient tray cards or bookmarks with information about sourcing from sustainable companies, vendors, and farmers;
- Promoting local, healthy, and sustainable items in cafeterias and on patient menus;
- Posting signed copies of the HFHC Pledge in cafeterias;
- Including local and sustainable food programs and issues in newsletters;
- Creating wellness committees and green teams;
- Hosting employee and community engagement classes and educational events; and
- Upon hire, teaching foodservice staff about the facility's sustainability initiatives.

---

5. The results of the full survey, conducted and analyzed by HCWH staff, will be published in the 2017 HCWH *Menus of Change* report. Previous reports are available online: https://noharm-uscanada.org/documents/menu-change-2015-program-report-highlights-and-survey-results.

Overall, the HFHC movement holds a twofold promise for advancing the alternative agro-food movement by shifting hospital food dollars to alternative foods and supply chain infrastructure and by shifting public discourse and policy around healthy food.

SUPPLY CHAIN CHALLENGES AND OPPORTUNITIES FOR FARM-TO-HOSPITAL

Ostensibly, institutions such as hospitals provide an opportunity to combine the market power of large-scale purchasing with the moral concerns of alternative consumers. But the leap in scale from an individual buying three onions at a farmers' market to a hospital buying 300 cases is not simply one of numbers, it presents an entirely different set of challenges, opportunities, and relationships.

Hospital buyers are rarely able to "shake the hand" of the farmers who grow their food due to the logistical constraints of their foodservice operations. When hospitals have to prepare tens to hundreds of gallons of soup at a time, the ability to source large and consistent volumes and ready-made products are concerns of no small consequence. While some hospitals are shifting to scratch cooking from whole ingredients, others source up to 90 percent of their fresh fruits and vegetables in processed form, such as green beans sliced by the half inch or butternut squash cut into one-inch cubes. This means that hospitals rely on the efficiency and standardization that is currently more available through industrial food supply chains.

For example, one Food and Nutrition Services Director reported, "Local, organic chicken was a real challenge" (Interview #38, 2011). He went on to describe that under the conventional purchasing system, he could place an order on his food distributor's website, and the next day hundreds of conventionally raised, uniform four-ounce chicken breasts would show up on his loading dock, shrink-wrapped, and stacked by the case. Procuring local, organic chicken first required weeks of working through bureaucratic purchasing and legal systems to set up a relationship with a new vendor. When the hospital finally received its first delivery of organic chicken, it was an ice-packed box of whole chickens with the heads and feet still on. "My cooks almost died," the Director reports. Having to chop off chicken heads was significantly more time- and labor-intensive than lining up a row of boneless, skinless meat parts in the griller. Most institutional kitchens no longer have the equipment or staff with the knowledge necessary to deal with such whole foods.

A number of structural and economic barriers within hospital food procurement systems push back on the underlying values and goals embodied by an environment nutrition approach. These include set and limited budgets; logistical issues related to preparing and serving mass quantities of food daily; public policy related to food safety and nutrition; the need for large and consistent product volumes; dependence on efficiencies in ordering, delivery, and billing systems; and contracted relationships with existing vendors. To address these challenges, leading hospitals are fostering the development of alternative supply chains and influencing traditional supply chain actors to adopt systems of greater transparency and practices that support more regionalized, sustainable food systems.

Distributors

As Fig. 2 shows, both national and regional food distributors help hospitals meet the logistical needs of their foodservice operations. They offer extensive aggregation and distribution systems, storage and refrigeration capacity, and the ability to respond to shortages in targeted regional and sustainable products with other products from their warehouses. They provide food safety assurances, insurance coverage, and information technology and invoicing systems. Many hospitals also prefer to work with established vendors, because setting up new vendor relationships can take weeks of working through bureaucratic systems and ongoing labor required to manage ordering systems for multiple vendors.

For most hospitals, nearly all foods with long shelf lives (processed, canned, and prepared foods, grains, and frozen meat and poultry) come through national broadline distributors (Pritchard, 2012; Klein, 2012). Broadline distributors such as US Foods and Sysco offer the ultimate in one-stop-shopping by carrying everything from hamburger patties to spaghetti noodles to paper plates. Mid- to large-scale farmers have a comparative advantage over small farmers in terms of supplying products through broadline distributors because they are more likely to be able to meet required volume, pack and grade standards, and food safety criteria, as well as to cover the high insurance premium distributors require them to carry (Klein and Michas, 2016; Klein, 2015).

Regional distributors play a major role in hospital foodservice, particularly for perishable goods such as produce, bakery, dairy, and fresh meat and poultry. Supply chains for perishable goods have remained more regionalized in the

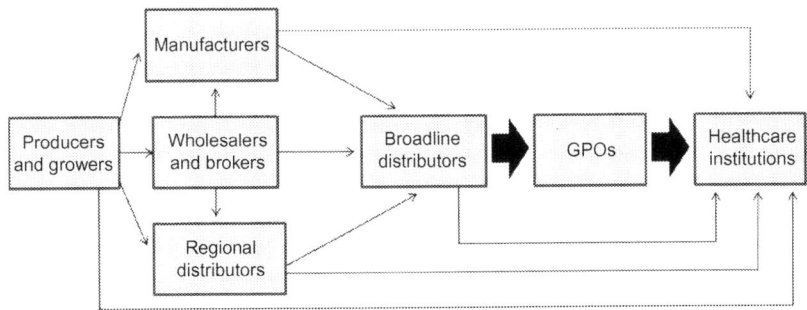

FIG. 2 Generic outline of a hospital food supply chain.

United States because perishables cannot be stored as easily, making them less amenable to the highly centralized model of distribution preferred by broadline distributors (Kaufman, 2000; Pritchard, 2012). Hospitals, therefore, have greater potential to connect with small- and mid-scale regional growers and ranchers for these product categories.

A 2013 survey of California hospitals participating in the HFHC program found that the majority of respondents (85 total), purchase local and sustainable food through their distributors. Eighty one percent of these hospitals sourced sustainable foods through their regional distributors (e.g., produce distributors), 71 percent through their broadline distributors, and only 8 percent purchased directly from farms and 5 percent from food hubs.[6] Alternative supply channels are typically not capable of adhering to the same or similar levels of price competition, efficiency, or technical and quality standards.

Group Purchasing Organizations

While efficiency is one of the top reasons hospitals rely on broadline distributors, contracts with Group Purchasing Organizations also drive buying decisions (Klein, 2015). GPOs act as gatekeepers to the health care market by negotiating transactions with the manufacturers, wholesalers, and distributors that supply the products a hospital needs to function, from medical equipment such as catheters and sonogram machines, to bulk supplies such as cotton swabs and latex gloves, to food and foodservice supplies. GPOs aggregate member hospitals' purchasing power to obtain lower prices and to eliminate duplicative transaction costs. While hospital foodservice departments may negotiate purchases directly with producers, distributors, or manufacturers, 80–90 percent of their procurement comes through GPO channels in keeping with contract terms. In foodservice, the largest GPOs have moved to two to three-year sole-source contracts with two national broadline distributors, US Foods and Sysco. Thus, in accordance with GPO contracts, most hospitals spend the majority of their food budget through a single distributor.

Hospitals report that the sustainable food options available to them through their GPOs are extremely limited, thus, the vast majority of local and sustainable food procurement is occurring outside of the GPO-hospital relationship (Klein, 2015). Currently, most local and sustainable food procurement is part of the allowed 10–20 percent off-contract purchasing. Often, these efforts are dependent on hospitals seeking out and developing contracts directly with local

---

6. When hospitals do create direct relationships with farmers, these happen most often through on-site farmers' markets or CSA programs. Of 85 hospitals surveyed in California in 2013, 43 reported hosting farmers' markets or farm stands, and fourteen reported CSA programs, often as part of their employee wellness programs. Fourteen hospitals had created their own gardens or farms on-site. Although it is less common, some hospitals are willing to create direct farm-to-hospital relationships with growers as a way to meet their commitment to regional and sustainable food systems.

and regional producers and distributors. As one hospital nutrition director notes, "[GPOs] are starting to take heed. But quite honestly, they are really slow and there isn't enough. That's why we are going off the trail to suit our needs" (Ramsey and Schilling, 2011).

Supply Chain Innovation

Some of the most promising opportunities to support the scaling up of local supply chains that embody environmental, health, and social goals while also meeting hospitals' needs for efficiency and standardization are hybrid supply chains that incorporate both conventional and alternative resources, infrastructure, and markets. For example, some hospitals have had partial success in pushing their existing regional distributors to take on the transparency and values-based purchasing role of food hubs (Klein and Michas, 2016).

In other innovative models, alliances between broadline distributors and regional companies or cooperatives are emerging. In a pilot project aimed at increasing their local food offerings, Sysco partnered with regional produce company Walsma & Lyons (Barham et al., 2012; Cantrell, 2010). As another example, Fifth Season Cooperative in southwestern Wisconsin formed an innovative alliance with Reinhardt Foodservice, a Midwestern broadline distributor, in order to serve their large university and hospital clients (Chapeta, 2012).

CASE STUDY: UNIVERSITY OF WASHINGTON MEDICAL CENTER PURCHASING LOCAL FOOD

Despite the aforementioned challenges, hospitals are finding creative ways to purchase local and sustainable food. In 2011, the University of Washington Medical Center (UWMC) signed the Healthy Food in Health Care Pledge, propelling the development of progressive purchasing policies. One way UWMC is enacting the Pledge is by purchasing regionally produced organic foods. This case study provides insight on the supply chain innovation required for UWMC to meet its goals. Details were compiled by HCWH staff during three interviews in the summer of 2017 with UWMC staff—the Food and Nutrition Services Director, Technology and Wellness Manager, and Café Supervisor. UWMC's foodservice purchasing pathways have brought the hospital national recognition and multiple HFHC awards.

UWMC is a nationally renowned teaching hospital affiliated with the University of Washington. UWMC is a 500-bed hospital providing highly specialized care to the greater Seattle area and beyond, with many patients traveling from outside of Washington State. Consistently ranked among the top hospitals in the country for patient wellness, the organization has more than 6000 employees, 63,000 patients annually, and serves more than 200,000 meals each year.

UWMC food purchasing is guided by the Healthy Food in Health Care Pledge as well as an internal purchasing policy that states: "The food we eat is critical to our wellbeing, and in times of sickness and recovery. Similarly, the increasing production of local, sustainable and organic food products improves the strength of local economies and environments" (UWMC, n.d.). UWMC foodservice staff envision a host of benefits from their regional organic purchasing efforts:

- Increasing patient, staff, student, and community access to healthy foods;
- Encouraging patrons to increase intake of organic produce;
- Creating healthier regional farming communities with reduced use of toxic synthetic chemicals;
- Supporting the growth of the regional organic sector through the introduction of a new market for crops; and
- Encouraging creative purchasing partnerships through new and diverse pathways.

The UWMC Food and Nutrition Services Director, Charles Zelinksi, and his staff were accountable for the implementation of the regional organic purchasing strategy development and roll out. According to staff, the department began by slowly introducing fresh produce one product at a time to allow kitchen staff time to acclimate themselves to the additional processing time needed for washing, chopping, and preparation. As staff became more accustomed to working with fresh produce, they simultaneously developed relationships to facilitate two local purchasing pathways—one working with a local food hub/aggregation center and one through direct purchasing from the University of Washington (UW) Student Farm and a CSA program through Full Circle Farms, a local organic grower.

Purchasing Pathway 1: Northwest Agricultural Business Center

Although UWMC does purchase organic produce through its existing supply chain vendors, it chose to take its commitment to regional organics to another level through a partnership with the Northwest Agricultural Business Center (NABC). The NABC is a local non-profit, more commonly referred to as a "food hub," that is focused on sourcing regional organic produce for local hospitals, restaurants, and institutions. NABC charges growers lower fees to keep more profit in the hands of the farmers.

Making a commitment to working with a small local program has its challenges. Robert Whittaker, UWMC's Café Supervisor, stated, "Partnering with a small, alternative distributor was sometimes difficult due to lack of consistent supply, increased staff time dedicated to labor for unprocessed produce, and inconsistent delivery schedules." Despite those difficulties, Mr. Whittaker recognized UWMC's contribution to local food programs was vital to the success of NABC and says that he was pleasantly surprised with how this relationship has matured over the past few years. The institution's support helped NABC grow and expand.

NABC greatly expanded its offerings, which allowed the UWMC cafeteria to introduce new produce items on a regular basis. Whittaker received positive feedback from customers, saying, "They often ask for recipes and comment on the quality of our produce. Customers notice that these products are superior, and we hope that this is a gateway to encouraging our communities to increase their vegetable intake."

Purchasing Pathway 2: The University of Washington Student Farm

Health care institutions have traditionally avoided direct purchasing relationships due to the time commitment, complicated ordering and distribution process, and difficulty in meeting HACCP food safety guidelines. The UWMC is breaking down the barriers that typically prevent direct buying by working closely with the UW Student Farm on campus to source herbs, salad greens, beets, and rainbow chard. This relationship further connects students, foodservice staff, and the community with UWMC's commitment to support local agriculture. Items sourced from the student farm are marketed to cafeteria patrons twice a month during a farmers' market accompanied by education and signage promoting farm fresh produce. In addition to working with the student farm, UWMC also started a CSA program with Full Circle Farms, a local organic grower.

Challenges and Lessons Learned

According to Whittaker, there were two main challenges to implementing the regional organic procurement policy—lack of consistent supply and training in the kitchen. Whittaker explained, "Kitchen staff were frustrated with the increased labor time necessary to process the fresh produce, which was frustrating for both staff and management." He cited that finding an advocate in the kitchen was crucial to turning morale around. The "change advocate" rallied the kitchen staff around the changes that included increased labor time for washing and chopping produce and coaching of the team on the benefits of fresh produce for the institution's eaters in addition to the positive impact on the local economy and environment.

The institution's second challenge in working with smaller local food programs was the lack of consistent supply. UWMC worked closely with these smaller partners to help them understand exactly what was needed out of the relationship and to assist them in better understanding institutional needs through ongoing communication. The hospital found that it needed to be very clear about product specifications and volume needs in order to operate. And when it was unsure about a partner's ability to meet volume needs, it started small with one product and had a back-up distributor that could step in when supply fell short.

UWMC aims to continue increasing purchasing of regional organic food. The foodservice team has discussed seeking out a local farm to grow staple items such as organic greens and carrots in an effort to increase direct purchasing relationships. UWMC is also developing a partnership with a grower that acts as a food hub, Cloud Mountain Farms, to begin sourcing processed organic items such as precut produce, salsas, and coleslaws, which the hospital typically purchases from a broadline distributor.

This brief case study demonstrates that as hospitals within the HFHC movement attempt to move from environmental nutrition ideals to the institutionalization of alternative procurement practices, there is a need for rapid development of infrastructure and capacity within alternative supply chains. Hospitals are struggling to navigate the tensions between their new food commitments and their reliance on the efficiency, affordability, and standardization provided by the conventional, industrial food system. Whether hospitals and intermediaries like HCWH are able to successfully resolve these tensions will largely determine whether the alternative agro-food movement can effectively scale up to meet institutional demand without losing sight of the social, health, and environmental values that brought it into being.

CONCLUSION

The efforts of leading hospitals to institutionalize environmental nutrition ideals through food procurement strategies provide an alternative model of new supply chain relationships that incorporate both industrial values of efficiency and standardization with the environmental, health, and social values that are driving new purchasing goals. However, these models often require hospital foodservice directors and staff to go above and beyond in terms of planning, participation, and at times, cost. While leading hospitals can play an important role in pointing the direction toward a reformed food system rooted in environmental nutrition ideals, many supply side barriers to the development of new food initiatives are not simply obstacles to overcome, they are indications of powerful economic and political forces aligned, either actively or passively, in support of the status quo. Without strong intermediaries to help build effective infrastructure, capacity, and messaging in alternative food value chains and more importantly, without public policy to set the conditions for food systems change, the movement will have only partial success. Efforts to institutionalize environmental nutrition values in hospital foodservice are swimming upstream against powerful norms and highly concentrated and sophisticated industrial food value chains.

Hospitals and health care professionals can leverage their moral authority to continue to make the case for food system change, and, by extension, foster a more holistic sense of "health" in health care. Polls consistently show that health professionals rank as some of the most trusted experts in the United States (Gallup, 2014). Health professionals can be particularly strong advocates for policy and regulatory changes that drive greater sustainability in the food system. For example, in 2013, the HFHC submitted a letter to the Food and Drug Administration (FDA) signed by nearly 800 clinicians demanding a ban on the use of medically relevant antibiotics in animal agriculture. And in 2016, HFHC launched the Clinician Champions in Comprehensive Antibiotic Stewardship (CCCAS) Collaborative to promote policy action that supports judicious use of antibiotics in agriculture. Through policy advocacy, health care professionals can help create the conditions for a food system that guarantees environmental stewardship, maintenance of local economies, animal welfare, and protection of public health for all eaters, now and into the future.

REFERENCES

AHA. n.d. Fast facts on US hospitals: American Hospital Association, online.

AMA, 2008. Report 8 of the Council on Science and Public Health (A-09): sustainable food, Resolution 405. Reference Committee D.

APHA, 2012. Prevention Provisions in the Affordable Care Act. American Public Health Association, Washington, DC.

Barham, J., Tropp, D., Enterline, K., Farbman, J., Fisk, J., Kiraly, S., 2012. Regional Food Hub Resource Guide. USDA: Agricultural Marketing Service.

Calvert, G.M., Karnik, J., Mehler, L., Beckman, J., Morrissey, B., Sievert, J., Barrett, R., Lackovic, M., Mabee, L., Schwartz, A., 2008. Acute pesticide poisoning among agricultural workers in the United States, 1998–2005. Am. J. Ind. Med. 51, 883–898.

Cantrell, P., 2010. Sysco's journey from supply chain to value chain: 2008–2009 final report. Wallace Center, Winrock International.

Chapeta, D., 2012. Phone interview with operations manager of fifth season cooperative. Westby, WI.

Das, R., Steege, A., Baron, S., Beckman, J., Harrison, R., 2001. Pesticide-related illness among migrant farm workers in the United States. Int. J. Occup. Environ. Health 7, 303–312.

DGAC, 2015. Scientific Report of the 2015 Dietary Guidelines Advisory Committee. Dietary Guidelines Advisory Committee, USDA and US Department of Health and Human Services, Washington, DC.

Diaz, R.J., Rosenberg, R., 2008. Spreading dead zones and consequences for marine ecosystems. Science 321, 926–929.

Dignity Health. n.d.. If it's good for the planet, it's good for the patient, online.

Divisi, D., Di Tommaso, S., Salvemini, S., 2006. Diet and cancer. Acta Biomed. 77, 118–123.

Eskenazi, B., Harley, K., Bradman, A., Weltzien, E., Jewell, N.P., Barr, D.B., Furlong, C.E., Holland, N.T., 2004. Association of in utero organophosphate pesticide exposure and fetal growth and length of gestation in an agricultural population. Environ. Health Perspect. 112, 1116.

FSD, 2011. 2011 hospital census. FoodService Director.

FSD, 2012. 2012 hospital census. FoodService Director.

Gallup. 2014. Honesty/ethics in professions 2013.

Gibbs, S.G., Green, C.F., Tarwater, P.M., Mota, L.C., Mena, K.D., Scarpino, P.V., 2006. Isolation of antibiotic-resistant bacteria from the air plume downwind of a swine confined or concentrated animal feeding operation. Environ. Health Perspect. 114, 1032.

Gilchrist, M.J., Greko, C., Wallinga, D.B., Beran, G.W., Riley, D.G., Thorne, P.S., 2007. The potential role of concentrated animal feeding operations in infectious disease epidemics and antibiotic resistance. Environ. Health Perspect. 115, 313.

Gussow, J.D., Clancy, K.L., 1986. Dietary guidelines for sustainability. J. Nutr. Educ. 18, 1–5.

Harrison, J., 2011. Pesticide Drift and the Pursuit of Environmental Justice. The MIT Press, Boston.

Harvie, J., 2006. In: Redefining Healthy Food: An Ecological Health Approach to Food Production, Distribution, and Procurement. Designing the 21st Century Hospital, Hackensack, NJ.

Harvie, J., Moore, D., Brook, L., 2008. Menu of change: healthy food in health care. Health Care Without Harm.

Harvie, J., Mikkelsen, L., Shak, L., 2009. A new health care prevention agenda: sustainable food procurement and agricultural policy. J. Hunger Environ. Nutr. 4, 409–429.

HCWH. 2013. Personal email with Health Care Without Harm *Healthy Food in Health Care* organizer. January 28.

HCWH, 2016. Improving community health and building community wealth in the bay area. Health Care Without Harm.

HCWH. n.d.-a. Health Care Without Harm healthy food in health care program, online.

HCWH. n.d.-b. Healthy food in health care pledge: Health Care Without Harm, online.

Heederik, D., Sigsgaard, T., Thorne, P.S., Kline, J.N., Avery, R., Bønløkke, J.H., Chrischilles, E.A., Dosman, J.A., Duchaine, C., Kirkhorn, S.R., 2007. Health effects of airborne exposures from concentrated animal feeding operations. Environ. Health Perspect. 115, 298.

HEN, n.d.. Hunger and Environmental Nutrition Dietetics Practice Group. Academy of Nutrition and Dietetics. online.

Horrigan, L., Lawrence, R.S., Walker, P., 2002. How sustainable agriculture can address the environmental and human health harms of industrial agriculture. Environ. Health Perspect. 110, 445.

Iles, A., 2007. Making the seafood industry more sustainable: creating production chain transparency and accountability. J. Clean. Prod. 15, 577–589.

Interview #38. 2011. Phone interview with food and nutrition services director. March 15.

Interview #61. 2011. Phone interview with hospital CEO. November 16.

Kaufman, P.R., 2000. Understanding the Dynamics of Produce Markets: Consumption and Consolidation Growth. USDA Economic Research Service, Washington, DC. Agriculture Information Bulletin Number 758.

Klein, K., 2012. A new prescription for the local food movement. The Nation.

Klein, K., 2015. Values-based food procurement in hospitals: the role of health care group purchasing organizations. Agric. Hum. Values 32, 635–648.

Klein, K., Michas, A., 2016. The Farm Fresh Healthcare Project: analysis of a hybrid values-based supply chain. J. Agric. Food Syst. Community Dev. 5, 57–74.

Klein, K., Thottathil, S., Clinton, S., 2014. Environmental Nutrition: Redefining Healthy Food in the Health Care Sector. Health Care Without Harm, Reston, VA.

KP. 2006. Kaiser Permanente Comprehensive Food Policy.

Majumdar, D., 2003. The blue baby syndrome. Resonance 8, 20–30.

McIntyre, B., Herren, H., Wakhungu, J., Watson, R., 2009. Agriculture at a Crossroads. International Assessment of Agricultural Knowledge, Science and Technology for Development, U.N.D. Program, Washington, DC.

MichiganGreen. 2009. Michigan Health & Hospital Association Green Health Care Committee newsletter.

Mudry, J.J., 2009. Measured Meals: Nutrition in America. State University of New York Press, Albany, NY.

Pfeiffer, D.A., 2006. Eating Fossil Fuels: Oil, Food and the Coming Crisis in Agriculture. New Society Publishers, Vancouver.
Pritchard, J., 2012. Muddy Waters: Making Sense of the Healthcare Supply Chain in the Era of Reform. Medical Distribution Solutions, Inc., Lawrenceville, GA.
Ramsey and Schilling, 2011. Solving the purchasing puzzle. FoodService Director. December 15.
Sachs, E., Feenstra, G., 2008. Emerging local food purchasing initiatives in Northern California Hospitals. Agricultural Sustainability Institute: UC Davis.
Salas-Salvadó, J., Martinez-Gonzalez, M., Bullo, M., Ros, E., 2011. The role of diet in the prevention of type 2 diabetes. Nutr. Metab. Cardiovasc. Dis. 21, B32–B48.
Schettler, T., 2002. The case for ecological medicine. In: The Networker. Science and Environmental Health Network.
Schettler, T., 2004. Nutrition and food production systems: a role for health care. In: The Networker. Science and Environmental Health Network.
Scrinis, G., 2008. On the ideology of nutritionism. Gastronomica 8, 39–48.
Silbergeld, E.K., Graham, J., Price, L.B., 2008. Industrial food animal production, antimicrobial resistance, and human health. Annu. Rev. Public Health 29, 151–169.
Sirois, E., Pryor, K., Thottathil, S., 2013. Menu of change: healthy food in health care. Health Care Without Harm.
Stein, J.; T. Schettler; B. Rohrer; and M. Valenti. 2011. Environmental threats to healthy aging: with a closer look at Alzheimer's and Parkinson's diseases. 2008.
Sutton, P., Wallinga, D., Perron, J., Gottlieb, M., Sayre, L., Woodruff, T., 2011. Reproductive health and the industrialized food system: a point of intervention for health policy. Health Aff. 30, 888–897.
Thorne, P.S., 2007. Environmental health impacts of concentrated animal feeding operations: anticipating hazards, searching for solutions. Environ. Health Perspect. 115, 296.
Thu, K.M., 2002. Public health concerns for neighbors of large-scale swine production operations. J. Agric. Saf. Health 8, 175–184.
UWMC. n.d. Internal Food Purchasing Policy. Seattle: University of Washington Medical Center.
WHO. n.d.. Health topics: nutrition: World Health Organization, online.
WWF. n.d.. Unsustainable fishing: World Wildlife Fund, online.

Kendra Klein is a senior staff scientist at Friends of the Earth (FOE). She is a writer, researcher, and advocate with over fifteen years of experience in environmental sustainability, food, agriculture, and environmental health. Prior to joining FOE, she coordinated the California Healthy Food in Health Care campaign and worked on toxics at Breast Cancer Action. Kendra has apprenticed on organic farms, is a Switzer Environmental Fellow, and has written for The Nation, Gastronomica, Civil Eats, Food Tank, and EcoWatch. She holds a BA from Miami University of Ohio and a PhD in Environmental Science, Policy & Management from the University of California, Berkeley.

Jenna Newbrey MBA, RD, CD is the Northwest Regional Coordinator for Health Care Without Harm's Healthy Food in Health Care program, and manages programs throughout the region that focus on strengthening regional food systems and increasing healthy food access by leveraging the purchasing power of health care Institutions. Jenna is a registered dietitian with a background in public health, health care, and food procurement with an MBA is Sustainable Systems. She sits on the advisory board for the Northwest Food Buyers' Alliance, Food Alliance, King County's Local Institutional Food Team, and the Washington Chapter of the Association for Healthcare Foodservice. Jenna works on a number of local food purchasing projects in the region with a large network of health care organizations throughout both Washington and Oregon, leveraging the power of institutions to transform our food system.

Emma Sirois is the National Associate Director of Health Care Without Harm's Healthy Food in Health Care program. In this capacity Emma coordinates a team of content experts and organizers to work with health care institutions and health professionals across the country to promote sustainable food systems by leveraging the purchasing power and health authority of the health care sector. Emma has 20 years of experience working toward healthy, sustainable, and vibrant food systems through education, advocacy, and public policy work with non-profit and government entities nationally and in Oregon and Arizona. Emma holds a Master's degree in Urban and Environmental Planning from Arizona State University.

Chapter 12

Bringing School Foodservice Staff Back in: Accounting for Changes in Workloads and Mindsets in K-12 Values-Based Procurement

Amy Rosenthal*, Christine C. Caruso[†]
*Rutgers University, New Brunswick, NJ, United States, [†]University of Saint Joseph, West Hartford, CT, United States

Chapter Outline

Introduction	261	Findings	267
Background	263	The Cafeteria Worker Experience	268
Purchasing Changes in the National School Lunch Program	263	Procurement Changes	271
Context of Institutional Foodservice Work	266	Conclusion	277
		References	279

INTRODUCTION

A growing body of research has documented a trend of values-based institutional purchasing as foodservice managers within institutions such as schools and hospitals increasingly consider criteria such as sustainability, healthfulness, or other social and environmental concerns when they purchase food (Friedmann, 2007; Klein, 2015; Kimmons et al., 2012; Harris et al., 2012; Fitch and Santo, 2016; de Schutter, 2014; Barlett, 2011). This literature tends to highlight (1) difficulties related to building supply chains to move these foods from farm-to-institution (Benson and Fleury, 2017; Vogt and Kaiser, 2008; Feenstra et al., 2011; Conner et al., 2011); (2) attitudes of would-be purchasers and the in-house constraints they face, such as limited budgets, inadequate kitchen infrastructure, and restrictive policies (Motta and Sharma, 2016;

Colasanti, Matts and Hamm, 2012; Gregoire and Strohbehn, 2002; Landry et al., 2015); and (3) potential impacts of these types of purchases upstream in the food chain as well as on consumers (Thilmany McFadden et al., 2016; Tuck et al., 2010; Fitch and Santo, 2016; Cluss et al., 2014; Joshi and Misako Azuma, 2008; Lo and Delwiche, 2015). Thus far, researchers have paid insufficient attention to the impacts of food procurement changes on the daily experiences and responsibilities of the frontline workers tasked with preparing these foods for the consumers (Vancil-Leap, 2016b; Tsui et al., 2013). Even researchers and advocates focused on labor issues tend to emphasize the impacts of values-based purchasing upstream in the food chain, namely on agricultural workers and food processors, with less attention paid to those preparing food.[1]

This chapter begins addressing this gap by examining the perspectives of frontline staff in the context of values-based procurement shifts, specifically to more healthful and regionally sourced foods. We focus on cafeteria workers in the National School Lunch Program (NSLP), drawing on data from a research study funded by the Robert Wood Johnson Foundation and conducted by School Food Focus, a national non-profit organization that worked with school districts to leverage their purchasing power for healthier school meals.[2] The "PreK-12 School Food: Making It Healthier, Making It Regional" (MHMR) project has examined meal program operations in six school districts, all of which are actively attempting to purchase and serve foods they consider to be healthier and regionally sourced. The purpose of the project is to investigate the changes these school food authorities (SFAs) make to their meal program operations as they purchase and serve these foods.

A missing element of research on such procurement initiatives in schools has been the impact on cafeteria staff of these menu changes. We find that foodservice workers are a crucial link in the chain to move these foods from production to consumption. Values-based procurement changes cannot be successful without the labor of cafeteria staff; however, these workers experience daily constraints that limit both their effectiveness and their own wellbeing. Such challenges may be magnified by the introduction of healthier and regionally sourced foods. We argue that values-based initiatives currently do not sufficiently account for the effects of increased workloads and complexity of tasks on foodservice labor, and that those implementing these initiatives should do more to incorporate the perspectives of frontline foodservice workers and managers.

These findings are based on data from meal program operations in school districts in Virginia, Iowa, Florida, Georgia, Kentucky, and South Carolina,

---

[1]. See, for example, Feldstein et al. (2017). Similarly, the winter 2015–16 issue of the *Journal of Agriculture, Food Systems and Community Development* devoted to "Labor in the Food System from Farm to Table" featured ten articles about agricultural labor and one on frontline (restaurant) workers.

[2]. As of January 2018 School Food Focus merged with FoodCorps, a national service organization which connects children to healthy food in school.

collected in school year (SY) 2016–17 and 2017–18. The research team conducted interviews with one to six district-level staff in each district, as well as three to six cafeteria managers. Two members of the team observed meal preparation and/or meal service in three to six schools in each district and administered a written survey to school cafeteria staff. Findings represent analysis of interviews with 21 cafeteria managers and 147 completed staff surveys in 23 schools, as well as field notes.

The chapter begins with background on the NSLP and current issues related to values-based school food procurement as well as the broader context of the U.S. labor landscape of which foodservice workers are a part. Next, findings from the MHMR study are presented, first with a narrative composite vignette that conveys the daily experience of a school cafeteria manager. This narrative is followed by a brief discussion, leading to an analysis of survey and interview data that address how cafeteria staff have been impacted as SFAs procure more healthful and local foods, namely in the areas of workload, training and equipment needs, and attitudes. We conclude with a discussion of recommendations to navigate emerging challenges related to values-based procurement in school cafeterias, with relevance across institutional food settings.

BACKGROUND

Purchasing Changes in the National School Lunch Program

The National School Lunch Program (NSLP) provides meals, often free or at a reduced price, to more than 30 million students in public and nonprofit schools across the country (U.S. Department of Agriculture Food and Nutrition Service, 2018b). Combined, school district meal programs buy more than six billion dollars of food per year, and they receive more than one billion dollars more in donated commodities from the federal government (Young et al., 2012). Most school meal programs, officially known as school food authorities (SFAs), are organized as a unit of the school district administration (which usually aligns with a city or county, for example, Minneapolis Public Schools Culinary and Wellness Services or Henrico County Public Schools Nutrition Services). The central SFA office oversees all schools in the district, including setting a district-wide menu, contracting with vendors, equipping cafeterias, and hiring and training foodservice staff. Responsibility for executing the menu and getting meals to students rests with the cafeteria workers and the cafeteria manager; however, central office SFA staff make most decisions about what to purchase and serve (Young et al., 2012). As in other institutional settings, these procurement decisions are influenced by many factors, including price, quality, and ease of logistics. In some school districts, all or some elements of meal program operations may be contracted out to foodservice management companies. In SY 2009–10, approximately 13.5 percent of all school districts used a foodservice management company (Young et al., 2012).

Over the past several years, school meal purchasers have had to heed new federal nutrition guidelines for the NSLP. Since initiation of the first school feeding programs in the early 20th century, public health advocates have seen schools as a potential site of intervention in children's health (Levine, 2008; Ruis, 2017). Though at first nutritionists' and social workers' efforts centered on student undernutrition and provision of adequate calories, since the 1980s changes in nutritional advice and rising rates of childhood obesity have caused those concerned with children's health to advocate for low-sodium, low-fat meals with more fruits, vegetables, and whole grains (Levine, 2008; Poppendieck, 2010). In 2010, policymakers, encouraged by these advocates as well as a sympathetic White House, passed the Healthy, Hunger-Free Kids Act (HHFKA), authorizing a major revision to federal nutrition guidelines for school meals (Harrington, 2017; The Kids' Safe and Healthful Foods Project, 2016). As of SY 2012–13, the new standards restricted sodium, saturated fat, and calories, and increased required amounts of whole grains and produce.

In response to the new regulations, as well as part of their ongoing efforts to improve the health and quality of menu offerings generally, there has been increasing interest in many school districts to serve more freshly-prepared foods (Schober et al., 2016). Most school cafeterias have come to be dominated by pre-cooked, "heat and serve" items and the infrastructure they require, namely freezers and convection ovens (Poppendieck, 2010; Levine, 2008; Gaddis, 2014b). Doing food preparation in the district or school instead gives the school food authority staff more control over macronutrient (e.g., calorie and sodium) levels as well as exclusion of additives such as preservatives, artificial flavors, and food dyes, many of which may have negative effects on children's health (National Toxicology Program, 2014; Kobylewski and Jacobson, 2010).

Efforts to improve the healthfulness of school meals often overlap with initiatives to leverage the purchasing power of school districts for upstream improvements in the food supply chain. The ability to prepare whole foods, as opposed to only heat-and-serve items, allows more flexibility for districts to purchase foods outside of conventional supply chains, for example, prioritizing local farmers who might not offer pre-processed produce or pre-cooked meat. "Farm-to-school" programs, which help schools source produce and other foods from nearby farms, are particularly popular, especially in light of the 2010 regulations that set a higher baseline for the amount and types of fruits and vegetables cafeterias must serve. In fact, the HHFKA instructed the U.S. Department of Agriculture (USDA) to provide support for farm-to-school, leading to the inauguration of the USDA Farm to School Program in 2012, and set aside five million dollars per year in mandatory funding for a Farm to School Grant Program to "improve access to local foods in eligible schools" (U.S. Department of Agriculture Food and Nutrition Service, 2018a; U.S. Department of Agriculture, n.d.; Joshi et al., 2014). Local food purchasing helps schools source fresh, healthy items to meet the guidelines; in addition, school foodservice personnel, USDA officials, and food systems advocates emphasize the economic impact of local food purchasing, which initial evidence has shown supports the farmers and processors that schools procure from (Benson and Fleury, 2017; Pinchot, 2014; Tuck et al., 2010; Colasanti, Matts and Hamm, 2012). As of the 2013–14 school year, school districts surveyed by the Farm to School Program spent almost $790 million on local food items, up from $386 million in purchase in the 2011–12 school year (U.S. Department of Agriculture, 2016).

However, recent literature has documented several barriers that SFAs face in these attempts to improve healthfulness and quality in school meals. Many school food professionals opposed the 2010 nutrition reforms, their main concerns being increased costs, difficulties in procuring appropriate foods, and student dislike for the new offerings (Harrington, 2017; The Kids' Safe and Healthful Foods Project, 2016; Donze Black, 2015; Cornish, Askelson and Golembiewski, 2016). By December 2015, USDA records showed 98.5 percent of schools in compliance with the guidelines; however, 62 percent of school foodservice directors still reported some or major difficulty in meeting the lunch regulations in school year 2014–15 (The Kids' Safe and Healthful Foods Project, 2016).[3] Studies have found that SFAs have encountered challenges in meeting the new regulations or otherwise serving healthier foods, such as budget constraints and difficulty sourcing compliant products (Askelson, Lubker Cornish and Golembiewski, 2015; Cornish, Askelson and Golembiewski, 2016; Ohri-Vachaspati et al., 2015). School foodservice staff face similar challenges in purchasing local foods, for example, finding appropriate products supplied in adequate quantity at an affordable price (Gregoire and Strohbehn, 2002; Kloppenburg, Wubben and Grunes, 2008; Lyson, 2016; Motta and Sharma, 2016; Landry et al., 2015; Conner et al., 2011). Meal program staff have also highlighted the need for equipment and facilities that allow them to receive, store and prepare fresh produce, whole grains, and other more healthful foods (The Kids' Safe and Healthful Foods Project, 2013).

While researchers agree on the general importance of various school staff to the success of health initiatives at school (Cornish, Askelson and Golembiewski, 2016; Greaney et al., 2014; Nollen et al., 2007), few studies explicitly examine the role of cafeteria staff in executing school meal programs and how their activities may affect the outcomes of efforts to improve meal healthfulness and quality. The few related findings suggest that increased training of foodservice staff could contribute to healthier food environments and better nutrition outcomes (Tsui et al., 2013; Thomson et al., 2012). Similarly, little research explores the experience of cafeteria staff as workers. Notable exceptions are Vancil-Leap's exploration of gendered expectations of school "lunch ladies," (2016a,b) and

---

3. In early 2017, the USDA announced that schools would have more time to reach compliance with the HHFKA regulations and that certain restrictions would be lifted. This action indicates the possibility of further changes to the 2010 guidelines and their enforcement (Greenberg 2017).

Gaddis on culinary capacity and sustainability (2014a,b), who both highlight the physical and emotional labor expected from the school cafeteria worker.

Overall, frontline staff are particularly absent in research on school food, as participants or even as informants. Cafeteria managers may act as school-level informants (Nollen et al., 2007; Odum et al., 2013; Haesly et al., 2014); however, as of publishing, we could find only a single study in the United States that surveyed the frontline cafeteria workers themselves, asking their perspective on involvement in nutrition education (Perera et al., 2015). There is clearly a lack of literature that explores cafeteria staff reactions to meal program initiatives and their ensuing effect on the job of cafeteria workers.

Context of Institutional Foodservice Work

NSLP cafeteria workers could be more broadly categorized as institutional foodservice labor, itself a subset of the foodservice sector. While comprehensive data on institutional foodservice work is limited (Tsui et al., 2013), examining the labor landscape, especially the foodservice sector, clarifies the broader context of school cafeteria work. Across the workforce, service sector staff, including foodservice and preparation workers, are particularly saddled with the burden of low wages and poor work conditions (U.S. Department of Labor Bureau of Labor Statistics, 2016). The majority of foodservice jobs are of low pay and low quality, frequently with no benefits, and characterized by high turnover rates (Tsui et al., 2013). At the same time, foodservice jobs have seen some of the greatest growth across employment sectors since the recession (Cohen, Poppendieck and Freudenberg, 2017).

The average school cafeteria staff member working in the United States earns $9.85 per hour (Payscale, 2017). While this is somewhat higher than the current federal minimum wage of $7.25 per hour, the more likely challenge with income in this profession is the limited hours available. Nonstandard employment, which describes all jobs that are not permanent full-time positions, including part-time employment, has remained steady at 18 to 20 percent of the total workforce for the past several decades (Bernhardt, 2014; Kalleberg, 2000). However, among the foodservice sector, rates of part-time employment hover at around 40 percent (Workforce Strategies Initiative, n.d.), twice the rate of part-time employment across all sectors.[4]

Low wages and nonstandard employment have contributed to the rise of the working poor (Appelbaum, Bernhardt and Murnane, 2003; Fine, 2005), those formally employed who do not earn enough to support their families. This phenomenon is likely due to many determinants, and Appelbaum and colleagues emphasize the role of wages, which have not kept pace with inflation, as well as

---

4. That said, part-time work can also be voluntary (Booth and van Ours 2013; Boeri and van Ours 2013). For example, Booth and van Ours (2013) found that Dutch women preferred part-time labor, especially in households with a male partner working full-time and a highly gendered division of labor.

the decline of labor unions. According to a report by the United States Bureau of Labor Statistics, in 2014 there were 9.5 million individuals classified as working poor, with women and people of color more likely to be among them (U.S. Department of Labor Bureau of Labor Statistics, 2016). This is consistent with Bronfenbrenner's observation of the intersectionality of U.S. labor: "Women are in the workforce today in numbers nearly equal to men. Still, a combination of outright discrimination and gender-based occupational segregation has left the majority of women trapped in low-paying jobs with few or no benefits or opportunities for advancement" (2005, p. 4). School foodservice work is no exception: it continues to be a highly-gendered occupation, with an approximately 86 percent female workforce (U.S. Department of Labor Bureau of Labor Statistics, 2018).

Understanding the shift to lower wages and worsening working conditions requires an examination of the role of labor unions in the United States. Since the mid-20th century, union participation has steadily declined by more than half (Fine, 2005). Lack of participation in labor unions and lower rates of union organization can be directly linked to low-wage and precarious work in many sectors, including foodservice and manufacturing. For women, union participation is associated with more than 30 percent higher wages (Bronfenbrenner, 2005). Appelbaum et al. (2003) note specific impacts of the decline in union participation and organizing for foodservice workers: wage freezes; benefits cuts; workload increases; increased use of temporary workers and subcontracting; replacing native-born workers with recent immigrants to facilitate cost-cutting; and use of machinery to automate food production. All of these factors impact the already strained conditions of school cafeteria workers and present an important backdrop to understanding the challenges of programmatic changes, including values-based procurement, that shape their job site, work responsibilities, and day-to-day experiences.

FINDINGS

This chapter's findings are based on data from meal program operations in six school districts, located in Virginia, Iowa, Florida, Georgia, Kentucky, and South Carolina, collected in the 2016–17 and 2017–18 school years. The research team selected school food authorities based on criteria including procurement changes previously made, student enrollment, NSLP free and reduced price meal eligibility, and location (Table 1). Due to the project's exploratory nature and for reasons of convenience, no SFAs resistant to values-based purchasing were included in the sample.[5] The SFAs are all in generally sound financial health, and none contract out operations to a foodservice management

---

5. Including a true "control" SFA would be difficult because almost all school food authorities are changing procurement practices to some extent to be in compliance with the 2010 Healthy, Hunger-Free Kids Act.

TABLE 1 Making It Healthier, Making It Regional Participating SFA Details, SY 2016–17

SFA Located In:	Total Student Enrollment	Students Qualifying for Free or Reduced Lunch	Lunch Participation Rate	Total Cafeteria Staff
Florida	186,332	68 percent	60 percent	1081
Georgia	180,000	42 percent	68 percent	1300
Kentucky	100,063	68 percent	68 percent	913
Iowa	32,979	74 percent	66 percent	419
South Carolina	17,301	35 percent	68 percent	160
Virginia	89,901	40 percent	62 percent	854

company. These may not be representative of SFAs across the country, so these findings may not be applicable in other contexts.

The research team first interviewed the director of each SFA (or his or her designated replacement) via phone. Two researchers visited each district, observing lunch and breakfast preparation and/or service in three to five schools suggested by the SFA director. In each school, the researchers interviewed the cafeteria manager and administered a written survey to the cafeteria staff. Findings of this chapter are based on analysis of interviews with 21 cafeteria managers and 147 completed staff surveys in 23 schools, as well as field notes. The survey sample was approximately 92 percent female, with an average age of 48 and a little more than eight years of experience working in their SFA.

The Cafeteria Worker Experience

What follows is a narrative composite vignette (Wertz et al., 2011) of the daily experience of a foodservice manager at a school-based cafeteria, drawn from the authors' field observations and interpretations, as well as surveys and semi-structured interviews with kitchen staff and managers. This composite vignette aims to offer a sense of the lived—and felt—experience of a frontline worker in school foodservice in general, as well as one confronting procurement changes related to an SFA's efforts serve healthier and more regionally sourced foods. The intention of this passage is to provide the reader with a deeper appreciation of the character of school food work by walking through the activities that take place in a "typical" school kitchen, while examining common themes found across the SFAs in the study. These themes include: (1) context and challenges of school food preparation; (2) the motivations of school foodservice frontline staff; and (3) the experience and impacts of introducing healthier and more regionally sourced menu items.

> *I woke up today at around 4:30 a.m. to get to work before 6. I sometimes try to arrive even earlier to get a handle on the day; no matter how long I've been doing this job, you just never know what's going to get thrown at you.*
>
> *Right now two of the ladies – really good workers, too – are out. It's tough even when we're all here, and now here we are all trying to fill in. Some days I just feel like I'm running around and don't get a minute to stop and take a breath until well after lunch is over. At the end of most days I'm bone tired; it's just really hard work. It's more than cooking; it's cleaning, it's stocking, it's serving and just running around and being on your feet for hours every day. I'm used to it though, and all the ladies too – you don't really ever hear anyone complain about it or anything.*
>
> *It would be great if we had one or two more staff, though. Especially now that we're doing so much more fresh fruit and vegetables. Sure, we still open some cans, but we have fresh items every day, so that's more washing, cutting, and don't get me started on about peeling those little stickers off every single apple or pear....*
>
> *I've been in this cafeteria for maybe four years now, but with the district for over ten – my, where does the time go? I got started as a general worker and liked the hours because my kids were in school here in the district. Now that they've graduated I'm glad to be a manager. And honestly, with the cost of living going up around here I really couldn't be working part-time anymore. Some of the ladies even have two more jobs on top of this one to make ends meet.*
>
> *Anyway, since we serve free breakfast for everybody in this middle school the ladies come in around 6:30 a.m. Even though we're used to it, it takes a lot to get those 350 breakfasts out to the kids – we've got about fifteen minutes between the buses coming in and them getting to class. Today one of the buses came in late so there was a big rush right as the bell was ringing; it was crazy. It feels like going into battle sometimes but you know, we all just want to make sure that they're getting good, healthy stuff and starting their day right.*
>
> *After breakfast service, most of the ladies move back to prepping lunch, while a couple of us sort out what's left, temp the milk, throw out what we have to. We work really hard to make sure we have the right amount so that we don't waste anything, but sometimes the temperatures are off, so we have to toss it. We have really strict standards that we follow, and it's drilled into all of us through our training. It's lot to juggle sometimes, but the job's not really hard. It's funny, though; I think we do a lot more by lunch time than some people do all day.*
>
> *Since we're a middle school we have a lot of options for lunch. I like to make sure that the students have a few different options for sides and that at least a couple of them are fresh. The last couple of years we've been getting a lot more fresh vegetables from the nearby farms. I think it's good, but yeah, that's some more work for us. I think it's exciting to try new recipes, too, but it does take time for us to work out the kinks and get it right. Sometimes it's a lot of pressure;*

I mean if it doesn't come out well, doesn't look good then of course the kids aren't going to take it.

So, like today the corn came from a can, but we are baking the local sweet potatoes. Oh, and I still need to figure out the fruit. We were supposed to have bananas today, but they just aren't ready, so I need to go back and see what we have enough of. We always say that the last child in the last lunch period should have all the same options as the first child coming through the first lunch period.

We have five lunch periods here, about 25 minutes each, give or take. We only have enough staff for two lines because we need some people in back to refill, so that means we've got to get all these kids through two lines – serve them, make sure they have all the components they're supposed to, check them out. It breaks my heart but some of the kids are still in line when the lunch period is just about over. I mean it's hard on the ladies, too – it's real fast paced. But even with all of this going on, I see that they're checking in with the kids, telling them they like their new hair styles or shoes or whatever, or asking them what's going on if it seems like they're down.

Before the last lunch period is over we already start clean-up back in the kitchen. Then we can do some prep for the next day and sometimes for some events we help out with for other parts of the school. Or maybe I do a quick training with them or show them a new recipe. And then for me it's the paperwork – review our numbers, do the ordering, rotate the stock, process the shipments. Yeah, it's not over for us when the bell rings!

    The narrative above recruits the voices of the many women (and a few men) interviewed for this study and weaves them together with the investigators' observations at the field sites. While no experiences are absolutely universal, the sentiments illustrated here represent common themes and findings across the six SFAs. Cafeteria staff work very hard and at a constant pace, in order to prepare and serve hundreds of meals under many constraints, including time, and often space, equipment, and staffing. Workers are charged with guaranteeing the safety of the food served and kitchen cleanliness, both to meet regulations and protect the health of the students that are their charges. Given the heavy workload and time pressure that is a regular part of this job, any changes, including menu changes that diverge significantly from the common practices and skills of the team, further impact the workload in unexpected and significant ways.

    An important sentiment observed and communicated across the field sites relates to the care workers have in approaching their jobs, motivated by the children they serve. Care for and a desire to feed students is the driver of this work, with a nearly universal emphasis on ensuring that all students on a daily basis have access to the same options. This task seems all the more challenging as there are many practical reasons to avoid cooking too much and ending up with waste. By extension, further efforts to reduce waste, also shaped by care for the students, inspire workers to make food look and taste appealing.

    As our narrator suggests, the scope of duties undertaken is quite broad and must be done against strict time constraints in order to prepare and serve the food to students during the appropriate windows in the school day. In addition to their regular tasks cafeteria staff must also deal with daily, unforeseen issues that arise, from unexpected field trips that require packing bagged lunches to missed deliveries that cause last-minute menu changes.

These time pressures are frequently exacerbated by understaffing in the kitchens, prevalent in most of the SFAs in the study. Some SFAs have several permanent positions unfilled, impacting the shared responsibilities of existing staff to complete the daily tasks. These conditions are intensified by the inevitable absences kitchen teams will encounter, some of which may be long term. While substitute workers may be assigned to kitchens with absent staff to fill the gap, their inexperience and lack of integration into the team means they cannot entirely make up for the worker being replaced. School food positions typically require no experience, so even when new permanent workers are hired, many of their skills come from on-the-job training. Managers must take the time to help them, and staff must perform these new tasks while learning and refining them in the fast-paced context of meal preparation.

Procurement Changes

It is into this context of a constantly busy and stressful cafeteria work environment that values-based procurement initiatives, in the form of new menu offerings and resulting changes to food preparation and service activities, are introduced to cafeteria staff. The SFAs included in this study all operate in compliance with the 2010 Healthy, Hunger-Free Kids Act regulations and are explicitly committed to improving the healthfulness and the quality of the foods that they serve. As one of several strategies to achieving those goals, these meal programs all focus on procuring more fresh, whole, and (when possible) local ingredients to prepare within their own facilities. While in many cases SFAs swap out pre-made items for ones that meet the new nutritional requirements (e.g., less sodium), using their own recipes allows meal programs greater control over macronutrient profiles. Also, the 2010 regulation requiring all students to take a fruit or a vegetable with each meal led SFAs to significantly increase the amount of produce they purchase and offer, especially fresh fruits and vegetables.

All of the SFAs in our sample prioritized values-based purchasing in terms of foods grown "locally,"[6] and these initiatives overlap with efforts to serve more produce, provide fresher foods and do more scratch-cooking. The SFAs most often purchase local produce, though some also buy local chicken, grains, and manufactured items. Particularly when items come directly from a smaller

---

6. There is no federally recognized definition of "local foods" (U.S. Department of Agriculture Economic Research Service, 2017). For four of the SFAs, "local" is defined as produced within the state. For one, "local" means within 100 miles of the foodservice central office, and for another, from within the state or a neighboring state.

farm, instead of through the mainstream processing and manufacturing supply chains, they are often not pre-processed. As such, these items must be washed and prepped, or even cooked. For example, one SFA in the sample wanted to purchase local wheat. Because that item requires cooking, the cafeteria staff had to bake from scratch in order to include it in the menu.

The result of the combination of these various priorities—children's health, meal quality, and local purchasing—is a change in the activities of the cafeteria staff as SFAs try to buy fewer heat-and-serve or pre-packaged items and instead prepare more foods from basic ingredients. These efforts include, for example, preparing sauces and dressings from scratch, baking muffins, or putting together wraps and salads. Further, the district-level staff may also add special items to the menu, such as a scratch-baked fruit dessert, to entice students to purchase (or continue to purchase) lunch. These changes to staff activities raise concerns about whether values-based procurement initiatives adequately account for the increased staff workload, training and equipment needs, and changes in staff attitudes that are necessary for implementation. The following sections use survey and interview data to investigate the impact of purchasing changes, as experienced by cafeteria staff, in these areas.

Workload

As a result of procurement changes, cafeteria staff must do more complicated activities to produce the same number of meals. Preparing fresh produce options, such as boxing grab-and-go salads or chopping cantaloupe, requires more labor than opening a can of green beans. Workers may need to get used to looking for dirt or bugs on fresher produce, or switch preparation methods, like the SFA that had to dedicate more time to tearing a delicate, local Bibb lettuce by hand. As new items are introduced, the cafeteria staff must adjust to different types of tasks. Many have not previously (or not in many years) been asked to make a smoothie in the blender or follow a recipe to bake a cake from scratch. These tasks not only take more time to do but also require time to adjust to and learn.

However, number of staff or hours for current staff do not tend to increase in proportion to new tasks or offerings. As such, the already heavy cafeteria workload intensifies as workers are asked to do more in the same amount of time. In response to a survey question that asked staff whether their jobs had become harder, easier, or stayed the same over the past five years, many responses noted increased workloads. When asked to explain how her job has gotten harder, one staff member responded simply, "By doing all the cooking." Another noted, "More students; more responsibilities of other jobs; more time getting done with normal duties; more job duties if other staff is [sic] ill." Some comments pointed specifically to the time-consuming nature of prepping fruits and vegetables. The burden of this workload intensification is particularly difficult in the context of meal programs that are already understaffed. As one staff member put it, "more work, less help." It should be noted that some staff observed a decrease in cooking over the past years, noting that they no longer bake rolls or make certain items from scratch. However, the majority of comments indicated a sense of increasing and more involved work requirements.

Training and Equipment Needs

Cooking skills are not a prerequisite for cafeteria staff positions, so when food preparation tasks beyond opening boxes and reheating are required, employees may need supplemental training. As more fresh produce enters the cafeteria, knife skills become a particular concern. One SFA has responded by requiring all staff to wear knife guards while cutting produce; most others do occasional district-wide training and rely on cafeteria managers to monitor safety and do on-the-job training.

These adjustments are more challenging if the foods are not purchased regularly, which can happen particularly with local foods, due to their expense and seasonality. For example, one SFA purchases local chicken but can only afford to do so about two times per year. It must purchase this product raw, whereas all of its other chicken comes into the cafeteria pre-cooked. Working with raw meat requires careful attention to food safety protocols, and while the staff are able to safely prepare the product, they find it very stressful. Staff commented that it would not be so difficult if they served the product regularly and became accustomed to the process, but in serving it so rarely, they never become truly practiced and comfortable.

More complicated food preparation also may require equipment different from that with which most cafeterias are equipped. Two SFAs have recently fitted all their kitchens with combi-ovens, which allow for more precise cooking. Three have also purchased or refurbished equipment to quickly cut produce, saving staff time and energy when it comes to certain common items, such as tomatoes or oranges.

Most of the nonmanagerial staff surveyed seem to feel they have adequate equipment and training. When asked if they have equipment and training necessary to do their jobs, staff on average agreed with both statements, with slightly more agreement that they have adequate training. Fewer staff agreed that they would like more culinary training (Table 2).

TABLE 2 Cafeteria Worker Opinions on Equipment and Training

	"I have the equipment I need to do my job well."	"I have the training I need to do my job well."	"I would like more training in culinary skills."
Average level of agreement[a]	3.9	4.1	3.5

[a] 1, strongly disagree; 5, strongly agree.

From the managers' perspectives, formal training and accessing the proper equipment seem to be less of a concern than finding time and opportunities to work closely with staff as they adjust to new tasks.[7] In general, cafeteria managers seem to be expected to offer support and on-the-job training to staff who may need it, for the variety of new tasks that might be expected. They highlighted the day-to-day process of adapting to change as the most difficult part of serving new types of foods. As one cafeteria manager put it:

> *That, I think, was a challenge for my staff in the beginning, change. Compared to having them already made and just popping them on a pan in the oven.*

For the already busy managers, leading this process requires extra time and attention to troubleshoot any problems that arise:

> *What it requires of me as a manager to study it ... So it takes me sitting in my office quiet and then I have to make a schedule to meet it. Sometimes the first time we do it, it fails.*

As another said:

> *When there's a change you have to figure it out, there's no book or help, you learn over time.*

Ultimately, managers must spend time and energy to incorporate new foods and preparation styles into the kitchen. In order to most effectively implement all that is asked of them, they must adjust work schedules and find efficiencies so they can do more with less time. Sometimes they "fail" at new recipes; they also may sacrifice effective preparation of current menu items when they add new ones. One SFA that had been adding several new, scratch-cooked items decided to minimize these additions in order to focus more explicitly on "the basics." The consequences of serving burned pizza or forgetting to season the green beans may not seem high, but, as noted above, cafeteria managers operate under intense time and monetary pressure, with little room to waste hours or product. Also, negative experiences with cafeteria foods color student expectations of school meals, making students reluctant to try certain foods again or to eat in the cafeteria at all.

Attitudes

Despite the challenges, cafeteria managers largely agree that their workers do ultimately adjust to changes. Speaking about increased washing and chopping of produce, one manager said, "Once they get used to it, they don't really have any issue." Similarly, describing working with new recipes, another manager

---

7. This finding may be particular to the SFAs in this study given the length of time they have been incorporating healthy and local foods into their menus, as well as their relative financial health. These factors may indicate that these SFAs are ahead of others in offering necessary training and equipment.

noted, "You start doing it, and then after you do it, you're like 'Okay, that wasn't too hard.'" Many managers expressed a stoic, practical attitude in the face of new requirements and requests handed down to them from above; as one manager said, "It's something we have to do, we get it done."

They often rationalized the extra effort these adaptations require with a sense of purpose and motivated themselves with their responsibility to students. Cafeteria managers and frontline workers agreed on an overall mission to serve healthy and local meals to students. One cafeteria manager said:

Some [staff] still say, "This is ridiculous, more work," but for the most part we've tried to realize it's a positive thing for the kids. It might cause us a little more work but in the end we're taking care of the kids.

Almost all managers noted the importance of feeding children healthy foods and thus responded positively to efforts to procure and prepare fresher foods, which they felt contributed to this end:

I'm teaching my kids to eat healthy and going in that route. So for me, it makes me happy we're serving these types of foods to them.

Cafeteria workers recognized this as well: almost all agreed with the statement "It is important to serve healthy foods." One staff member noted her job has gotten harder but wrote, "It has been exciting [to] get to learn more about cooking healthier foods. It is great working with fresh ingredients." Staff also, on average, agreed with the statement "It is important for us to serve local foods." (Table 3).

However, there does not seem to be much concern among workers about making the meals even more healthy or adding additional scratch-cooked foods; most staff seem satisfied with the current menus they serve and the types of food preparation they do. Almost 90 percent of workers indicated that they serve enough or too many healthy foods. Workers were slightly more interested in increasing scratch cooking than healthy foods. While about the same number

TABLE 3 Cafeteria Staff Opinion on Importance of Healthy, Scratch and Local Foods

	"It is important for us to serve healthy foods."	"We should serve more foods that we cook from scratch."	"It is important for us to serve local foods."	"It is important for us to tell the students when the food is local."
Level of agreement[a]	4.3	3.1	4.0	3.6

[a] 1, strongly disagree; 5, strongly agree.

TABLE 4 Cafeteria Staff Opinion on Amounts of Healthy, Scratch and Local Foods

Total	Amount of Healthy Food (N = 140)	Amount of Scratch (N = 94)	Amount of Local Food (N = 130)
Not enough	12 percent	23 percent	22 percent
Right amount	79 percent	68 percent	63 percent
Too much	9 percent	9 percent	15 percent

thought their SFA served too many scratch foods as those who thought they served too many healthy foods, about twice as many wanted more scratch-cooking (Table 4). Attitudes are more divided on local foods. More than half of staff thought they serve the right amount of local food, while 22 percent want more, and 15 percent want less.

These divided attitudes may be related to the varied ways that workers think students react to these initiatives. In comments in response to the question "Please explain what students like about the food," eight staff members highlighted healthy options (specific items such as fruit and vegetables, or in general); for example, "More healthy food is what they like." However, responding to "Please explain what students do *not* like about the food" (emphasis added), five staff members noted healthy options or the removal of unhealthy options such as French fries. "They say there is no flavor or too healthy," according to one staff member. Both workers and managers expressed that the students are not interested in whether food was locally produced. One worker wrote, "[The students] don't care if it is local or not." Most managers also acknowledged that, while they themselves may see why buying local is important, they don't think the students pay attention to the origin of the produce: "they don't care where it came from."

So while cafeteria staff may acknowledge the importance of healthy and local foods on the one hand, they also see on a daily basis which foods students choose or reject, often hear students' opinions firsthand, and may even see what they tend to throw away. As such, each day they work hard to prepare food that they then may see students pass over, call "gross," or discard untouched. These behaviors are not only in reaction to healthier foods; however, as previously noted, some staff feel students react less favorably toward such foods, especially some produce and whole-grain items. Such student reactions could influence staff opinions on purchasing changes. "Healthy" and "local" foods have been mandated by their superiors, for reasons that may not always be fully communicated, especially for local purchasing. While some districts do attempt to communicate the importance of serving healthy and local foods, these messages may not always make it through the chain of communication from district-level staff to worker, or may become less salient over time.

Because cafeteria workers are the meal program staff in most direct contact with students, the abstract motivation for serving healthy and local foods remains in tension with daily evidence of food waste and the increased effort of preparation. One staff member summed up on a survey what we heard from others informally: "I just and never will think you should make a child take a piece of fruit or veggie that they are just going to throw away." Although almost all cafeteria managers and district-level staff noted that they think, over time, students are eating more produce and wasting less, cafeteria staff still experience students snubbing or discarding items on a regular basis. This phenomenon is especially difficult for the many cafeteria staff who care deeply about whether students like the food and are concerned with whether or not they leave the cafeteria hungry.

CONCLUSION

Values-based purchasing initiatives correlate with more labor-intensive and complex food preparation techniques. Our research findings suggest that these add to an already heavy cafeteria staff workload. Even though managers and staff often recognize the importance of serving local and healthy foods and may be motivated to make new initiatives work for the students' wellbeing, they still face daily constraints that affect their own wellbeing as well as their ultimate effectiveness in and continued dedication to implementing changes. These findings point to a few recommendations that may help values-based purchasing initiatives better account for the workload they add for foodservice laborers.

Ideally, staff would be compensated appropriately and given increased hours or additional help as needed for more complex tasks, as well as any needed training and equipment. Such training could include more support for cafeteria managers to increase their ability to institute more efficient work processes and help staff maximize their skills. In our research, we noted how a skilled cafeteria manager can ease the friction of changes as they are rolled out and incorporated into daily routines. He or she may be able to adjust schedules, find shortcuts, or otherwise adapt the workflow to ease the burden on staff. This training could also include more regularized communication about why the SFA is making these changes, so both managers and staff are reminded of the value of serving healthy and local foods.

The importance of finding strategies to promote buy-in from foodservice staff was also a theme that arose in this research. Peer reactions appeared to influence the process of adapting to change. One SFA has had greater success in introducing new items and initiatives after launching a program that pilots them in several schools, which are paid for their effort in working out any kinks. These staff then introduced the new processes to their peers at other schools, and district-level staff said the frontline workers react more positively than when the changes are introduced by superiors.

In addition, greater emphasis on ongoing evaluation of implementation could make staff feel more supported, both in their daily tasks and overall professional development, as well as ensure higher quality food production. Staff in our study received little feedback or assistance in evaluating or improving their cafeteria practices. A regularized, nonpunitive process could also include more feedback from students, which would allow space for the opinions of another key segment of stakeholders. While referred to and treated as customers, students don't often get to offer their opinion on their experience of school lunch. Such a process could not only make school meals more acceptable to students but also help educate them as to why they are served certain foods in the cafeteria and the benefits of the healthy and local options.

In sum, we suggest that:

- SFA purchasers and other district-level staff (as well as other institutional buyers) consider the impact of changes in procurement on foodservice labor. This should include soliciting cafeteria staff input; piloting new items and recipes to fix issues and develop best practices before rolling them out broadly; facilitating peer-to-peer roll out of changes; and supporting and training cafeteria managers to develop workflow processes and messaging strategies.
- School district administrators and policymakers at the state and national level ensure sufficient funding for values-based purchasing initiatives. This includes sufficient resources for school food authorities to hire, train, and fairly compensate an adequate number of cafeteria staff and managers; to install and maintain necessary kitchen equipment; and to continue to purchase high-quality foods.
- Researchers and advocates strive to include more input from foodservice laborers in their efforts to study, report on, and develop values-based purchasing initiatives. This includes exploring the experiences of frontline foodservice staff in different institutional settings, with different types of work arrangements (part-time, full-time, temporary, etc.), and varying levels of access to resources such as training and equipment.

In considering the implementation and impact of values-based institutional purchasing initiatives, we cannot neglect the perspectives of the frontline workers ultimately responsible for preparing and serving the foods purchased. Values-based purchasing should recognize the contributions of labor across the food supply chain, including cafeteria staff; otherwise, we would be implicitly prioritizing one set of "values" at the expense of another. Also, we would miss opportunities to more smoothly embed these types of purchases into institutional operations, and large-scale changes to the food system will require changes made in the procurement office to be tenable all the way to the plate.

REFERENCES

Appelbaum, E., Bernhardt, A., Murnane, R.J., 2003. Low-Wage America: an overview. In: Appelbaum, E., Bernhardt, A., Murnane, R. (Eds.), Low-Wage America: How Employers Are Reshaping Opportunity in the Workplace. Russell Sage Foundation, New York, NY, pp. 1–29.

Askelson, N.M., Lubker Cornish, D., Golembiewski, E., 2015. Rural school foodservice director perceptions on voluntary school meal reforms. J. Agric. Food Syst. Community Dev. 6, 65–75.

Barlett, P.F., 2011. Campus sustainable food projects: Critique and engagement. Am. Anthropol. 113, 101–115.

Benson, M., Fleury, D., 2017. Institutions: An emerging market for local and regional Foods. In: Dumont, A., Davis, D., Wascalus, J., Wilson, T.C., Barham, J., Tropp, D. (Eds.), Harvesting Opportunity: The Power of Regional Food System Investments to Transform Communities. Federal Reserve Bank of St. Louis, pp. 189–208. Available from: https://www.stlouisfed.org/~/media/Files/PDFs/Community-Development/Harvesting-Opportunity/Harvesting_Opportunity.pdf (Accessed November 28, 2017).

Bernhardt, A., 2014. Labor standards and the reorganization of work: Gaps in data and research. IRLE working paper no. 100-14, University of California, Berkeley. Available from: http://escholarship.org/uc/item/3hc6t3d5 (Accessed November 28, 2017).

Boeri, T., van Ours, J., 2013. The Economics of Imperfect Labor Markets. Princeton University Press, Princeton, NJ.

Booth, A.L., van Ours, J.C., 2013. Part-time jobs: What women want? J. Popul. Econ. 26, 263–283.

Bronfenbrenner, K., 2005. Organizing women: the nature and process of union organizing efforts among U.S. women workers since the mid-1990s. Work. Occup. 32, 441–463.

Cluss, P.A., Fee, L., Culyba, R.J., Bhat, K.B., Owen, K., 2014. Effect of foodservice nutrition improvements on elementary school cafeteria lunch purchase patterns. J. Sch. Health 84, 355–362.

Cohen, N., Poppendieck, J., Freudenberg, N., 2017. Food Justice in the Trump Age: Priorities for Urban Food Advocates. CUNY Academic Works. Available from: http://academicworks.cuny.edu/sph_pubs/161 (Accessed November 28, 2017).

Colasanti, K.J., Matts, C., Hamm, M.W., 2012. Results from the 2009 Michigan farm to school survey: participation grows from 2004. J. Nutr. Educ. Behav. 44, 343–349.

Conner, D., Nowak, A., Berkenkamp, J., Feenstra, G., Van Soelen Kim, J., Liquori, T., Hamm, M., 2011. Value chains for sustainable procurement in large school districts: fostering partnerships. J. Agric. Food Syst. Community Dev. 1, 1–14.

Cornish, D., Askelson, N., Golembiewski, E., 2016. "Reforms looked really good on paper": Rural foodservice responses to the healthy, hunger-free kids act of 2010. J. Sch. Health 86, 113–120.

de Schutter, O., 2014. The power of procurement: Public purchasing in the Service of Realizing the Right to Food. UN Human Rights Council. Available from: http://www.srfood.org/images/stories/pdf/otherdocuments/20140514_procurement_en.pdf (Accessed November 28, 2017).

Donze Black, J., 2015. A changing landscape for school food. Stanford Social Innov. Rev. 13 (4), 17.

Feenstra, G., Allen, P., Hardesty, S., Ohmart, J., Perez, J., 2011. Using a supply chain analysis to assess the sustainability of farm-to-institution programs. J. Agric. Food Syst. Community Dev. 1, 1–16.

Feldstein, S., Lo, J., Spach, C., 2017. The importance of inclusion in local and regional food system efforts. In: Dumont, A., Davis, D., Wascalus, J., Wilson, T.C., Barham, J., Troff, D. (Eds.), Harvesting Opportunity: The Power of Regional Food System Investments to Transform Communities. Federal Research Bank of St. Louis, pp. 189–208. Available from: https://www.stlouisfed.org/~/media/Files/PDFs/Community-Development/Harvesting-Opportunity/Harvesting_Opportunity.pdf (Accessed 28 November 2017).

Fine, J., 2005. Community unions and the revival of the American labor movement. Polit. Soc. 33 (1), 153–199.

Fitch, C., Santo, R., 2016. Instituting Change: An Overview of Institutional Food Procurement and Recommendations for Improvement. John Hopkins Center for a Livable Future. Available from: https://www.jhsph.edu/research/centers-and-institutes/johns-hopkins-center-for-a-livable-future/_pdf/research/Instituting-change.pdf (Accessed November 28, 2017).

Friedmann, H., 2007. Scaling up: bringing public institutions and foodservice corporations into the project for a local, sustainable food system in Ontario. Agric. Hum. Values 24 (3), 389–398.

Gaddis, J., 2014a. Mobilizing to re-value and re-skill foodservice labor in U.S. school lunchrooms: a pathway to community-level food sovereignty? Radical Teacher 98, 15–21.

Gaddis, J., 2014b. Fit to Feed: Labor, ecology, and the remaking of the National School Lunch Program. Ph.D thesis, Yale University.

Greaney, M.L., Hardwick, C.K., Spadano-Gasbarro, J.L., Mezgebu, S., Horan, C.M., Schlotterbeck, S., Austin, S.B., Peterson, K.E., 2014. Implementing a multicomponent school-based obesity prevention intervention: a qualitative study. J. Nutr. Educ. Behav. 46, 576–582.

Greenberg, J., 2017. Sonny Perdue changes school lunch rules, but says Obama standards for milk, grains remain. Politifact, 4 May. Available from: http://www.politifact.com/georgia/statements/2017/may/04/sonny-perdue/agsec-perdue-changes-school-lunch-rules-says-stan/ (Accessed 17 November 2017).

Gregoire, M.B., Strohbehn, C., 2002. Benefits and obstacles to purchasing food from local growers and producers. J. Child Nutr. Manag. 26 (2).

Haesly, B., Nanney, M.S., Coulter, S., Fong, S., Pratt, R.J., 2014. Impact on staff of improving access to the school breakfast program: a qualitative study. J. Sch. Health 84, 267–274.

Harrington, C., 2017. Healthy Hunger-Free Kids? The US School Lunch Revolution. In: Ashbee, E., Dumbrell, J. (Eds.), The Obama Presidency and the Politics of Change. Palgrave Macmillan, Cham, Switzerland.

Harris, D., Lott, M., Lakins, V., Bowden, B., Kimmons, J., 2012. Farm to institution: Creating access to healthy local and regional foods. Adv. Nutr. 3, 343–349.

Joshi, A., Misako Azuma, A., 2008. Bearing Fruit: Farm to School Program Evaluation Resources and Recommendations. National Farm to School Network. Available from: http://www.farmtoschool.org/resources-main/bearing-fruit-farm-to-school-program-evaluation-resources-and-recommendations (Accessed November 28, 2017).

Joshi, A., Henderson, T., Ratcliffe, M.M., Feenstra, G., 2014. Evaluation for Transformation: A Cross-sectoral Evaluation Framework for Farm to School. National Farm to School Network. Available from: http://www.farmtoschool.org/resources-main/bearing-fruit-farm-to-school-program-evaluation-resources-and-recommendations (Accessed November 28, 2017).

Kalleberg, A.L., 2000. Nonstandard employment relations: Part-time Temporary and Contract Work. Annu. Rev. Sociol. 26, 341–365.

Kimmons, J., Jones, S., McPeak, H.H., Bowden, B., 2012. Developing and implementing health and sustainability guidelines for institutional foodservice. Adv. Nutr. 3, 337–342.

Klein, K., 2015. Values-based food procurement in hospitals: the role of health care group purchasing organizations. Agric. Hum. Values 32 (4), 635–648.

Kloppenburg, J., Wubben, D., Grunes, M., 2008. Linking the land and the lunchroom: Lessons from the Wisconsin homegrown lunch project. J. Hunger Environ. Nutr. 3, 440–455.

Kobylewski, S., Jacobson, M.F., 2010. Food Dyes: A Rainbow of Risks. Center for Science in the Public Interest, Washington, DC. Available from: https://cspinet.org/sites/default/files/attachment/food-dyes-rainbow-of-risks.pdf (Accessed November 28, 2017).

Landry, A.S., Lingsch, K.J., Weiss, C., Connell, C.L., Yadrick, K., 2015. Barriers and possible facilitators to participation in farm to school week. J. Child Nutr. Manag. 39, 1–6.

Levine, S., 2008. School Lunch Politics: The Surprising History of America's Favorite Welfare Program. Princeton University Press, Princeton, NJ.

Lo, J., Delwiche, A., 2015. The good food purchasing policy: A tool to intertwine worker justice with a sustainable food system. J. Agric. Food Syst. Community Dev. 6, 185–194.

Lyson, H.C., 2016. National policy and state dynamics: a state-level analysis of the factors influencing the prevalence of farm to school programs in the United States. Food Policy 63, 23–35.

Motta, V., Sharma, A., 2016. Benefits and transaction costs of purchasing local foods in school districts. Int. J. Hosp. Manag. 55, 81–87.

National Toxicology Program, 2014. 13th Report on Carcinogens. Research Triangle Park, NC. Available from: https://ntp.niehs.nih.gov/annualreport/2015/analysis/roc/index.html (Accessed November 28, 2017).

Nollen, N.L., Befort, C.A., Snow, P., Daley, C.M., Ellerbeck, E.F., Ahluwalia, J.S., 2007. The school food environment and adolescent obesity: qualitative insights from high school principals and foodservice personnel. Int. J. Behav. Nutr. Phys. Act. 4 (18).

Odum, M., McKyer, E.L.J., Tisone, C.A., Outley, C.W., 2013. Elementary school personnel's perceptions on childhood obesity: pervasiveness and facilitating factors. J. Sch. Health 83 (3), 206–212.

Ohri-Vachaspati, P., Turner, L., Adams, M.A., Bruening, M., Chaloupka, F.J., 2015. School resources and engagement in technical assistance programs is associated with higher prevalence of salad bars in elementary school lunches in the United States. J. Acad. Nutr. Diet. 116, 417–426.

Payscale, 2017. School Cafeteria Worker Salary. Available from: http://www.payscale.com/research/US/Job=School_Cafeteria_Worker/Hourly_Rate.

Perera, T., Frei, S., Frei, B., Wong, S.S., Bobe, G., 2015. The Role of School Foodservice Personnel in Nutrition Education: challenges and opportunities at U.S. Elementary Schools. J. Health Educ. Res. Dev. 3.

Pinchot, A., 2014. The Economics of Local Food Systems: A Literature Review of the Production, Distribution and Consumption of Local Food. University of Minnesota Extension. Available from: https://www.extension.umn.edu/community/research/reports/docs/2014-Economics-of-Local-Food-Systems.pdf (Accessed November 28, 2017).

Poppendieck, J., 2010. Free for all: Fixing School Food in America. University of California Press, Berkeley, CA.

Ruis, A.R., 2017. Eating to Learn, Learning to Eat. Rutgers University Press, New Brunswick, NJ.

Schober, D., Carpenter, L., Currie, V., Yaroch, A., 2016. Evaluation of the LiveWell@school food initiative shows increases in scratch cooking and improvement in nutritional content. J. Sch. Health 86, 604–611.

The Kids' Safe and Healthful Foods Project, 2013. Serving Healthy School Meals, Robert Wood Johnson Foundation and The PEW Charitable Trusts. Available from: http://www.pewhealth.org/reports-analysis/reports/serving-healthy-school-meals-kitchen-equipment-85899527489 (Accessed November 28, 2017).

The Kids' Safe and Healthful Foods Project, 2016. School Meal Programs Innovate to Improve Student Nutrition. Robert Wood Johnson Foundation and The PEW Charitable Trusts. Available from: http://www.pewtrusts.org/~/media/assets/2016/12/school_meal_programs_innovate_to_improve_student_nutrition.pdf (Accessed November 28, 2017).

Thilmany McFadden, D., Conner, D., Deller, S., Hughes, D., Meter, K., Morales, A., Schmit, T., Swenson, D., Bauman, A., Phillips Goldenberg, M., Hill, R., Jablonski, B.B.R., Tropp, D., 2016. The Economics of Local Food Systems: A Toolkit to Guide Community Discussions, Assessments, and Choices. Agricultural Marketing Service. Available at: https://www.rd.usda.gov/files/ILAMSToolkit.pdf (Accessed November 28, 2017).

Thomson, J.L., Tussing-Humphreys, L.M., Martin, C.K., LeBlanc, M.M., Onufrak, S.J., 2012. Associations among school characteristics and foodservice practices in a nationally representative sample of United States schools. J. Nutr. Educ. Behav. 44, 423–431.

Tsui, E.K., Deutsch, J., Patinella, S., Freudenberg, N., 2013. Missed opportunities for improving nutrition through institutional food: the case for food worker training. Am. J. Public Health 103 (9), e14–e20.

Tuck, B., Haynes, M., King, R., Pesch, R., 2010. The Economic Impact of Farm-to-School Lunch Programs: A Central Minnesota Example. University of Minnesota Extension Center for Community Vitality and University of Minnesota Department of Applied Economics. Available from: https://www.extension.umn.edu/community/economic-impact-analysis/reports/docs/2010-EIA-Farm-School-Programs.pdf (Accessed November 28, 2017).

U.S. Department of Agriculture, 2016. Research Shows Farm to School Works. Available from: http://agrilife.org/ipm/files/2016/07/FactSheet_Research_Shows_F2S_Works.pdf (Accessed November 28, 2017).

U.S. Department of Agriculture, n.d. The Farm to School Program 2012–2015: Four Years in Review. Available from: https://fns-prod.azureedge.net/sites/default/files/f2s/Farm-to-School-at-USDA--4-Years-in-Review.pdf (Accessed 28 November 2017).

U.S. Department of Agriculture Economic Research Service, 2017. Local Foods: Overview. Available from: https://www.ers.usda.gov/topics/food-markets-prices/local-foods.aspx.

U.S. Department of Agriculture Food and Nutrition Service, 2018a. Farm to School Grant Program. Available from: https://www.fns.usda.gov/farmtoschool/farm-school-grant-program (Accessed 28 March 2018).

U.S. Department of Agriculture Food and Nutrition Service, 2018b. National School Lunch Program: Participation and Lunches Served. Available from: https://fns-prod.azureedge.net/sites/default/files/pd/slsummar.pdf (Accessed 28 March 2018).

U.S. Department of Labor Bureau of Labor Statistics, 2016. A Profile of the Working Poor, 2014. Available from: https://www.bls.gov/opub/reports/working-poor/2014/home.htm (Accessed November 28, 2017).

U.S. Department of Labor Bureau of Labor Statistics, 2018. Food and Beverage Serving and Related Workers. Available from: https://www.bls.gov/ooh/food-preparation-and-serving/food-and-beverage-serving-and-related-workers.htm (Accessed 28 March 2018).

Vancil-Leap, A., 2016a. The physical and emotional contours of feeding labor by school foodservice Employees. In: Demos, V., Segal, M. (Eds.), Gender and Food: From Production to Consumption. Emerald Insight, pp. 243–264.

Vancil-Leap, A., 2016b. Resistance and Adherence to the Gendered Representations of School Lunch Ladies. Gender Issues 34 (1), 67–85.

Vogt, R.A., Kaiser, L.L., 2008. Still a time to act: a review of institutional marketing of regionally-grown food. Agric. Hum. Values 25, 241–255.

Wertz, M.S., Nosek, M., McNiesh, S., Marlow, E., 2011. The composite first person narrative: Texture, structure, and meaning in writing phenomenological descriptions. Int. J. Qual. Stud. Health Well-being 6 (2).

Workforce Strategies Initiative, n.d. The Restaurant Workforce in the United States. Available from: http://www.aspenwsi.org/wordpress/wp-content/uploads/The-Restaurant-Workforce-in-the-United-States.pdf (Accessed 28 November 2017).

Young, N., Diakova, S., Earley, T., Carnagey, J., Krome, A., Root, C., 2012. School Food Purchase Study-III Final Report. U.S. Department of Agriculture, Food and Nutrition Service, Alexandria, VA.

Amy Rosenthal is doctoral candidate in the Bloustein School of Public Policy and Planning at Rutgers University. Her research focuses on food systems, social justice, and how the two do, do not, and could overlap. She is the Primary Investigator on the Making It Healthier, Making It Regional project and a Research Specialist for School Food Focus, where she was previously the Research and Data Manager. Amy has an M.A. in Food Studies from New York University and a B.A. in History from Stanford University.

Dr. Christine C. Caruso is a Public Health and Environmental Psychologist, currently serving as Research Fellow for School Food Focus/FoodCorps, where she is contributing to a study investigating values-based procurement practices in school food, funded by the Robert Wood Johnson Foundation. Her research interests focus on food systems, institutional food, social determinants of nutrition, participatory democracy, and health equity. In addition, she currently serves as a member of the Nutrition across the Lifespan working group for the City of Hartford's Food Policy Commission, as well as a steering committee member for Hartford Decide$, the participatory budgeting organization serving Hartford, Connecticut.

Chapter 13

Food Banks as Local Food Champions: How Hunger Relief Agencies Invest in Local and Regional Food Systems

Megan Bucknum*, Deborah Bentzel[†]
*Rowan University, Glassboro, NJ, United States, [†]The Food Trust, Philadelphia, PA, United States

Chapter Outline

Introduction	286
Food Banking 101	286
The Organization of the Food Bank Network	287
Aggregation and Distribution by Food Banks	288
Food Bank Value Chains: Donations and Procurement	289
Donations	290
USDA Foods	291
Food Bank-Led Purchasing	292
Making the Case for Farm-to-Food-Bank: Existing Food Bank Assets	293
Procurement Expertise	293
Strategic Partnerships Across Sectors	294
Infrastructure	294
Stimulating Demand for Fresh Food	295
Policy Support for Farm-to-Food-Bank Efforts	295
Farm-to-Food Bank Examples	296
Methodology	296
Greater Cleveland Food Bank	298
Community Food Bank of Southern Arizona	298
Food Bank of South Jersey	298
Houston Food Bank and Feeding Texas	299
Northern Illinois Food Bank	299
Second Harvest Food Bank of Greater New Orleans and Acadiana	299
Emerging Themes	299
Current Challenges to the Growth of Farm-to-Food-Bank	301
Coordination Around Perishables	301
Price	302
Finding Suppliers	302
Conclusion	302
References	303
Further Reading	304

INTRODUCTION

Food banks in the United States play a pivotal role in closing gaps in food access to the nearly 42.2 million Americans who do not have reliable access to food (Weinfield, et al., 2014). One common misconception about food banks is that they strictly distribute donated food to people experiencing food insecurity. While food banks do connect food insecure individuals to free emergency food (in collaboration with pantries and soup kitchens), they also purchase high volumes of food and coordinate a broader range of programs and activities to serve their target populations.

This chapter examines the role that food banks play in strengthening local and regional food systems, and presents vignettes of food banks serving as regional food system drivers. We define "local and regional food systems" to be the "place-specific clusters of agricultural producers of all kinds – farmers, ranchers, fishers – along with consumers and institutions engaged in producing, processing, distributing and selling foods" (Low et al., 2015). This chapter draws on interviews with nineteen individuals representing food banks and other hunger-related and food policy organizations, as well as secondary research of U.S. food banks to examine how food banks act as key institutions in food systems change. The authors interviewed employees at the Cleveland Food Bank, Community Food Bank of Southern Arizona, Food Bank of South Jersey, Houston Food Bank, Second Harvest Food Bank of Greater New Orleans and Acadiana, and representatives from Feeding America, National Sustainable Agriculture Coalition, and Feeding Texas for this chapter.

We contend that food bank agencies play a vital role in supporting and sustaining vibrant local food systems, and can leverage their purchasing volumes to support regional food security and regional economies. The chapter begins with an overview of how food banks procure foods for distribution to partner agencies and communities at large. Next is an exploration of how food banks are unique with respect to their farm-to-institution capacity, resources, and collaborations. The last section presents selected vignettes to exemplify how food banks are identifying opportunities to innovate in their efforts, particularly in the face of challenges, to use regional food resources to feed regional consumers.

FOOD BANKING 101

As a regional food institution, the goal of a food bank is to ensure that *all* regional consumers can be fed, while also redistributing foods that would otherwise go to waste. According to Feeding America (2017a,b,c), when the first food bank was launched in Arizona in the late 1960s, its founder, John Van Hengel, began gleaning fruits and vegetables for a feeding program. This gleaning effort, that originated in Phoenix neighborhoods that were constructed on former orchard land, expanded into soliciting donations from local grocery stores to be redistributed to smaller food pantries (Poppendieck, 1998). This charity-based approach to hunger emerged as the most viable solution to cuts in federal public assistance that were included in the Omnibus Budget Reconciliation Act of 1981 (Poppendieck, 1998). The creation of America's Second Harvest, the first national food bank network (now called Feeding America), was a response to this shift of federal funding and the resulting increase in food insecurity (Winne, 2008). As these early food banks initially formed to quickly respond to hunger, early procurement efforts targeted diverting potential food waste into food donations. As food banks became more established, emphasis shifted from being a depository for food surpluses into institutions aiming not only to address hunger by providing people with their daily caloric needs, but also by addressing the physical health of individuals and the economic health of the community they serve (Winne, 2008).

This shift in goals increased the amount of fresh produce being procured by food banks, beginning with surplus from produce distributors and then extending directly to farmers and other food producers. Two University of Southern California professors outlined food bank best practices for procuring and distributing produce to member agencies, and they highlighted produce procurement programs among their training materials (Weis, 2011). Food bank procurement continues to provide local producers an outlet for their misshapen, surplus, or otherwise unsaleable goods. Through these acquisitions, food banks are able to provide low-income consumers fresher, minimally processed, local food.

According to Feeding America (2017a,b,c), 13 percent of households throughout the United States experience food insecurity. The power of food banks are their ability to help level the nutritional playing field for children and adults alike, by providing supplemental nutrition to those in need. Adequate access to healthy foods equates to more time spent in school, fewer days lost on the job, and overall better health outcomes (Center for Health Law and Policy Innovation, 2015). The emergency food system's infrastructure can be hidden from plain view—from warehouses in commercial districts to food pantries tucked into churches in low-income neighborhoods and backpack programs in elementary schools serving underserved, food insecure families. The following sections aim to reveal this infrastructure.

THE ORGANIZATION OF THE FOOD BANK NETWORK

For the most part, food banks within the United States are organized through the food banking network, Feeding America. This network of nearly 200 food banks and 60,000 food pantries works together to address hunger and food insecurity in every state, the District of Columbia, and Puerto Rico. Other food banks operate as independent entities, without a fiscal sponsor or other umbrella agency, and they independently create their own guiding policies and programs, procurement strategies, and advocacy efforts.

Food banks receive food from a vast donor network of "national food and grocery manufacturers, retailers, shippers, packers and growers…government agencies and other organizations" who can supply food banks at a scale appropriate to the food bank's service reach and warehouse size (Feeding America, 2017a,b,c). Food banks secure funding for operations and programming, like many non-profit organizations: through private, family, and corporate foundations; through local, state, and federal government grants and contracts; federated funds, charitable giving, and corporate sponsorship.

The two most common governmental programs in which food banks participate are The Emergency Food Assistance Program (TEFAP) and the Commodity Supplemental Food Program (CSFP), both of which are administered by the United States Department of Agriculture (USDA) Food and Nutrition Services (FNS) in conjunction with state governments (USDA, FNS, 2016a). TEFAP is a federal commodity surplus program. Under TEFAP, the USDA purchases food and makes it available to state distributing agencies (e.g., state departments of agriculture, health, or welfare) who then distribute on to food banks (USDA, FNS, 2016b).

Aggregation and Distribution by Food Banks

Regardless of the source of the goods entering into the food bank supply chain, food is typically donated to food banks in pallet, case, and container quantities, but distributed to individuals by the unit. Breaking down quantities into this individual unit highlights a key role food banks play in sorting, repacking, and redistributing foods in the quantities and varieties most applicable to the communities being served on the ground. Food banks distribute to partner agencies (food pantries, cupboards, soup kitchens, and schools), which provide direct services by distributing food directly to individuals (Feeding America, 2017a,b,c). These agencies range in size, populations served, organizational structure, and capacity. Many of these agencies are volunteer run, or do not have the resources to have a dedicated procurement person, so many find that working with a food bank that can organize food and monetary donations from businesses and individuals is a necessity.

Many partner agencies are only open one or a few days a week and may not have access to facilities with adequate cold storage infrastructure (such as refrigerators). Because of these agency constraints, most regionally based emergency food value chain networks are centered around a food bank. This network architecture allows upstream and downstream value chain stakeholders to leverage the food bank's infrastructural economy of scale, improving the overall efficiency and food safety of a food bank's distribution system.

Individual food banks have their own processes for organizing food orders and distribution to their partner agencies that are based upon both the food banks' capacity and the collective needs of the partner agencies. Food banks typically operate as independent non-profit organizations, and therefore have their own internal organizational processes and partner agency policies. Even though food banks may internally operate somewhat differently from one another, the following commonalities of the relationships between food banks and their partner agencies can be used to provide a high-level overview of food bank operations:

- *Coordinated ordering.* In addition to food donations, food banks also purchase food to diversify and enhance the nutritive value of their offerings. Partner agencies, such as food pantries and emergency meal sites, typically work with the food bank to order the foods that will best serve the communities they reach. For example, food pantries may work with their regional food bank to help ensure that culturally appropriate foods are available for specific communities. How partner agencies order food and provide feedback on food options varies; however, pantries receive an array of food items available at any one time. Because the food bank purchases this food, as opposed to accepting a donation, partner agencies are charged by food banks the cost of the food plus a nominal mark-up for the food bank's role in procurement and storage. Depending on the items chosen by the partner agency, and the food bank's funding streams, a food bank may or may not charge a partner agency a fee for the food.
- *Shared maintenance fees.* The most typical fee charged by food banks of partner agencies is a "shared maintenance fee." Food banks within the Feeding America network have the autonomy to determine what fee scale to use for partner agency food orders, but are not permitted to charge more than the Feeding America network cap of $0.19 per pound (Long Island Cares, Inc., n.d.). This fee is based upon the number of pounds of food that a partner agency orders from the food bank. Shared maintenance fees help support the cost of the food bank infrastructure maintenance such as coolers, pallet jacks, loading docks, and other warehouse essentials.
- *Delivery.* Many food banks offer delivery to partner agencies if a minimum quantity or order size is obtained. Arrangements pertaining to delivery vary greatly between food banks.

FOOD BANK VALUE CHAINS: DONATIONS AND PROCUREMENT

Food banks are large-scale purchasers of food, placing them in the same institutional food procurement category as hospitals, colleges and universities, K-12 schools, and correctional facilities. This institutional nature of food bank procurement makes it essential to take a critical look at the way that food moves across this specific value chain, the set of activities from inputs to producers to processors and distributors. Any changes or adjustments to a procurement strategy must take into account the existing suppliers and volume needs of the organization. Like other institutions, food banks purchase at a scale with the potential to support local and regional food systems and economies. They also have opportunities to procure healthier and more sustainable food to help address health problems such as obesity, diabetes, malnutrition, and exposure to toxins that disproportionately affect low-income communities and communities of color. The next sections dissect how food moves from the farm to the food bank, both directly and more often, via several intermediaries.

Food banks across the country typically stock their inventory using three methods: by accepting donations, through federally funded programs (USDA Foods via TEFAP and CSFP), and through food bank-led purchasing. Opportunities for food banks to procure locally and regionally grown foods, as well as locally manufactured goods, exist in each of these scenarios.

Donations

Donated foods from food processors, manufacturers, retailers, and shippers comprise a significant amount of the goods food banks receive and turn to distribute. Food donations made to food banks may be made directly by a food manufacturer or producer (such as a bread baker or citrus grower) or through grocery stores and other retailers as they off-load excess goods. The type of donated food varies greatly and can include all types of food. Individual food banks set the specific requirements about the types of food donations they will accept; however, many have stopped accepting high sugar food and drink donations. In 2015, Feeding America published a Healthy Food Donation List to help set guidelines for individual and corporate food donations (Feeding America, 2015). Some food banks have hired nutritionists that evaluate the nutritional value of food donations for the purposes of determining whether to accept the donation (The Reinvestment Fund, 2016).

Transportation of these foods from grocery stores may be provided by the retailer, or may be coordinated by the food bank. Typically, all forms of food are accepted for donation—frozen, dried, shelf-stable, or fresh—and represent all parts of the plate, from proteins to fluid milk, produce, and grains. Feeding America member food banks receive donation support to streamline deliveries, and to support inventory control and supply needs. Feeding America coordinates directly with food manufacturers, processors, packers, and shippers to organize donations on behalf of individual food banks and in alignment with individual food bank procurement goals (e.g., an individual food bank may choose to focus on increasing offerings of foods that are culturally appropriate for Latino communities).

Individual donations also play an important role in food banks' ability to obtain enough foods for distribution to those in need (Feeding America, 2017a,b,c). Food drives provide a means by which individual donations are aggregated. Once these mixed deliveries reach the food bank warehouses, they are sorted into food categories (soups, pasta and other grains, canned fruits and vegetables, etc.) and inventoried for future distribution.

Locally and regionally grown foods can enter the food bank system through donations. For example, food banks procure locally grown foods via "gleaning" programs, which refers to the harvest of fruits and vegetables that would have otherwise been left in the field due to either surplus or cosmetic irregularity (New Jersey Agricultural Society, n.d.). Many farmers allow volunteer groups, organized by a local food bank, to glean their fields, to save the farmers labor costs otherwise associated with harvest and packing. Alternatively, farmers with surplus may simply donate part of their harvest that was not sold through a for-profit distribution channel.

Gleaning is a widespread procurement practice, with many programs organized at the state level. California is a leader in total gleaned produce volume, with 127 million pounds of fresh produce being distributed throughout the state's food banks by the California Association of Food Banks' (CAFB) Farm to Family program in 2012 (California Association of Food Banks, 2012). CAFB worked with California producers (and producer associations) to facilitate donations, as well as payments for the food and necessary packaging, to divert and capture this food to help feed people throughout the state. According to Vitiello et al. (2015): "Similar gleaning programs operate in Arizona (18.5 million pounds in 2011), Texas (13 million), Ohio (26 million, though much of this is from wholesalers), and smaller programs in Arkansas (1.2 million), Colorado (1 million), and Kentucky (almost 1 million)." These volumes serve as evidence of the capacity, interest, and ability for food banks to support local and regional food in respect to gleaning and donations.

USDA Foods

TEFAP acts as a supplemental nutrition program to support the diets of low-income and elderly at-risk individuals (USDA, FNS, 2016b). The flow of food through TEFAP starts at the federal level, whereby USDA and the Farm Service Bureau purchases domestic agricultural products that become known as "USDA Foods," (formerly known as "commodity foods," and sometimes referred to as, "entitlement foods") to distribute to states at a volume determined by need. In addition to food, state distributing agencies also receive funds to support storage, transportation, and administrative expenses.

The type of state agency acting as the distributor of USDA foods vary widely across the country, and include, but are not limited to: state departments of agriculture, education, health, human and social services, and sovereign native and tribal communities through the Food Distribution Program on Indian Reservations (FDPIR.) On average, about 20 percent of foods flowing through a local or regional food bank originate through TEFAP, and each year, Congress appropriates about $350 million in TEFAP funds for spending on domestically grown foods, and another $50 million to support administrative functions of the program (Cabili et al., 2013).

The Commodity Supplemental Food Program (CSFP) provides food packages to low-income individuals aged 60 or older, through the distribution of USDA Foods to state agencies, who may, from there, distribute foods to delivery to the homes of qualifying elderly individuals (U.S. Department of Agriculture, Food and Nutrition Services, 2016c). Food banks frequently play a role with the CSFP as they receive deliveries and repack USDA Foods into specific food kits, and further distribute these packages to partner agencies on the local level.

USDA Foods must be domestically grown, and these foods are procured by this federal agency through open, competitive Invitation for Bid (IFB) processes or through Requests for Purchase (RFP). For example, USDA Agricultural Marketing Service may publish an IFB or RFP if it seeks to purchase blueberries for distribution through one of USDA's nutrition programs. This bid would include information pertaining to product specifications, packing, distribution, and other service requirements. A blueberry farmer who responds to this solicitation with the best proposal and pricing would win the bid, a contract would be established, and the blueberries would enter the USDA Foods system.

USDA Foods include meats, poultry, fish, eggs, dairy products, grains, oils, fruits, vegetables, and other food items. Depending upon a food bank's location in the country, the facility may already be receiving locally or regionally grown foods through TEFAP or CSFP. USDA Foods provide those they serve with state of origin on all products. Over the past ten years, TEFAP has refined its portfolio to include a wider variety of lower-sugar, low-sodium, and whole grain foods, as well as increased variety and quality of fresh fruits and vegetables.

Food Bank-Led Purchasing

In addition to receiving USDA Foods and donations, food banks also purchase foods for distribution to partner agencies and for use in meal and other food assistance programs. Funds for purchasing foods may be acquired through grants and donations, or through state programs designed to incentivize food banks to purchase surplus agricultural products. Food banks sometimes issue a competitive bidding process to procure foods of all kinds, frequently securing at least three bids, as is often required by government regulations, before choosing a vendor.

With respect to procurement of locally or regionally grown foods, food banks often work closely with farmers to plan for needs for the coming year. This can be done through formal forward contracts, which are agreements between an agricultural producer and a buyer that are made prior to planting and that establish the volume and type of crop the buyer intends to purchase from the producer (as well as other specifications such as pricing, packing, and transportation details). However, our food bank informants did not report that they used forward contracts frequently, potentially due to the philanthropic nature of their business model, making it difficult to predict their budget. Instead, the food bank informants preferred establishing long-term informal relationships with their growers to help ensure that supply, demand, and pricing expectations are aligned.

MAKING THE CASE FOR FARM-TO-FOOD-BANK: EXISTING FOOD BANK ASSETS

[Produce industry growers] look at food banks as simply wanting a hand out. Not necessarily—we can also become a customer.
                                    Don Harris, Former Director, Produce for Feeding America

In 2016, Feeding America spent roughly $2.4 billion on food procurement for their member network, demonstrating how food banks are not merely passive recipients of aid, but rather active participants in agro-food value chains (Feeding America, 2017a,b,c). The combined purchasing power of both Feeding America and the individual buying power of food banks represents a large market opportunity for regional agricultural producers. It also has potential to help re-direct food that otherwise would be considered postharvest "waste" to provide more nutritious food options to people who cannot easily access them. Rather than only focusing on donations, there are a number of compelling reasons why food banks are already positioning themselves to be major players in local and regional food systems in meaningful and profitable ways.

Food banks can serve a key role in supporting the expansion of regional food systems by contributing infrastructure, procurement and logistical expertise, innovative partnerships, and stimulating demand for fresh and healthier food. The following section outlines how existing food bank operational practices provide an ideal foundation for a "farm-to-food-bank" program: a procurement strategy for food banks interested in purchasing food from regional agricultural producers.

Procurement Expertise

Procurement, a process by which there is research into vendors and supply needs, solicitation, contracting, and review of successes and challenges, is a set of activities at which food banks are experts. In 2016, approximately 900 million pounds of produce were aggregated and distributed across the country to local hunger relief agencies through food banks (Feeding America, 2017a,b,c). Like many other cost-sensitive institutions serving vulnerable populations, food banks also know how to balance their nutritional requirements with their budgets by working with a wide variety of vendors, including farmers, retailers, and wholesalers. Additionally, many food banks may already have relationships with regional producers through both gleaning, and other donation programs, as well as direct purchasing. Food banks' ability to communicate their supply needs, to build and maintain relationship with local farmers, food manufacturers, and other suppliers supports the argument that food banks are a vital and growing part of the Farm to Institution movement.

Strategic Partnerships Across Sectors

Food banks must have responsive operations and innovative strategies to increase their reach to those facing food insecurity and hunger. Whether coordinating with local organizations to provide on-the-ground programming around nutrition education, collaborating with state agencies for large-scale procurement efforts, or working with a community-based agency for enhanced outreach to those in need, food banks demonstrate how leveraging partnerships can expand impact and reach. The success of a food bank relies on their ability to forge successful partnerships to help with the needs of their organization—such as finding individual and corporate donors (for food and money) and attracting volunteer labor—as well as building community awareness of the services and programs they offer to those in need. Because food banks incorporate partnership building in all aspects of their work, many food banks already possess the collaborative skills needed to create and manage a farm-to-institution program. These programs are collaborative in nature, as they aim to leverage the economic benefits of geographically focused purchasing to address regional food security. When successful, farm-to-institution programs are able to simultaneously achieve upstream and downstream supply chain benefits.

For example, the Director of Food Procurement at one food bank shared in an interview that their food bank strived to create a donor network that is regional in focus. This creation of a regional network of donors and clients allows for regional stakeholders to participate in solving a regional problem, such as food insecurity. A network need not only include regional growers of food, but also regional manufacturers of food and regional employers. This regionally focused approach demonstrates how a food bank can address the underlying issues of hunger, such as the economic health of a region, while still addressing food security directly.

Infrastructure

Food banks are positioned to engage in farm-to-institution practices because their infrastructure supports the aggregation, storage, and distribution of a wide array of foods, including perishable items. Food banks' infrastructure typically includes dry and cold storage, loading docks, fleet of trucks and/or vans, and other equipment and software for managing inventory and shipments. As such, food banks generally have the infrastructure at hand to support farm-to-institution procurement strategies in a manner that both upholds food safety and prolongs the shelf-life of foods.

Likewise, transportation is a costly endeavor when moving food through all aspects of a supply chain. Food banks have extensive experience with creating systems for moving food, and supporting their vendors and partners to deliver and distribute foods through cost-efficient means. With fleets of trucks and vans, precise delivery routes, plans for back-hauling (loading items to transport back to its point of origin to avoid wasted resources on an empty truck), and other logistics knowledge, food banks are primed to play a key role in aggregating and distributing locally and regionally grown foods from the farm, to the food bank, and to partner agencies. Exemplifying the benefit of existing infrastructure, the Community Food Bank of Southern Arizona utilized its warehouse and transportation to create a farm-to-institution program in their region. In this program, the food bank has facilitated relationships and transactions between producers and institutions within their region, acting as a broker and distributor. Motivated by the notion that improving the economic well-being of agricultural producers will help improve the overall regional economy, the food bank found a way to use their existing resources to expand market opportunities for the producers, while continuing to work toward their central goal of improving food accessibility for consumers in their region.

Stimulating Demand for Fresh Food

Food banks have been driving the demand for healthy and fresh foods through education efforts on the local level. Through cooking classes and nutrition education opportunities, and by working to procure culturally appropriate foods, food banks have been building the demand for fresh food. According to Feeding America, by 2025, the demand of meals for hungry families will reach about eight million, doubling from its current rate of four million meals per year. One informant at Feeding America estimated that nearly two billion pounds of fruits and vegetables will be needed in emergency food channels to meet that demand by 2025. To be able to fulfill growing food needs, food banks will need to find a way to procure more healthy and fresh foods that meet both favorable nutritional guidelines and their budgets.

POLICY SUPPORT FOR FARM-TO-FOOD-BANK EFFORTS

Farm-to-food-bank support exists in several states, facilitating procurement practices that bring regional farm items to local food banks. Many states have invested in farm-to-food-bank support, either through monetary incentives, or by providing administrative support and resources to coordinate state-level efforts. For example, some states provide payments to farmers in the form of tax credits or reimbursements to incentivize local and regional producers to donate or sell to their local food bank. "Farmers Feeding Florida," is a program funded through state mechanisms that helps farmers cover transportation costs for donations to food banks (Feeding Florida, n.d.). Another common state-level strategy is to provide farmers with tax credits for donating or selling to local or regional food banks to help cover harvest, packing, and transportation costs. Finally, according to one informant, a handful of states (e.g., California, Pennsylvania, Texas, Ohio, and others) have funded and coordinated state agricultural surplus systems that distribute funds to food banks to purchase state-grown products that are in surplus quantities. Whether individual states offer any type of farmer support within their boundaries may be based on political will, resources, and funds available, and the need to incentivize local and regional food purchasing.

While there are no specific programs that address farm-to-food-bank incentives or guidelines on the federal level, discussions are taking place about how local food procurement could be supported through legislation such as the Farm Bill (King, 2017, personal communication, 22 March). Tax credit programs are generally considered to have bi-partisan appeal, and while there is nothing proposed for the 2018 Farm Bill, it is possible that draft language will be introduced to support farm to food bank programs. In addition, there are opportunities for the USDA's procurement programs to implement more specifications to prioritize the purchase of regional and healthier food when it is available.

FARM-TO-FOOD BANK EXAMPLES

The vignettes included here are intended to provide illustrative examples of how food banks, as institutional food purchasers, can tap into local and regional food supplies to the benefit of local farmers and local consumers. These brief examples reveal that food banks often take their own approach to regional food procurement, based on their existing infrastructure, geography, agricultural landscape, and the unique needs of the communities they serve. Almost every interviewer pointed out that their regional purchasing efforts are a result of place-based strategic partnerships. These food banks collaborated with regional partners to access regionally available resources to help support new endeavors and then tailored programs around the identified assets found within their community.

Methodology

The examples provided in the following vignettes were assembled through the collection and analysis of secondary data and interviews. The authors conducted interviews between March and April 2017 with nine procurement-focused employees at six food banks, as well as interviews with ten representatives from other hunger-related and food policy organizations, including Feeding America, National Sustainable Agriculture Coalition, and Feeding Texas. Interview candidates were identified by secondary research of food banks pursuing a regional procurement strategy. All interviews were administered over the phone, except for one that was conducted in person. The interview protocol focused on questions about food bank infrastructure, breadth and depth of services and programming, procurement strategies currently in use, including if and how food banks were able to source regionally grown foods and how partnerships contribute to their success with respect to healthy food procurement. Table 1 highlights information gleaned from the food banks interviews and provides an overview of their current regional food procurement efforts.

TABLE 1 Farm-to-Food-Bank Interview Findings

Food Bank	Region	Pounds of Food Distributed (2016 Estimate)	Infrastructure	Current Regional Food Procurement Efforts
Cleveland Food Bank	Six counties clustered near Cleveland metro area in Ohio	44,000,000	128,000 ft²	Beneficiary of the Ohio Association of Food Banks' purchases of surplus fruits and vegetables from Ohio producers, which is supported through state-funded Ohio Agricultural Clearance Program
Community Food Bank of Southern Arizona	Five counties within Southern Arizona	27,000,000	97,932 ft²	Created a Farm to Institution program to support local economic activity by enabling new sales outlets for local farm products
Food Bank of South Jersey	Four counties in Southern New Jersey	11,000,000	65,000 ft²	Leverages agricultural funding from the state to purchase produce from regional producers. Additionally, collaborating with a peach association and the company Campbell's to produce a peach salsa, whose proceeds go to the food bank
Houston Food Bank	Eighteen counties in Texas within a 100 mile radius of the food bank	98,990,393	308,000 ft²	Working with Feeding Texas and individual farms to secure food produced within Texas. Currently 89 percent of its food is produced in the state of Texas
Northern Illinois Food Bank	Thirteen counties within Northern Illinois		140,000 ft²	Collaborating with three regional producers that have planted fields dedicated to the food bank, which are harvested by food bank volunteer labor
Second Harvest Food Bank of Greater New Orleans and Acadiana	23 parishes in Southern Louisiana	31,000,000	100,000 ft²	Purchases off-grade produce from regional producers

Greater Cleveland Food Bank

By utilizing state funding, the Greater Cleveland Food Bank has been able to buy roughly eight million pounds of Ohio grown produce each year, through a collaboration with Ohio Association of Food Banks (OAFB). In 1999, OAFB created the state program for local purchasing to help subsidize Ohio-based food banks purchase from Ohio-based farmers, called the Ohio Agricultural Clearance Program. To maximize this funding opportunity, the Greater Cleveland Food Bank has been working with agricultural producers in the beginning of the growing season to coordinate product types and quantities that it would like to purchase. Although not a formal forward contract, this coordination is beneficial to the business planning undertaken by the food bank and farmer, as prices can be stabilized without being vulnerable to negative market forces.

Community Food Bank of Southern Arizona

The Community Food Bank of Southern Arizona not only purchases food from regional agricultural producers, but also, according to staff, it helps facilitate relationships and transactions for these producers with other regional institutions through its Farm to Institution program. The food bank has utilized its assets, such as its warehouse and its fleet of trucks, to distribute regionally produced food to other regional institutions (such as hospitals and schools), which enables it to assist in increasing the amount healthy fresh foods available in its region, while supporting regional agricultural businesses. While the total volume of product purchased by the food bank from regional producers is less than $10,000 a year, the organization places value on the facilitation of relationships that are developed through the transactions between producers and other institutions. In addition to acting like a local food distributor, the Community Food Bank of Southern Arizona also offers several gardening opportunities to its clients, empowering them to be directly involved in their food security.

Food Bank of South Jersey

Serving four counties outside of the Philadelphia Metropolitan area, the Food Bank of South Jersey leverages state dollars from the State of New Jersey's State Agricultural Development Committee's State Food Purchasing Program to purchase fresh foods that have been produced regionally within the state. The food bank has also been able to work with many agricultural producers, and producer organizations, to glean products that do not meet market standards. An exemplar of this strategy is its relationship with both a large peach packhouse and Campbell's Corporation in Camden, New Jersey, which supplies and processes the ingredients for a peach salsa that is marketed throughout the state as a fundraiser for the food bank.

Houston Food Bank and Feeding Texas

Through a partnership with Feeding Texas (the state-wide food bank association of Texas), the Houston Food Bank was able to procure 83 percent of its annual produce from Texas farmers in 2016, the equivalent of 31,800,311 pounds. By streamlining procurement at the state level to serve all of the food banks under its purview, Feeding Texas is able to work with larger farmers and capture better prices, given the volume at which it purchases. The crux of this coordination is having access to both the produce rich region of South Texas (including produce imported from Mexico) and a central facility in Edinburg, Texas, where full tractor trailer shipments can be delivered and "mixed" according to the orders of food banks throughout the state. This allows food banks to receive tractor trailer shipments of multiple products, not just one full trailer of one product, which would be more difficult for the individual food bank to distribute throughout its network.

Northern Illinois Food Bank

This food bank's regional procurement efforts emphasize working with willing producers as part of a dedicated planting strategy. Currently, three producers have planted their fields with food specifically destined for the food bank. When it comes time to harvest, the food bank arranges for volunteers to harvest the crops. The food bank mentioned that the opportunity for harvesting is beneficial to the diversity of volunteer experiences that that food bank provides. The vast volunteer network that Northern Illinois Food Bank has allows for this arrangement to be feasible, both from a harvesting perspective, as well as sorting and repacking efforts that need to be completed back at their warehouse.

Second Harvest Food Bank of Greater New Orleans and Acadiana

By tapping into the resources of this region, the Second Harvest Food Bank of Greater New Orleans and Acadiana has been able to forge relationships with the many food distribution centers, manufacturers, and producers located in its catchment area. These connections include many regional farmers who view the food bank as a customer to whom they can sell their excess inventory and off-grade products. The food bank pays around $0.10 a pound for these items, and offers pick up and boxes to offset the farmer's cost of production given the below market rate that they can pay. According to informants from the food bank, this arrangement has proven to be mutually beneficial to both the food bank and the farmer, as it is a tactical way to address a hunger problem with an inventory surplus issue.

EMERGING THEMES

Although each food bank is developing its own specific regional food procurement program that is appropriate for its operation and clientele, several themes around policy and supply chain infrastructure emerged from these highlighted food bank efforts. First, several food banks are leveraging partnerships with regional producers to organize regional food hubs by tapping into one of their greatest intrinsic assets: their infrastructure. A food hub, as defined by the USDA, "is a centrally located facility with a business management structure facilitating the aggregation, storage, processing, distribution, and/or marketing of locally/regionally produced food products" (Barham, 2010). For example, the Food Bank of Northeast Georgia has been working with the producers in their area to organize a food hub using excess warehouse capacity. Not only will this wholesale and retail facing aggregation and distribution entity strengthen the regional agricultural economy, it also can create the stage for additional entrepreneurial activities. Food Link in New York has a similar approach, becoming the first food bank to become a food hub. They did this to address excess capacity in their operation and to leverage opportunities for regional economic development. Food Link's food hub operation aggregates food from regional producers, wholesalers and retailers and distributes roughly sixteen million pounds of food annually to about 200,000 people in their catchment area (Kako, 2014). Similarly, the Houston Food Bank's large cooler space (29,400 square feet) allows for large-scale procurement of highly perishable goods, including Texas-grown fruits and vegetables. This infrastructure is leveraged on behalf of Feeding Texas-affiliated food banks, as the Houston site procures these perishable items on behalf of Feeding Texas members. Without such ample cooler space, this type of cooperative purchasing would not be possible.

Second, an increasing number of states throughout the country have begun to support local and regional food procurement incentives for farmers and food banks to connect a variety of funding mechanisms. These state-funded programs support economic development, public health, and hunger initiatives and build development. Food banks, and other institutions, can leverage these funding streams to help assist with their regional procurement strategies. Leveraging funds from multiple sources can help food banks increase their purchasing power and can enable them to pay market value (or closer to) for products produced within their state. Examples of these include the Ohio Agricultural Clearance Program.

Third, food banks are partnering with local producers and aggregators around "off-grade produce," which are fruit and vegetable items that may be considered "ugly," "cosmetically imperfect," "commercial grade," or "seconds." These are items that will not be sold to the retail market because they may be the wrong size or shape, or there may be a surplus. Food banks can purchase these products for below the market rate of retail-ready product. Food banks can act as distribution centers for off-grade products, creating an outlet for agricultural producers and a supply of fresh foods that can be sold slightly below market rate. Because of standards and grades set by the market and overseen through governmental regulation, food that does not meet these standards is typically considered "off-grade," non-retail ready, or cosmetically imperfect, and is either discarded or sold for a significant discount. The scale of food bank purchasing can often make these institutions a compatible outlet, which is not only beneficial to the food producer and the food bank, but can also decrease food waste.

Following are several examples from our interviews that highlight where food banks have successfully purchased off-grade food items from regional producers:

- *Second Harvest Food Bank of Greater New Orleans and Acadiana.* This food bank is currently buying sweet potatoes that are too small for the ConAgra packing house to which the regional producer sells the majority of his crop. The food bank purchases sweet potatoes that are under an inch-and-a-half for a discounted price because these products would otherwise be discarded.
- *Northern Illinois Food Bank.* A recently opened chicken slaughter facility has opened in Illinois to specifically harvest retired hens from the egg industry. Typically discarded on the farm, these birds represent an affordable source of protein for the food bank to purchase, when the birds are harvested in a certified facility.
- *Kentucky Catfish Study.* A 2007 USDA report entitled "Delivering the Goods: Lessons Learned from Direct Delivery of Kentucky Catfish" describes how an agricultural cooperative that was marketing catfish worked with regional food banks in Kentucky to sell excess inventory and off-grade products at a discounted price, allowing the cooperative to recover the cost of storage and/or having to take a complete financial loss.

CURRENT CHALLENGES TO THE GROWTH OF FARM-TO-FOOD-BANK

As this chapter suggests, many food banks across the country are attempting to create regionally based supply chains that integrate more regionally produced fresh foods into their product offering. However, there are several challenges that farm-to-food-bank initiatives in the United States must consider before growing in scope and efficacy.

Coordination Around Perishables

Sourcing and distributing fresh food also requires significant investments in infrastructure (refrigerated trucks and storage facilities), as well as new ways to get these items into the hands of clients before they spoil (The Reinvestment Fund, 2016). As food banks continue to increase the percentage of fresh product within their inventory, food banks will need to coordinate clearly with partner agencies to be able to adapt their current customer distribution setup to accommodate more perishable products. Additionally, food banks must also explicitly train their partner agencies in the correct food handling processes and operational standards for handling fresh and perishable foods.

Price

Even with additional government and foundational funding streams, food banks are challenged to compete with other institutions such as hospitals and universities on wholesale food pricing. Highlighting the benefits of working with a food bank—inventory management and ability to accept off-grade products—could be successful marketing messages that food banks can use when reaching out to potential regional suppliers.

Finding Suppliers

The linchpin to any successful farm-to-bank program is finding farmers that are interested in and capable of selling to and working with these institutions that require high volumes of food. The success stories highlighted in this chapter address the need for food banks interested in starting a farm-to-food-bank program to use their assets to forge mutually beneficial relationships with producers, processors, and distributors in their region who share values with the coordinating food bank; namely of food security and reducing food waste. Food Banks could consider reaching out to their local Cooperative Extension office, regional food hubs or other food aggregators, and/or any local food advocacy non-profit organizations that exist in their region to find suppliers. In addition, they could consider partnering with other institutions in the region that are actively building regional foodservice purchasing infrastructure, such as universities, hospitals, or large school systems.

CONCLUSION

In conclusion, while challenges, food banks are well positioned to leverage their assets to address food security within their region while providing fresh and regional food to their consumers and purchasing food regionally from producers. As large-scale purchasers of many types of foods, institutions such as food banks have the potential to support local and regional food economies through their procurement tactics. Through partnerships with farmers and food producers, collaborative food purchasing models and utilization of off-grade produce and other strategies, farm to food bank efforts are taking root across the country. Pricing and availability of regionally produced foods may prove challenging, given the large volumes in which food banks purchase goods. In other words, a challenge to initiating a farm to food bank effort may be a mismatch in economies of scale between the buyer (food bank) and the supplier (a farmer or producer). For food banks interested in expanding their local food offerings, understanding the demand for fresh, regional foods from partner agencies would be useful, as would working with a local partner or state agency to understand more about supply available for donation or purchase.

This chapter highlights existing practices of food banks acting as influential institutional purchasers to enhance regional food systems, and it is the authors' hope that these examples will inspire other food banks to establish new farm to food bank programs, as well as encourage regional food advocates and practitioners to view food banks as integral parts of a food supply chain. Consumers of all income levels deserve equitable access to healthy, affordable, culturally appropriate, and regionally sourced foods. Now more than ever, food banks are developing innovating models, partnerships, and practices to link the hungriest in our country to fresh, delicious foods produced by local agricultural producers. As institutions with significant purchasing power, food banks have the power to influence and disrupt conventional supply chains by infusing value into new ways of working with food producers so that farmers win, communities win, and the complicated supply chain players stretching from the farm to food bank win as well.

REFERENCES

Barham, J., 2010. Getting to scale with regional food hubs. Available from: https://www.usda.gov/media/blog/2010/12/14/getting-scale-regional-food-hubs. (Accessed April 28, 2017).

Cabili, C., Eslami, E., Briefel, R., 2013. White paper on The Emergency Food Assistance Program (TEFAP). U.S. Department of Agriculture, Alexandria, VA.

California Association of Food Banks, 2012. Farm to Family Out the Door: A Food Banks' Guide to Produce Distribution in California. California Association of Food Banks, Oakland, CA.

Center for Health Law and Policy Innovation of Harvard Law School, Feeding America, 2015. Food banks as partners in health promotion: creating connections for client and community health report. Available from: Center for Health Law and Policy Innovation. July 2015.

Feeding America, 2015. Healthy food donation list. Available from: http://hungerandhealth.feedingamerica.org/wp-content/uploads/legacy/mp/files/tool_and_resources/files/healthy-food-donation-list.pdf.

Feeding America, 2017a. Nourishing healthy futures: 2016 annual report. Available from: Feeding America.

Feeding America, 2017b. Our history. Available from: http://www.feedingamerica.org/about-us/about-feeding-america/our-history/.

Feeding America, 2017c. Food bank network. Available from: http://www.feedingamerica.org/about-us/how-we-work/food-bank-network/.

Feeding Florida, n.d. Farmers feeding florida. Available from: <https://www.feedingflorida.org/programs/farmers-feeding-florida>. (Accessed 21 April 2017).

Kako, N., 2014. Fighting against hunger, a New York food hub helps food banks across the state. Available from: http://www.csmonitor.com/Business/The-Bite/2014/0707/Fighting-against-hunger-a-New-York-food-hub-helps-food-banks-across-the-state. (Accessed April 28, 2017).

King, W., 2017. Policy Specialist [Interview] (22 March 2017).

Long Island Cares, Inc., n.d. Why shared maintenance fees for food banks? [Online] Available from: <http://www.licares.com/staff/financial_legal/Shared_Maintenance_Fees.pdf>. (Accessed 25 April 2017).

Low, S.A., et al., January 2015. Trends in U.S. Local and Regional Food Systems. US Department of Agriculture Economic Research Service, Washington, D.C.

New Jersey Agricultural Society, n.d. Food insecurity and food waste. Available from: <http://www.njagsociety.org/food-insecurity-and-food-waste>. (Accessed 22 April 2017).

Poppendieck, J., 1998. Sweet Charity: Emergency Food and the End of Entitlement. Penguin Putnam, New York.

The Reinvestment Fund, 2016. Feeding the line, or ending the line? Innovations among food banks in the United States report. Available from: The Reinvestment Fund [2016].

U.S. Department of Agriculture, Food and Nutrition Services, 2016a. Programs and Services. Available from: https://www.fns.usda.gov/programs-and-services.

U.S. Department of Agriculture, Food and Nutrition Services, 2016b. Nutritional Program Fact Sheet: The Emergency Food Assistance Program. U.S. Department of Agriculture, Food and Nutrition Services, Washington, DC.

U.S. Department of Agriculture, Food and Nutrition Services, 2016c. Nutrition Program Fact Sheet: Commodity Supplemental Food Program. U.S. Department of Agriculture, Food and Nutrition Services, Washington, DC.

Vitiello, D., Grisso, J.A., Whiteside, K.L., Fischman, R., 2015. From commodity surplus to food justice: food banks and local agriculture in the United States. Agric. Human Values 32 (3), 419–430.

Weinfield, N.S., et al., 2014. Hunger in America 2014: National Report. Westat and the Urban Institute, Washington, DC.

Weis, E., 2011. More Food Banks Offer Fresh Fruits, Vegetables. USA TODAY. 31 January. Available from: http://www.usatoday.com.

Winne, M., 2008. Closing the food gap: resetting the table in the land of plenty. Beacon Press, Boston.

FURTHER READING

Berney, G., Tropp, D., Clifton, K., McKenna, L., 2007. Delivering the Goods: Lessons Learned From Direct Delivery of Kentucky Catfish. U.S. Department of Agriculture, Agricultural Marketing Service, Washington, DC.

Campuzano, J. & Cole, K., 2017. Director of Supply Chain Procurement and Supply Chain Manager, Feeding Texas [Interview] (13 April 2017).

Eilmann, P., 2017. VP of Food Resources and Product Development, Greater Cleveland Food Bank [Interview] (31 March 2017).

Ericson, S., 2017. Director of Food Procurement, Northern Illinois Food Bank [Interview] (11 April 2017).

Food Bank of Northeast Georgia, n.d. Our history. Available from: <http://www.foodbanknega.org/about>. (Accessed 28 April 2017).

Harris, D., 2017. Director of Produce for Feeding America [Interview] (31 March 2017).

Houston Food Bank, n.d. National school lunch program distributor. Available from: <http://www.houstonfoodbank.org/programs/nslp/>. (Accessed 28 April 2017).

Jones, K., 2017. Program Manager, Local Food Pathways at Community Food Bank of Southern Arizona [Interview] (28 April 2017).

Macpherson, A. & McClay, J., 2017. Senior Manager of Strategic Operations Projects and Senior Manager, Sourcing & Procurement at Food Bank of South Jersey [Interview] (8 March 2017).

Maskill, M. & Massad, M., 2014. Associate Director of Operations and Traning and Communications Coordinator at the Houston Food Bank [Interview] (4 April 2014).

Nuismer, M., 2017. Food Sourcing Specialist, Second Harvest Food Bank of New Orleans [Interview] (10 April 2017).

Megan Bucknum is ¾ Faculty in the Department of Geography, Planning and Sustainability at Rowan University, in addition to being an independent consultant. As a food systems planner, she has experience facilitating public meetings, conducting interviews, designing participatory research projects, and co-authoring several practitioner focused publications within the field. Motivated by the power of narrative, her research projects use interviews and oral histories to explore barriers and opportunities along our food supply chains and to understand how people connect to the built and natural environment. Megan has a Master of Urban and Environmental Planning from the University of Virginia.

Deb Bentzel serves as The Food Trust's Associate Director for Community Food Systems, and leads the agency's regional food systems initiatives, including farm-to-institution and farmers' market programming. She provides training and technical assistance to schools and childcare centers, suppliers and farmers, and other groups interested in implementing local foods initiatives intended to address equitable access to nutritious foods and food education. Deb earned her Master of Public Health degree in 2003 from Boston University and prior to joining The Food Trust, managed research projects and community programs at Fair Food Philly, Harvard University, and the Veterans Administration.

Chapter 14

Plant Proteins Move to Center-Plate at Colleges and Universities

Kristie Middleton, Elise Littler
Farm Animal Protection, The Humane Society of the United States, Washington, DC, United States

Chapter Outline

Introduction	307	Foodservice Management	
Understanding the Plant-Based		Companies	314
Food Trend	**309**	**Common Themes**	**317**
Animal Welfare	309	The Role of Non-Profit	
Health	309	Intermediaries and Peer Networks	317
The Environment	310	Emergence of New Meatless	
Cost Savings	311	Product Lines Among Private	
Case Studies: College and		Manufacturers	319
University Foodservice	**312**	**Conclusion**	**321**
Self-Operating Foodservice		**References**	**321**
Facilities	312	**Further Reading**	**324**

INTRODUCTION

In 2011, the University of North Texas (UNT) did what no other university had done before: The 36,000-student university opened the nation's first all-vegan dining hall. Dining officials at the school, 30 miles north of the famous Fort Worth Stockyards—a national livestock hub—had been hearing from students that they wanted more vegan options. Dining staff took a dining hall that was struggling to sell many meals, revamped the menu, and redesigned the space, creating the Mean Greens dining hall.

Students responded with enthusiasm: Use of the dining hall steadily rose from 175 transactions per day, to more than 1700. The school saw a 35 percent increase in voluntary meal plans, and the dining hall became a favorite of faculty and staff. The university even started using Mean Greens as a retention and recruitment tool (Botts, 2017).[1] The Director of Special Projects on campus at the time, Ken Botts, observed, "People were voting with their dollars, and their votes indicated [that] they really wanted to support the plant-based options."

Although Mean Greens was the first all-vegan university dining hall, plant-based food has become trendy in popular culture on a national scale.[2] For example, vegetarian comfort food reached first place in *Forbes* magazine's "5 Top Food Trends You'll See in 2017" (Huen, 2016). In its 2017 food trend predictions, *People* magazine predicted, "The habit of cutting back on meat will continue to rise in popularity" (Lillien, 2017). The reasons that consumers frequently cite for switching to plant-based diets are that it is better for the environment, healthier, and cheaper. For institutions as well, reducing the amount of meat served can help reduce environmental impacts, promote better health, and enable the purchasing of higher quality foods using the money saved by purchasing less meat (Tanyeri, 2016a).

This chapter examines the movement toward plant-based diets among university students. It briefly reviews why demand for plant-based foods among students is increasing and how meat production impacts animal welfare, health, the environment, and food budgets. Then, we present case studies of several colleges and universities that have implemented plant-based foodservice initiatives, nearly all of which have leveraged the expertise and resources of the Humane Society of the United States' (HSUS) Food and Nutrition team to develop, strengthen, and promote these programs on campus.

The authors, the staff of HSUS, rely on communication materials and interviews with those involved, from students to foodservice staff, to provide profiles of each institution's efforts. The firsthand accounts of institutional implementation of plant-based menus were derived primarily from interviews with students, foodservice professionals, and manufacturers. We conducted three phone interviews, five in-person interviews, and two email interviews over the course of two years (2016 and 2017) to gather the information used in this chapter's case studies, and gathered newspaper articles and data from the universities themselves. We also discuss examples of external organizations that are supporting these initiatives on campuses, including non-profit organizations such as HSUS and private manufacturers. Ultimately, we believe that the movement toward plant-centric eating, which typically begins with either strong student interest or HSUS initiatives, has seen an overwhelmingly positive reception from participating colleges and universities and has changed what schools are purchasing and how they are designing their menu offerings.

---

1. Voluntary meal plans are those that students who are not required to be on a meal plan can elect to purchase.
2. A plant-based diet is one based on fruits, vegetables, legumes, and grains, which seeks to minimize or omit meat and other animal products.

UNDERSTANDING THE PLANT-BASED FOOD TREND

Many consumers cite ethical concerns surrounding the poor treatment of farm animals in industrial operations and the very practice of using animals for food when viable alternatives exist as reasons behind their desire to consume more plant-based foods. Individuals are also choosing to eat less meat or choosing to eliminate it from their diets completely for health, environmental, and budgetary reasons.

Animal Welfare

For decades, concerns about the treatment of animals in the agricultural sector have been growing. The vast majority of land animals raised for food in the United States—approximately nine billion—are raised in intensive confinement on concentrated animal feeding operations (CAFOs), which are large-scale farms that often house tens of thousands of animals in cramped conditions. CAFOs are used for many types of food animal operations, from poultry to beef. Long-term confinement, artificial inputs (such as antibiotics), and even physical abuse of animals are common in these operations. Animals are typically kept in dusty, crowded feedlots, crammed into wire cages, or kept in crates in which their natural movements are restricted for their entire lives. Further, despite federal regulations against inhumane slaughter of farm animals, oversight is nearly nonexistent (Humane Society of the United States, 2016).

Many students cite concerns about farm animal welfare or even the practice of keeping animals for human benefit and consumption as a reason for adopting a meatless or flexitarian diet.[3] "I'm committed to social justice. Animal liberation is part of that," said one student, who graduated from New York University (NYU)'s Gallatin School of Individualized Study in December 2016. She is a member of NYU's Animal Welfare Collective, which gathered petitions from more than 1000 students in support of a vegan dining hall on campus (Wetlaufer, 2017).

Health

The long-term benefits that a diet lower in meat can have on health and well-being have been compelling reasons to eat consciously. According to Gallup, obesity rates among the U.S. college-aged population (ages eighteen to 23) went from 13.9 to 14.4 percent between 2008 and 2012 (Mendes, 2012), which has a significant impact on overall health. The Centers for Disease Control and Prevention found that, "Obesity-related conditions include heart disease, stroke,

---

3. A mostly plant-based diet with a more flexible approach toward consumption of animal products; typically, flexitarians will occasionally eat meat, eggs, and/or dairy, but do not consider it necessary to or a focal point of their diet.

type 2 diabetes, and certain types of cancer, some of the leading causes of preventable death," (Centers for Disease Control and Prevention, 2016).

Evidence suggests that diets lower in meat can reduce risk of obesity. The American Heart Association found that people who eat meat-free are at a lower risk of obesity than those who eat meat (Vergnaud, 2010). This is in part due to reduced saturated fat and cholesterol consumption. A study presented at the Obesity Society's 2013 annual conference found that participants who had consumed vegan or vegetarian diets over an eight-week period lost an average of 8.2–9.9 pounds, while those consuming meat lost 5.1 pounds (Turner-McGrievy et al., 2017). Students interviewed for this chapter shared similar experiences. For example, one student explained that after struggling with weight gain early during her freshman year, she started tracking her calories. When she realized that eating animal products used more of her "calorie budget," she began to examine the impacts of her eating habits on her health and adopt a plant-based diet.

The Environment

Other students are concerned about the environmental consequences of meat-heavy diets. One student wrote in an op-ed in *Washington Square News,* her school's student-published newspaper, "Eating a more plant-based diet is the most environmentally friendly diet...a plant-based dining hall would allow many NYU students to actively live out the widely held values of sustainability, health and compassion," (Wetlaufer, 2016).

Meat production has multiple negative impacts on the environment, including greenhouse gas emissions, animal waste, high water use, and land degradation. A report from Chatham House, a London-based think tank, implicated animal agribusiness as a bigger greenhouse gas emitter than the entire transportation sector (Wellesley, 2017). Since 1990, greenhouse gas emissions from livestock, which include carbon dioxide, methane, and nitrous oxide, have increased 16.4 percent. As of 2015, 9 percent of greenhouse gas emissions in the United States in 2015 were produced by the agriculture industry, with about 44 percent of the industry's greenhouse gas emissions (250.51 million metric tons) coming directly from livestock (USDA, 2015).[4]

Compounding on the issue of greenhouse gasses are the other byproducts of factory farming, including farm animal waste, which can pollute nearby soil and groundwater and affect the health of local communities and wildlife (Steinfeld et al., 2006). Large-scale factory farming is also an environmental justice problem, because low-income communities and communities of color

---

4. Forty four percent is a modest estimate, because this figure discounts the emissions produced by the facilities that feed and house those livestock, the fuel combustion required to grow and harvest these crops, and the emissions produced by the transport of livestock and of meat and other animal products to stores across the country.

are disproportionately affected by the impacts of animal waste. For example, in North Carolina, the concentration of hog farms near low-income African American communities exposes that population to contaminated water and air pollution, because crops are sprayed with animal waste and runoff seeps into groundwater supplies (Nicole, 2013).

The production of meat, eggs, and dairy also requires vast amounts of water and land resources to sustain animals, using resources that could otherwise be used for the cultivation of crops (Weis, 2013). Corn accounts for more than 95 percent of total feed grain production and use, meaning that a significant majority of the crops grown in this country are being used to feed the animals we raise for consumption (USDA, 2017). Just over 70 percent of the 76 million acres of soy grown in the United States is used for animal feed, an overwhelming majority of which (94 percent) are genetically modified (USDA, 2015).

Supporting the livestock industry also takes a lot of water, and not just to quench the animals' thirst. The feed conversion ratios for beef, pork, and poultry are 7:1, 5:1, and 2.5:1 respectively, and all of those crops must be irrigated, according to Dr. Robert Lawrence, a professor at the Johns Hopkins Bloomberg School of Public Health (Eichenseher, 2013). Accounting for the amount of water required to produce meat, citing research from Arjen Hoekstra, environmental nutritionist Helen Harwatt, PhD calculated that it typically takes an estimated 13,812 gal of water to produce one pound of cooked beef, 5365 to produce one pound of cooked pork, and 3998 to produce a pound of cooked chicken, while a pound of cooked pulses only requires 908 gal of water (Harwatt, 2018, Hoekstra, 2014).

Cost Savings

The growth in interest in plant-based meals is also supported by research that shows that eating a plant-based diet can be friendly for the wallet, too. A study comparing a government-recommended meal plan containing meat with comparable plant-based meal plans found that the plant-based diets saved an average of $14.36 per week, amounting to roughly $750 less spent on food annually (Flynn and Schiff, 2012). The plant-based diets saw lower costs, even when consumers splurged on "luxury" goods, such as olive oil rather than canola oil, and contained significantly more servings of vegetables, fruits, and whole grains. With tuition, housing, textbooks, and myriad additional costs of being a student, these savings associated with a plant-based diet have made it an appealing option on college campuses.

A leader of the Yale foodservice team, Rafi Taherian, notes that universities save money by offering plant-based meals as well. He said, "We're now seeing a significant reduction in consumption of animal-based proteins on campus. That helps us offset the cost of increasing the quality, and we also see a tremendous amount of student satisfaction in terms of what they eat and how they experience our environment" (Tanyeri, 2016b).

CASE STUDIES: COLLEGE AND UNIVERSITY FOODSERVICE

College-aged students are interested in eating more plant-based food and less meat. According to 2011 research from food industry firm Technomic (2016), 21 percent of college students eat meat only occasionally, or eat vegetarian or vegan exclusively. Research also shows that campuses of many kinds, from small to large, and with foodservice that is self-operated or contracted out, all have populations of students who prefer plant-based foods.[5] For example, about 10 percent of students at the University of California, Berkeley, which self-operates its foodservice, report to be vegetarian (Lubow, 2017). Eight percent of students at Arizona State University, which has more than 70,000 students and contracts out its foodservice operations, identify as vegan or vegetarian (Technomic, 2016).

According to Dana Tanyeri, Contributing Editor for *Foodservice Equipment and Supplies* (2016a) magazine, "Those leading the [college and university foodservice] segment's evolution, which they characterize as being driven by a generation of students whose habits, tastes and expectations have changed faster and more dramatically than any in history, say it has impacted everything from product sourcing, equipment specifications and staffing to dining hall design." To meet this growing interest that students have in plant-based foods, institutions across the country are experimenting with offering more plant-based options.

Self-Operating Foodservice Facilities

The next section lists examples of this interest in college and university institutional foodservice operations that are self-operated.

University of California, Berkeley

The University of California, Berkeley's foodservice is self-operated (and called "Cal Dining"). The dining officials at Cal Dining are making an effort to serve more sustainable and healthy options for their student population. Cal Dining opened Brown's California Café in 2015. According to its website, the operation "uses scratch-cooking techniques to produce delicious vegetable-forward menu items as well as 'better meat' options such as grass-fed burgers and rotisserie organic chicken" (Cal Dining, n.d.). For example, guests can enjoy the "Flipped Plate," which includes choices such as brown rice; caramelized onion and parsley pilaf; roasted broccoli with cumin; house-made hummus with paprika pita chips; and an entrée –including plant-based options such

---

5. University dining operations are either "self-operated" facilities or contracted to foodservice providers, which are outside companies that are responsible for every aspect of the food operations: creating menus, sourcing ingredients, cooking, cleaning, and disposing of waste.

as root vegetable gratin and grilled portabella mushroom caps, or a small serving of meat. The café won a 2016 Higher Education Energy Efficiency and Sustainability Best Practice Award for Sustainable Food Systems.

To keep the momentum going, in 2016, Cal Dining opened Café 3. Dubbed a "dining hall for all," Café 3 features a food station designed to accommodate Kosher and halal diets, and has a plant-centric focus (Eliahou, 2016). "At our 'plant-forward' dining hall, nine of the ten platforms serve strictly vegetarian or vegan foods. We have heard from many students that this dining hall has allowed them to realize they could comfortably eat a vegetarian diet on a regular basis," said the Environmental Initiatives Coordinator, Samantha Lubow (2017). Lubow also shared that one creative way Cal Dining is increasing uptake of plant-based foods is by working more directly with students:

> *Each year, students in the foodservice management class have a final project in which they work directly with one of our executive chefs to adjust one of our more 'traditional' menu options and create something aligned with [plant-based foods]. In the past, students' projects have adjusted our pork sliders to oyster mushroom sliders, our huli huli chicken to huli huli eggplant, and more. If the new recipe is financially feasible, we add it to our menu.*

(Lubow, 2017)

University of North Texas

Through traditional feedback forms such as comment cards and student advisory meetings, as well as through social media, during the 2010–11 school year, University of North Texas (UNT) solicited input from students on what food options they were looking for in student dining halls. UNT, also a self-operating campus, had five dining halls at the time, one of which was underperforming in that it was selling very few meals (measuring 175 swipes of meal cards per day). Given that many students were complaining over the dearth of healthy, flavorful plant-based options on campus, Director of Dining Bill McNeace commissioned then-Director of Special Projects for Dining Services, Ken Botts, to transform the underperforming dining hall into a vegan operation called *Mean Greens* Café, and work with a classically trained pastry chef, Wanda White, to create the menu (Botts, 2017).

Making the changes to the dining hall's menus required research. "One of the first things Wanda asked me was, 'Ken, how am I supposed to cook without milk and eggs?'" shared Botts, a thirty-year foodservice industry veteran. He sought out vegan cookbooks and scoured the internet, and over the next six weeks, the two tested recipes and created a cycle menu. Chef Wanda used many of the dining hall's existing recipes to create animal-free versions, such as cashew-based macaroni and cheese. White found that making the changes was much easier than she originally anticipated: "It's as simple as swapping out cow's milk with almond or soy milk, and butter with a dairy-free butter," she stated (2017).

University of Massachusetts, Amherst

University of Massachusetts, Amherst (UMass Amherst) is the largest self-operated college dining services operation in the country, serving 45,000 daily meals and a total of 5.5 million meals per year. *The Princeton Review* ranked UMass Amherst Dining as number one for best campus food in the nation (Fitzgibbons, 2017).

Seven percent of respondents to a fall 2015 survey of 4000 UMass students identified as vegan or vegetarian (1.4 percent vegan and 5.6 vegetarian) while in the school's fall 2016 survey, 8.1 percent of students reported being vegan or vegetarian (2.3 percent vegan and 5.8 percent vegetarian) (DiStefano, 2017). While many students may not identify as vegetarian or vegan, "consumption of fruits and vegetables continues to grow – [an] average UMass student consumes roughly 1,200 pounds of produce per capita annually" (DiStefano, 2017).

Plant-based options in the school's dining halls include a variety of fifteen Mediterranean, Southeast Asian, and Latin-inspired plant-based options daily on its menus. Some examples include chana masala, made-to-order stir-fry, zucchini cakes, green rice burrito bowls, and more. The average amount of plant-based protein consumption is five times what it was in 2012, and red meat consumption has declined by roughly 16 percent (DiStefano, 2017).

Foodservice Management Companies

The next section provides examples of plant-based work in colleges and universities who contract out their foodservices.

Compass Group North America

Compass Group North America ("Compass") serves 9.4 million meals each day in North America (Compass Group, 2017). Universities, schools, hospitals, corporate cafeterias, sporting arenas, private events, and other institutions contract the company to provide foodservices. In 2015, the company announced a new sustainability platform called Envision 2020. According to Christine Seitz, Vice President of Culinary Strategy and Innovations (2017): "The guiding principle in our Envision 2020 journey continues to focus on 'making it easy to do the right thing' by encouraging our customers to make better choices and minimize our eco-impact. Because of our scale, we can create change that is radical and transformational." The Envision 2020 goals include reducing red meat purchases by 10 percent by 2020, increasing produce and whole grain purchases by 5 percent every year, and making half of all grains offered whole grains.

The company is working to meet its goals through a variety of innovative approaches, including by offering more plant-based food items. For example,

Compass began serving the company Hampton Creek's products in 2014. Best known for its "Just Mayo" product, Hampton Creek is creating plant-based equivalents to animal-based foods, such as mayonnaise. Additional strategies for increasing plant-based options, and decreasing meat purchases, include serving meat-mushroom blends, in which ground meat is reduced in burgers and meatballs by being replaced with mushrooms. In 2015, Compass Group also began partnering with HSUS to create more plant-based dining concepts for colleges and universities.

Canisius College

One example of a current college for which Compass Group operates dining services is Canisius College in Buffalo, New York. Its dining services are operated by Chartwells, a subsidiary of Compass Group, and it serves a student body of 2900 undergraduates. The dining services team has a good relationship with the Canisius Veg Club, the student vegetarian organization. In response to the Veg Club's requests for more meatless options, dining services revamped its "On the Go" concept for students interested in quickly purchasing a ready-made meal to eat elsewhere in January 2016 to include more plant-based options. According to the head chef at Canisus, Jennifer DiFrancesco, sales went up by 76 percent after this change (DiFrancesco, 2016).

The success of the vegan options in the On the Go concept motivated dining services to incorporate more plant-based foods into all of its menus, according to DiFrancesco, and in the fall semester of 2016, Chartwells opened a 100 percent plant-based retail station called Pitchforks. Dining services incentivized participation at Pitchforks with a discount: the Chartwells team priced plant-based meals fifty cents less than the meat-based meals (DiFrancesco, 2016). According to John Tychinski, Director of Dining Services at Canisius College, "Chartwells has seen a 400 percent increase in sales year over year from the vegetarian station it replaced" (Scanlon, 2016).

To create the menus, DiFrancesco solicited feedback on the menu from the Veg Club for what options students would like to eat. Along with a couple of campus chefs, she visited popular local vegetarian and vegan restaurants and pored over vegan cookbooks to develop a 28-day cycle menu. She also used a large portion of the station's initial vegan recipes from Chartwells' database. The rotating lineup of vegan street and comfort foods includes options such as loaded sweet potato burgers with smashed avocados, tempeh Reubens, plant-based chicken enchiladas, kung pao cauliflower with zucchini lo mein noodles, creamy butternut squash linguine with fried sage, braised cauliflower tacos, "pulled pork" made from charred king oyster mushrooms, and sweet potato mac and cheese. Her team also created a rotating dessert menu with oatmeal banana chocolate cookies, tiramisu parfaits, pumpkin cheesecakes, and fresh fruit tarts with a vegan vanilla bean cream (DiFrancesco, 2016).

Aramark Corporation

Aramark is another company that provides food, facility, and other services to schools, hospitals, entertainment venues, and major institutions in nineteen countries worldwide, including 500 college campuses in the United States. In 2015, Aramark announced that it would be expanding its menu to offer a vegan option at every meal at some colleges (Sentenac, 2015). This was implemented based on student feedback from Aramark's proprietary platform and dining surveys, which indicated that the number of students interested in vegan options has steadily increased over the past few years. The company now offers more than 80 vegan entrees for chefs to incorporate in its residential menus, and helps provide an expansive plant-based menu to students at the two universities featured next in this chapter and many others (Sentenac, 2015).

Arizona State University

With more than 76,000 students, Arizona State University (ASU) is the largest public university in the United States by enrollment. In 2014, nearly 10,000 meal plans were purchased on ASU campuses, generating more than $29 million in annual program revenue. Aramark currently operates all seven dining halls and 47 retail outlets on ASU's four campuses.

In 2015, students at ASU established Veg Aware, a student organization promoting vegetarian eating on campus. One of the first campaigns of the organization was to collect more than one thousand signatures on petitions from students who desired getting more vegan options on campus. In response, in January 2015, resident dining opened the Daily Root, a station which featured plant-based options such as breakfast scrambles, a waffle bar, Thai curry plant-based "chicken" linguine, "beefy" enchiladas, blackened tofu with paella, and jalapeño vegan "chicken" paninis. The plant-based options at Daily Root became so popular that ASU added even more vegan and vegetarian options to other stations in the dining halls. After a remodel, four out of the six stations at the Tempe campus are 90 percent vegetarian (Plante, 2015).

American University

The Dining Services of American University (AU) in Washington, D.C. is presently run by Aramark and serves meals to approximately 1800 students annually. AU Dining holds focus groups each semester with vegetarian and vegan students and staff to solicit feedback for improving its program; and they received a lot of feedback about the need to serve more vegetarian and vegan option. As a result, in 2014, the AU dining team launched "Meatless Monday" in its Terrace Dining Hall in December 2014, and serves one plant-based entrée at every meal very Monday.

Dining services replaces the carved meat item with a carved vegetable or vegetable entrée on Mondays, such as mushroom Wellington or carved butternut squash. The school also features vegan and vegetarian specials at each station on Meatless Monday, and promotes the day with marketing materials, employee engagement, and community involvement to increase awareness of and participation in Meatless Monday (Johnson, 2016). Additionally, in the fall semester of 2015, AU Dining's "rotating retail" concept featured a vegan pop-up restaurant it called Food Forward. In partnership with the HSUS, Food Forward featured sandwiches such as mushroom bahn mis and not-meatball subs, a spicy kale Caesar salad, spinach waffles, ratatouille, fresh juice, and a rotating offering of vegan cupcakes. Presented on a digital display in the concept were infographics and images educating guests on the benefits of plant-based eating.

COMMON THEMES

In addition to the aforementioned health and environmental concerns, we believe two other factors have greatly influenced the adoption of plant-based meals by colleges and universities. The following sections elaborate on two: the involvement of non-profit intermediaries to provide educational tools for institutional foodservice staff in conjunction with the peer networks they are building for food foodservice professionals experimenting with plant-based food, and the emergence of new meatless product lines with private manufacturers.

The Role of Non-Profit Intermediaries and Peer Networks

One commonality in these case studies is that the plant-based initiatives on these campuses are supported by external organizations and campaigns that have provided students and the foodservice industry resources to increase vegetarian and vegan options. In our interviews, three non-profits were mentioned frequently: The Menus of Change University Research Collaborative (MCURC), Meatless Monday, and HSUS.

The MCURC research collaborative is a working group comprising 142 chefs, scholars, and foodservice leaders representing 44 colleges and universities and five ex-officio members working to shift their menus to be more plant-centric. MCURC emerged in 2014 out of an annual conference, called Menus of Change (MOC), hosted by the Culinary Institute of America and the Harvard T.H. Chan School of Public Health. At the conference, attendees share best practices and strategies for making their offerings more sustainable and healthy. Conference organizers also produce research-backed tools and reports to support culinary innovation and provide guidance to those in the foodservice industry who are interested in delivering healthier and more sustainably and ethically sourced food. One report, *Principles of Healthy, Sustainable Menus*, recommends plant-forward strategies such as moving nuts and legumes to the center of the plate and serving more globally inspired plant-based recipes and meal offerings.

MCURC brings together a network of scholars and experts from colleges and universities to collaborate on research and education supporting innovation in food production and diet (MOC Collaborative, 2017). It connects insight from academic programs, dining services, and athletics to catalyze the transformation of the current food system into one that is healthier and more sustainable. MCURC leverages the insight and research conducted by member universities by using campus dining halls as real-world testing grounds to evaluate best implementation strategies of its 24 Principles of Healthy, Sustainable Menus. Three of the six schools profiled in this chapter participate in MCURC: University of California, Berkeley, the University of Massachusetts, Amherst, and the University of North Texas.

Meatless Monday is a public health campaign started by the Johns Hopkins Bloomberg School of Public Health and its health initiative, the Monday Campaigns. Institutions that commit to Meatless Monday serve plant-based options at every meal on Mondays and highlight the environmental and health benefits of a reduced-meat or meat-free diet. The goal of Meatless Monday is to have individuals reduce their meat consumption by 15 percent. Several environmental non-profits have adopted this strategy in their meat-related work, and now, more than 150 colleges and universities are participating in Meatless Monday (The Monday Campaigns, Inc., 2017).

The Food and Nutrition team at HSUS has worked with 150 colleges and universities across the country to help them add more plant-based food options to their menus. Often, representatives reach out to foodservice professionals who provide meals to students and gauge their level of interest in putting some mode of plant-based program in place. Once they have the attention and interest of an institution, the team assists schools in the transition. This includes providing programming and marketing materials, running pop-up stations and offering free samples, delivering hands-on training on plant-based meal preparation to culinary staff, and creating reports to measure the sustainability impacts of the program.

Many of the events the HSUS holds are hosted at university campuses, which has played an integral role in informing students, staff, and administrators on the reasons to move toward a more plant-based diet and how to begin that change. The HSUS works with universities in two key ways: Forward Food Leadership Summits and Forward Food Culinary Experiences.

Forward Food Leadership Summits are a one-day event for anyone in the foodservice industry (foodservice directors, dietitians, chefs, etc.) to learn from industry leaders about meeting the demand for plant-based meals and marketing healthy foods. For example, HSUS held a summit in September 2017 at Rutgers University covering two key topics: meeting the growing demand for plant-based foods and successfully promoting healthy options. Speakers consisted of health and nutrition experts, members of the foodservice industry, and members of the HSUS food and nutrition team. As an institution already working with the HSUS on providing more plant-based options, Rutgers took a hands-on role in running the event; the summit agenda included a cooking demonstration by the university's executive chef followed by a plant-based lunch at Rutgers' plant-forward Harvest Café. Attendee feedback following the event was overwhelmingly positive, and many expressed new enthusiasm about plant-based eating and incorporating what they had learned into their workplaces.

Many universities that attend the HSUS summits are inspired to reach out to organizers for assistance on how to make the transition in their own cafeterias and other food stations. In response, the HSUS Food and Nutrition team provides a customized Forward Food Culinary Experience, a free hands-on training for culinary professionals at universities, hospitals, military bases, and other institutions to learn how to incorporate more plant-based foods into their menus. The agenda of these places a strong educational focus on sourcing, planning, and actively preparing plant-based meals on a large scale. Attendees are guided through new plant-based recipes by the HSUS's chefs before enjoying the fruits of their labor for lunch. The HSUS expects that these culinary experiences will influence about 3200 chefs and registered dieticians across the country in 2018 alone (PR Newswire, 2017).

Of the universities discussed in the previous section, all except the University of Arizona are involved with at least one of these programs. The HSUS and its Forward Food program have worked directly with Canisius College, the University of North Texas, and American University, and some of these universities' chefs to conduct plant-based culinary training sessions at campuses and other institutions nationwide to promote HSUS's plant-based initiatives.

At least 30 universities have made changes directly because of what they learned at these HSUS events, to varying degrees and in ways that make the transition uniquely their own. Carnegie Mellon University, for example, launched a 100 percent plant-based concept on campus beginning in the fall of 2017 called Garden Bistro, consisting of comfort style sandwiches and self-made sauté bowls, allowing a variety of students from across campus to experience plant-based meals. The University of California, Irvine offers a discount to meatless options at its retail locations for Meatless Monday, which incentivizes students to purchase a more affordable product that is simultaneously more humane and sustainable.

Emergence of New Meatless Product Lines Among Private Manufacturers

Colleges and universities are one of the fastest growing segments of the foodservice industry, making student demand a strong influencing factor on the types of food being produced (MAFSI, 2016). In a poll of 167 college and university dining operators nationwide, *Foodservice Director* found that vegetarian or vegan dishes comprise about half of the menu items served on campus, and that number appears to be increasing (Foodservice Director, 2015).

Those leading the segment's evolution, which they characterize as being driven by a generation of students whose habits, tastes and expectations have changed faster and more dramatically than any in history, say it has impacted everything from product sourcing, equipment specifications, and staffing to dining hall design. Universities are part of a greater national dining trend toward a more plant-based diet, which is driving change in food production at all levels. Ken Toong, the associate director of dining at UMass Amherst, believes that "ten years from now, Millennials and Gen Z will be the largest portion of the population, and animal welfare, vegetarianism, sustainability and social responsibility are important to them" (Tanyeri, 2016a). A 2012 report titled *Trouble in Aisle 5*, which studied the purchasing habits of Millennials, found that this age group has a stronger interest in specialty food than previous generations and are more likely to seek out natural and organic food (Hoffman, 2012). Food companies recognize that today's youth and students are an emerging market, and are updating their products to meet current and anticipated demand for healthy, sustainable, traceable, and more humane options.

Companies such as Hampton Creek (discussed herein), Beyond Meat, Hungry Planet, and US Foods are creating new food products to meet this growing demand for plant-based and/or socially conscious foods. For example, based in Los Angeles, Beyond Meat, is a plant-based meat producer that supplies schools, universities, restaurants, and retailers nationwide with plant-based "meats." The company's completely vegan burgers were developed to have the same texture and mouthfeel of a beef burger, and even "sizzled and smelled like a real burger while it cooked" according to a critic's review of the product (Price, 2016). Made with pea protein, Beyond Meat's strips, crumbles, and burgers contain up to 23 g of protein per serving, which is as much or more protein than a serving of animal meat (Beyond Meat, 2015).[6]

Another manufacturing company working directly with institutions is Hungry Planet, which produces one-to-one plant-based animal protein substitutes for use by culinary professionals. Hungry Planet's products include plant-based counterparts to beef, chicken, pork, Italian sausage, chorizo sausage, turkey, lamb, and crab, all of which can be used in place of meat in almost any recipe. Hungry Planet provides the meat substitutes for all the food stations offering plant-based alternatives at Canisius College, and Chef Jenn DiFrancesco credits its products in part for Pitchforks' success: "Our best-selling dish is the chicken enchilada, which is made with Hungry Planet's ground chicken. We had students coming back up to the station asking, 'Are you sure this isn't meat?'" DiFrancesco (2016) also notes that these products made transitioning to a plant-based menu and training chefs in vegan cooking significantly easier because the products can be used as a one-to-one substitute for meat.

---

6. Tyson, one of the world's largest meat companies, has an institutional foodservice division that recognizes how students are driving the move toward healthier, more sustainable food (Tyson, 2016). Its latest newsletter, *Focus*, states that "Gen Z is seeking out, and sometimes paying a premium for, food they consider healthier" (Tyson, 2016). The company has participated in past MOC conferences, invested in Beyond Meat, and launched a $150 million venture capital fund aimed at investing in alternatives to animal protein (McGrath, 2016).

The Culinary Institute of America has worked with industry groups and universities to develop meat-plant blends that reduce the amount of meat consumed at institutions and make for healthier meals. For example, it conducted one sensory study with the University of California, Davis that showed that customers preferred the flavor, texture, spice level, and salt level of a mushroom-beef blend over the 100 percent beef burger (Myrdal Miller et al., 2014). US Foods, one of the nation's largest foodservice distributors that serves institutions ranging from hospitals to universities, recently formulated a meat-mushroom blended burger to reduce the amount of meat offered in one serving. "Record-high beef trim prices inspired us to create a delicious alternative that's also significantly lower in fat and calories than a traditional all-beef burger. This recipe includes 33 percent mushrooms for a savory flavor, excellent texture, and a convincing health story as well," it claims (US Foods, 2015).

CONCLUSION

Increasingly, colleges and universities are serving more plant-based options. In many cases, the changes are originating from a surge in student demand for more plant-based food options. In others, university leaders and foodservice directors are finding innovative ways to market plant-based food options in order to meet sustainability goals or nudge students to more healthful choices. The transformative nature of the potential benefits to our food system and planetary health is difficult to exaggerate. Reducing the current demand for and consumption of meat and animal products is critical for preventing climate change and to conserve resources. Regardless of the impetus behind institutional foodservice increasing vegetarian and vegan options, several non-profit organizations and manufacturers are responding by offering resources and new products to support such changes on these campuses. Evidence to date suggests that student consumers are responding positively to the new plant-based food options.

The biggest barriers to date within the foodservice sector, according to Botts are a lack of tasty options, convenience, and cost of prepared plant-based meats (2017). However, the growth in demand is predicted to continue to surge and along with that a windfall of plant-based products as well as investment backing (CB Insights, 2017). *Foodservice Director Magazine,* in 2018, wrote, "plant-based dining for everyone," is what "operators will be talking about in 2018 and beyond" (FSD staff, 2017). Prices will likely fall and distribution will be greater should those predictions prove to be true. Overall, these shifts in university purchasing practices have the potential to stimulate broader changes in the food system and create a more humane, more sustainable, and healthier food supply.

REFERENCES

Botts, K., 2017. Pers. Comm., 17 January.
Cal Dining, Browns A California Café. Available from: http://caldining.berkeley.edu/locations/on-campus-retail/pat-browns. Accessed December 30, 2017.

CB Insights, 2017. Our Meatless Future: How The $90B Global Meat Market Gets Disrupted. 9 November 2017. Available from: https://www.cbinsights.com/research/future-of-meat-industrial-farming/#why. Accessed January 4, 2017.

Centers for Disease Control and Prevention, 2016. Adult Obesity Facts. Available from: https://www.cdc.gov/obesity/data/adult.html.

Compass Group, 2017. Compass Group USA Announces A Landmark Commitment To Reduce Food Waste By 25% By 2020, 9 March 2017. Available from: http://www.compass-usa.com/compass-groupusa-announces-landmark-commitment-reduce-food-waste-25-2020/.

DiFrancesco, J., 2016. Pers. Comm. 17 August 2016.

DiStefano, G., 2017. Pers. Comm., 13 February 2017.

Eichenseher, T., 2013. The Ripple Effect of Meatless Monday: Can it Extend to China? Discover Magazine. 03 June 2013. Available from: http://blogs.discovermagazine.com/waterworks/2013/06/03/the-ripple-effect-of-meatless-monday-can-it-extend-to-china.

Eliahou, M., 2016. Café 3 debuts 'Hall for All' food station aimed at improving inclusivity. Daily Cal. 24 August 2016. Available from: http://www.dailycal.org/2016/08/24/cafe-3-debuts-hall-for-all-food-station-aimed-at-improving-inclusivity/.

Fitzgibbons, D., 2017. UMass Amherst No. 1 in Campus Dining for Second Straight Year, Says Princeton Review, 31 July 2017. Available from: https://www.umass.edu/newsoffice/article/umass-amherstno-1-campus-dining-second.

Flynn, M., Schiff, A., 2012. Economical healthy diets: including lean animal protein costs more than using extra virgin olive oil. J. Hunger Environ. Nutr. 10 (4), 467–482. https://doi.org/10.1080/19320248.2015.1045675.

Foodservice Director, 2015. 2015 C&U Census: Higher learning. Available from: http://www.foodservicedirector.com/research/industry-census/articles/2015-cu-census-higher-learning. Accessed December 30, 2017.

Harwatt, H., 2018. Pers. Comm., 4 January 2018.

Hoekstra, A., 2014. Water for animal products: a blind spot in water policy. Environ. Res. Lett. 9. Available from: http://iopscience.iop.org/article/10.1088/1748-9326/9/9/091003/pdf.

Hoffman, B., 2012. How 'Millennials' Are Changing Food as We Know It. Forbes. 4 September. Available from: https://www.forbes.com/sites/bethhoffman/2012/09/04/how-millenials-are-changing-food-as-we-know-it/#51a598964041.

Huen, E., 2016. 5 Food Trends You'll See in 2017. Forbes. 30 November 2016. Available from: http://www.forbes.com/sites/eustaciahuen/2016/11/30/5-food-trends-youll-see-in-2017/#3e6d0647ad36.

Humane Society of the United States, 2016. Cruel Slaughter Practices. Available from: http://www.humanesociety.org/issues/slaughter/.

Johnson, K., 2016. American University. Available from: https://stars.aashe.org/institutions/american-university-dc/report/2016-03-30/OP/dining-services/OP-7.

Lillien, L., 2017. Hungry Girl: Seaweed, Mocktails and More of My Food Trend Predictions for 2017. People.com. 16 January 2017. Available from: http://people.com/food/hungry-girl-lisa-lillien-food-trends-2017/.

Lubow, S., 2017., Pers. Comm., 23 January 2017.

MAFSI, 2016. Commercial Foodservice Market Forecast Report. Available from: http://www.mafsi.org/assets/MarketForecast/2016%20mafsi%20report%20final.pdf. Accessed December 30, 2017.

McGrath, M., 2016. Tyson Launches $150 Million VC Fund That Could Help Hedge Against A Meatless Future. Forbes, 5 December 2016. Forbes. . Available from: https://www.forbes.com/sites/maggiemcgrath/2016/12/05/in-a-hedgeagainst-a-meatless-future-tyson-foods-launches-150-million-vc-fund/#18e71e8a435b.

Mendes, E., 2012. In U.S., Obesity Up in Nearly All Age Groups Since 2008. Gallup. 24 October. Available from: http://www.gallup.com/poll/158351/obesity-nearly-age-groups-2008.aspx.

Menus of Change University Research Collaborative, 2017. Overview of the Menus of Change University Research Collaborative. Available from: http://www.moccollaborative.org/images/uploads/pdf/MCURC_Overview_(10-9-17).pdf.

The Monday Campaigns, Inc, 2017. The Movement. Available from: http://www.meatlessmonday.com/the-global-movement/.

Myrdal Miller, A., et al., 2014. Flavor-enhancing properties of mushrooms in meat-based dishes in which sodium has been reduced and meat has been partially substituted with mushrooms. J. Food Sci. 79 (9), S1795–S1804. https://doi.org/10.1111/1750-3841.12549.

Nicole, W., 2013. CAFOs and environmental justice: the case of North Carolina. Environ. Health Perspect. 121 (6), A182–A189.

Plante, A., 2015. The Daily Root offers vegan options on all ASU campuses. State Press. 27 April. Available from: http://www.statepress.com/article/2015/04/the-daily-root-offers-vegan-options-on-all-asu-campuses.

Price, E., 2016. Plant Burgers are coming to a Fast Food Restaurant near You. Fast Company. Available from: https://www.fastcompany.com/3062085/plant-burgers-are-coming-to-a-fast-food-restaurant-near-you.

PR Newswire, 2017. School and Hospital Menus Add More Plant-Based Options to Meet Growing Demand. 18 November. Available from: https://www.prnewswire.com/news-releases/school-and-hospital-menus-add-more-plant-based-options-to-meet-growing-demand-300515891.html.

Scanlon, S., 2016. Canisius dining patrons dig in to hot new vegan food station. Buffalo News. 18 November. Available from: https://buffalonews.com/2016/11/18/canisius-dining-patrons-dig-hot-new-vegan-food-station/.

Sentenac, H., 2015. Aramark Offers New Vegan Options to Over 500 College Campuses. Latest Vegan News. Available from: http://latestvegannews.com/aramark-offers-new-vegan-options-to-over-500-college-campuses/.

Steinfeld, H., 2006. Livestock's Long Shadow. Food and Agriculture Organization of the United Nations, Rome. Available from: http://www.fao.org/docrep/010/a0701e/a0701e00.HTM.

Tanyeri, D., 2016a. What's Ahead for Campus Dining? Seven Industry Leaders Share Their Predictions. Foodservice Equipment and Supplies Magazine. 1 July. Available from: http://fesmag.com/features/foodservice-issues/13577-what%E2%80%99s-ahead-for-campus-dining-seven-industry-leaders-share-their-predictions.

Tanyeri, D., 2016b. One-on-One with: Rafi Taherian, Associate Vice President, Yale Hospitality. Foodservice Equipment and Supplies Magazine, 1 July. Available from: http://fesmag.com/features/foodservice-issues/13576-one-on-onewith-rafi-taherian,-associate-vice-president,-yale-hospitality,-2016-gold-plate-award-winner.

Technomic, 2016. Menu and Attitudes Assessment. Technomic, Chicago.

Turner-McGrievy, G., Mandes, T., Crimarco, A., 2017. A plant-based diet for overweight and obesity prevention and treatment. J. Geriatr. Cardiol. 14 (5), 369–374. https://doi.org/10.11909/j.issn.1671-5411.2017.05.002.

Tyson, 2016. Focus: Trends, Solutions, and More. Available from: https://tysonscore2.azureedge.net/assets/media/foodservice/documents/focus-cu-magazine-rev.ashx.

US Foods, 2015. Savory Burger: Harvest Value. Available from: https://usfoods.com/food/scoop/scoop-favorites/beef-pork-poultry/savory-burger.html. Accessed December 30, 2017.

USDA, 2015. Greenhouse Gas Inventory Data Explorer. Environmental Protection Agency. Available from: https://www3.epa.gov/climatechange/ghgemissions/inventoryexplorer/#agriculture/allgas/source/all.

USDA, 2017. Corn and Other Feed Grains: Background. The Economic Research Service. Available from: https://www.ers.usda.gov/topics/crops/corn/background.aspx.

Vergnaud, A., et al., 2010. Meat consumption and prospective weight change in participants of the EPIC-PANACEA study. Am. J. Clin. Nutr. 92 (2), 398–407. https://doi.org/10.3945/ajcn.2009.28713.

Weis, T., 2013. The Ecological Hoofprint: The Global Burden of Industrial Livestock. Zed Books, London.

Wellesley, L., 2017. Changing Climate, Changing Diets: Pathways to Lower Meat Consumption. Chatham House. 24 November 2017. Available from: https://www.chathamhouse.org/publication/changing-climate-changing-diets#sthash.sESJQiHj.dpuf.

Wetlaufer, E., 2016. Plant-Based Dining Halls Are a Win for All. Washington Square News, 31 October 2016. . Available from: http://www.nyunews.com/2016/10/31/plantbased-dining-halls-are-a-win-for-all/.

Wetlaufer, E., 2017. Pers. Comm., 23 January 2017.

FURTHER READING

American College Health Association, 2013. American College Health Association National College Health Assessment Spring 2013 Reference Group Executive Summary. Available from: http://www.acha-ncha.org/docs/ACHA-NCHA II_ReferenceGroup_ExecutiveSummary_Spring2013.pdf. Accessed December 30, 2017.

Baum & Whiteman, 2017. 13 Hottest Food & Beverage Trends in Restaurant & Hotel Dining for 2017. Baum & Whiteman, New York. Available from: http://www.baumwhiteman.com/2017TRENDS.pdf. Accessed January 1, 2018.

Boyman, J., Boyman, T., 2017. Pers. Comm. 28 July 2017.

Bucholtz, S., 2016. Pitchforks Satisfies Hunger and Vegans Alike. The Griffin. 9 September 2016. Available from: https://canisiusgriffin.wordpress.com/2016/09/09/pitchforks-satisfies-hunger-and-vegans-alike/. Accessed November 28, 2017.

Butler, S., et al., 2004. Change in diet, physical activity, and body weight in female college freshman. Am. J. Health Behav. 28 (1), 24–32. https://doi.org/10.5993/AJHB.28.1.3. 28 November 2017

Carter, S., 2012. The freshman 15. The Chicago Tribune. Available from: http://www.chicagotribune.com/lifestyles/health/sc-health-0725-school-college-obesity-20120725-story.html. Accessed 25 September 2017.

Chadwick, K., 2016. Pers. Comm., 9 June 2016.

ABC News, 2011. Adios A-Meat-Gos. Available from: http://abcnews.go.com/blogs/health/2011/09/01/adios-a-meat-gos-texas-college-caf-goes-vegan/. Accessed 16 January 2017.

Council for a Strong America, 2014. Retreat is Not an Option. Available from: https://www.strongnation.org/articles/14-retreat-is-not-an-option. Accessed 30 January 2017.

Fayet, F., et al., 2012. Prevalence and correlates of dieting in college women: a cross sectional study. Int. J. Women's Health 4, 405–411. https://doi.org/10.2147/IJWH.S33920. 1 January 2018

Foodservice Director staff, 2017. Trending Topics. Foodservice Director 30 (12), 25.

Gates, J., 2017. Pers. Comm., 25 January 2017.

Gross, K., 2017. Pers. Comm., 25 January 2017.

Grossman-Cohen, B., 2012. 'Meatless Monday' too hot a potato for USDA. CNN.com. 2 August. Available from: http://www.cnn.com/2012/08/02/opinion/grossman-cohen-meatless-monday/. Accessed 27 January 2017.

Ipsos MORI, 2016. Vegan Society Poll. Available from: https://www.ipsos.com/ipsos-mori/en-uk/vegan-society-poll?language_content_entity=en-uk, Accessed July 2010.

Kindy, K., 2013. USDA plan to speed up poultry-processing lines could increase risk of bird abuse. Washington Post. 29 October. Available from: https://www.washingtonpost.com/politics/usda-plan-to-speed-up-poultry-processing-lines-could-increase-risk-of-bird-abuse/2013/10/29/aeeffe1e-3b2e-11e3-b6a9-da62c264f40e_story.html?utm_term=.51a10a276d7e, Accessed 24 January 2017.

Klampe, M., 2011. Study: College students not eating enough fruits and veggies. 17 August 2011. Available from: http://oregonstate.edu/ua/ncs/archives/2011/aug/study-college-students-not-eating-enough-fruits-and-veggies, Accessed 30 January 2017.

Knowles, T., et al., 2008. Leg disorders in broiler chickens: prevalence, risk factors and prevention. PLoS One 3 (2), https://doi.org/10.1371/journal.pone.0001545. e1545.

Menus of Change University Research Collaborative, 2016. Overview of the Menus of Change University Research Collaborative. Available from: http://www.moccollaborative.org/images/uploads/pdf/MCURC-Overview-2016.pdf, Accessed 28 January 2017.

National Institute of Diabetes and Digestive and Kidney Diseases, 2012. Overweight and Obesity Statistics. Available from: https://www.niddk.nih.gov/health-information/health-statistics/Pages/overweight-obesity-statistics.aspx, Accessed 28 January 2017.

Oxfam America, 2017. Eat for Good recipes—Eat less meat, Squash blossom risotto by Chef Holly Smith, Café Juanita. Available from: https://www.oxfamamerica.org/take-action/campaign/food-farming-and-hunger/eat-for-good/eat-for-good-recipes-squash-blossom-risotto/, Accessed 28 January 2017.

Pomranz, M., 2017. Sonic to Bring 'Blended' Beef-Mushroom Burger to Fast Food World. Food Wine. Available from: http://www.foodandwine.com/news/sonic-bring-blended-beef-mushroom-burger-fast-food-world, Accessed 6 July 2017.

Technomic, 2011. Technomic Finds College Students Calling for Healthier Choices and Greater Say in Shaping Campus Dining Programs. 11 August 2011. Available from: http://www.prnewswire.com/news-releases/technomic-finds-college-students-calling-for-healthier-choices-and-greater-say-in-shaping-campus-dining-programs-127523548.html, Accessed 18 January 2017.

UC Berkeley Sustainability and Energy, 2016. Cal Wins Sustainability Best Practice Awards for Jacobs Hall and Brown's Café. UC Berkeley Department of Sustainability and Energy. 27 March 2016. Available from: http://sustainability.berkeley.edu/news/cal-wins-sustainability-best-practice-awards-jacobs-hall-and-browns-cafe, Accessed 24 January 2017.

White, W., 2017. Pers. Comm., 30 January 2017.

Wideman, R., et al., 2013. Pulmonary arterial hypertension (ascites syndrome) in broilers: a review. Poult. Sci. 92 (1), 64–83. https://doi.org/10.3382/ps.2012-02745, 1 January 2018.

Kristie Middleton is managing director of farm animal protection for The Humane Society of the United States and the author of *MeatLess: Transform the Way You Eat and Live–One Meal at a Time*. She is a sought-after speaker and thought leader on the topic. Middleton has partnered with the nation's biggest school districts including Los Angeles, Detroit, and Boston to implement plant-based initiatives such as Meatless Monday. Her work has been covered by national media, including The New York Times, Los Angeles Times, and CNN. She holds certificate in plant-based nutrition from T. Colin Campbell Center for Nutrition Studies.

Elise Littler is a former McGrath Public Policy Intern at the Humane Society of the United States. She received her BA in English from the University of British Columbia in Vancouver and her Master of Public Affairs degree from the Lyndon B. Johnson School of Public Affairs at the University of Texas, Austin. She currently works at the American Institutes for Research as a Proposal Development Coordinator in its educational assessments division.

Index

Note: Page numbers followed by *f* indicate figures and *t* indicate tables.

A

Agriculture, 47–48
 consolidation of, 26
 diversity, 150
 extension agency facilitation, 15, 160–163
 industrialization, 88
 pesticide, 241
 processing sectors, 6–7
 producers, 189–191
 production, 26
 supply chains, 223–224
Agro-food industry, 22
Agro-foods, 57, 239–240
 global value chain approach, 27
  governance, 54–55
  upgrading, 55–57
 institutional foodservice, 59–62
 intermediaries, 64–65
 market-driven, global food system, 50–52
 production, 24
 products, 58
 sectors, 55
 structure and governance, 57–59
 supply chains, 62
 value chains, 47–48, 88–89, 293
  agro-food intermediaries growth, 64–65
  financialization of, 48
  values-based procurement, 62–67
  limitations, 65–67
American Society for Prevention of Cruelty to Animals (ASPCA), 7
Anchors for resilient communities (ARC), 245–246
Animal agriculture, 185, 256–257
Animal-based proteins, 57, 112–113
Animal welfare, 309
Antibiotics, 241
Aramark Corporation, 316
Arizona State University, 316

B

Beef quality assurance (BQA) programs, 201
Beef supply chain, 12, 195–196, 201, 208, 212
Beef-to-school (B2S) programs, 195–196. *See also* Montana's beef-to-school (B2S) project
Broadline distributors, 251
Bronson Methodist Hospital, 168–169

C

Cafeteria worker experience, 268–271
 on equipment and training, 273*t*
 healthy, scratch and local foods, 275–276*t*
California Association of Food Banks (CAFB), 291
Canisius College, 315
Capacity-building intermediaries, 65
Carbon and water footprints, 25–26
Cattle production basics, 197–198
Center for Good Food Purchasing, 121–123
Center for Regional Food Systems (CRFS), 156, 160–161
Centers for Disease Control and Prevention (CDC), 128
Chain upgrading institutions, 56, 64
Climate models, 25
Commodity
 crop chains, 57
 foods, 291
Commodity supplemental food program (CSFP), 288, 291–292
Community based food procurement, 84–85
Community Food Bank of Southern Arizona, 298
Community supported agriculture (CSA) programs, 247
Community-supported fishery (CSF), 185–186
Compass Group North America, 314–315
Concentrated animal feeding operations (CAFOs), 309
Consumer-oriented approach, critiques of, 7–8

327

Consumers, 4–17
    activist groups, 64
    animal welfare, 88
    B2S supply chain partners, 202
    calories for, 23
    cost savings, 311
    low-income consumers, 30, 287
    markets, 55, 220
    pay-to-play consumer activism, 66
    waste, 58
Conventional institutional foodservice value chain, 60f

D

Dietary Guidelines for Sustainability, 243
Diet-related disease, 165
Double movement, 55
Drexel Food Lab (DFL), 128–129, 131, 135, 140t
    deli salads, 141t

E

Early childcare and education (ECE), 156, 162, 165–166
Environmental nutrition approach, 12–13, 240–242
    healthy food, redefining, 241–244
    history and uptake, 243–244
    hospitals
        foodservice, 247–250
        purchasing practices in, 249
    institutionalization, 256
    local food, 247–248
    modeling food choices, 249–250
    sustainable food, 247–248
    U.S. health care sector, 244
Environmental upgrading, defined, 57
Environmental Working Group (EWG), 7
Equity, social movements, 33–34

F

Farm-and-Sea-to-Campus (FSCN) program, 186
Farmers Feeding Florida programs, 295–296
Farm-to-college (FTC) programs, 186–188
Farm-to-food-bank, 293–295
    Community Food Bank of Southern Arizona, 298
    coordination around perishables, 301
    finding suppliers, 302
    fresh food, 295
    Greater Cleveland Food Bank, 298
    growth of, 301–302
    Houston Food Bank and Feeding Texas, 299
    infrastructure, 294–295
    interview findings, 297t
    methodology, 296
    Northern Illinois Food Bank, 299
    policy support for, 295–296
    price, 302
    procurement expertise, 293
    Second Harvest Food Bank of Greater New Orleans and Acadiana, 299
    South Jersey, 298
    strategic partnerships across sectors, 294
Farm-to-hospital (FTH) programs, 180, 185–186
    supply chain challenges
        distributors, 251–252
        group purchasing organizations, 252–253
        supply chain innovation, 253
Farm to Institution New England (FINE), 10, 16
    Boston Medical Center, 185–186
    core functions (convening, communication, and metrics), 180–182
    distributors, 189
    farm-to-college, 186–188
    farm-to-hospital, 185–186
    farm-to-school, 182–184
    history, 177–178
    Maine Food for UMaine System Project, 188
    network, 178–182
    Northeast Farm to School Institute, 184
    producers, 189–191
    regional FTS efforts, 183–184
    regional network, impact as, 191
    research into supply chain, 188–191
    strategic objectives, 191–192
    supply and demand, 177
Farm-to-institution programs, 12–13, 84–85, 94, 294
    food bank infrastructure, 294
    in Michigan, 150, 154–155t
        Baxter Child Development Center, 165–166
        Bronson Methodist Hospital, 168–169
        Calumet Center, 166–168
        commodity organizations, 164
        defined, 149–150
        growth in, 151
        history, 151–158
        non-profit and agriculture extension agency facilitation, 160–163

state government policy, 158–160
supply chain partnerships, 163–164
survey data, 152–153t
Farm-to-school (FTS) programs, 150, 182–184, 199
features of, 184
Northeast Farm to School Institute, 184
regional efforts, 183–184
Fast-food restaurants, 59
Federal food safety standards, 201
Federal nutrition standards, 199–200
Feeding Texas, 299
Financialization, 26
FINE. *See* Farm to Institution New England (FINE)
Food
activism, 55
deserts, 29
distribution companies, 6, 38–39, 59, 79–81
insecurity, 286
marketing system, 4–5
prices, 13
processing sectors, 6–7, 61–62
purchasing plans, 4
safety regulations, 28, 48
stamps, 33–34
transnational corporations, 50
Food and Nutrition Services (FNS), 288
Food Bank of South Jersey, 298
Food banks
aggregation, 288–289
charity-based approach, 286–287
coordinated ordering, 289
delivery, 289
distribution, 288–289
donations, 289–292
farm-to-food-bank, 293–295
Community Food Bank of Southern Arizona, 298
coordination around perishables, 301
finding suppliers, 302
fresh food, stimulating demand for, 295
Greater Cleveland Food Bank, 298
growth of, 301–302
Houston Food Bank and Feeding Texas, 299
infrastructure, 294–295
methodology, 296
Northern Illinois Food Bank, 299
policy support for, 295–296
price, 302
procurement expertise, 293

Second Harvest Food Bank of Greater New Orleans and Acadiana, 299
South Jersey, 298
strategic partnerships across sectors, 294
food purchasing, 292
goals, 286–287
organization of, 287–289
procurement, 289–292
shared maintenance fees, 289
USDA foods, 291–292
Food-borne illnesses, 3–4
Food Distribution Program on Indian Reservations (FDPIR), 291
Food Forward, 316–317
Food-purchasing decisions, 37
Foodservice institutions, 36–37
defined, 5
reforming industrialized agriculture, 9–10
Foodservice management companies (FSMCs), 61, 78, 90, 92, 186–187
American University, 316–317
Aramark Corporation, 316
Arizona State University, 316
Canisius College, 315
Compass Group North America, 314–315
foodservice contracts, 89–93
rebate pricing system, 89–93
Foodservice managers, 130–131
Foodservice operations, 3–4
Food systems
defined, 23
environmental sustainability, 25–26
livelihoods, 26–28
nutrition, 28–29
supply chain innovations and growing pains, Los Angeles, 114–117
values-based, data-driven purchasing practices, Oakland, CA, 112–114
workers, 87
Fresh food, 295, 301
Fresh fruit and vegetable chains, 57
Functional upgrading, 56

G

Get Healthy Philly (GHP), 129, 131, 135
Global agro-food value chains, 48
Global Animal Partnership (GAP), 88
Global food systems. *See also* Food systems
environmental sustainability, 25–26
equity, 33–34
food chain shortening, 32–33
food system transformation, 34–39

Global food systems *(Continued)*
 institutional foodservice, 38
 livelihoods, 26–28
 nutritional challenge, 28–29
 sustainability, 33–34
 third-party certifications, 31–32
Global industrialization, 47–48
Global value chains (GVC) approach
 agro-food value chains, 49
 governance, 54–55
 upgrading, 55–57
Good Food Purchasing Policy, 14–15, 86, 105, 107–108, 110t
 bidding processes, 122
 data quality and availability, 121–122
 defined, 106–107
 difficulties with contracts, 122
 food system change
  supply chain innovations and growing pains, Los Angeles, 114–117
  values-based, data-driven purchasing practices, Oakland, CA, 112–114
 fostering innovation in product development, 115–116
 implementation, 107–108
  and data collection, 110–111
 labor condition improvement, 116–117
 misguided perceptions, 121
 models, 113
 requests for proposals, 117–121
 transparency, 107
 values-driven procurement programs, 107
Greater Cleveland Food Bank, 298
Greenhouse gas emissions, 25
Group purchasing organizations (GPO), 6, 240, 252–253

H

Health and Human Services (HHS), 128
Health care sector, 168–169, 241–242
Health Care Without Harm (HCWH), 185, 240–241
Health professionals, 239–240, 256–257
Healthy food, 8, 239–240
 health care movement, 242, 244–247
 redefining, 241–244
Healthy Food in Health Care (HFHC) program, 245, 249
Healthy foodservice guidelines, 130
Healthy, Hunger-Free Kids Act (HHFKA), 264–265

Hospitals, environmental nutrition approach, 162
 foodservice, 247–250
 food supply chain, 251f
 purchasing practices in, 249
Houston Food Bank, 299
Hunger and Environmental Nutrition Dietetics Practice Group (HEN), 243

I

Import-substitution industrialization, 50
Individual quick freeze (IQF) patties, 213
Industrialized agriculture, 9–10
Institutional foodservice, 4, 34–35, 95–96
 challenges faced by, 13–14
 consumers, 127–128
 defined, 4–5
 food suppliers, 6
 food system challenges, in U.S., 79–88
 management, 78–79
 procurement policies, 82–83t
 value chains, 59–62, 65
 values-based purchasing, 14–15, 51–52
 work, 266–267
Institutional markets, 6, 226
Institutional value chains, 4
Institutional values-based procurement, 93–94
Invitation for Bid (IFB), 292
Iowa's Food-Buying Institutions, 225–227
Iowa State University (ISU), 221

J

Juvenile Justice System, 166–168

K

Kentucky Catfish Study, 301
K-12 school food programs, 6, 162–163

L

Large-scale commodity food production, 30
Large-scale factory farming, 310–311
Large-scale Iowa buyers, 228t
Legume-based proteins, 112–113
Livestock industry, 311
Livestock production, 241
Local and regional food systems, 93–95, 176, 226
 defined, 286
 farm-to-food-bank, 293–295
 food bank value chains, 289–292

Local food, 8, 63, 68, 229, 247–248
  consumption, 224–227
  procurement activity, 150
  sourcing, 164
  systems, 176, 286
Los Angeles Unified School District (LAUSD), 114–115

M

Maine Food for UMaine System Project, 177, 188
Making It Healthier, Making It Regional (MHMR), 262
  SFA details, SY 2016–17, 268t
Market concentration, 223–224
Meal participation, 213–214
Meat-heavy diets, 310
Meatless Monday, 318
Meat/meat alternates, 200
Meat production, 310
Menus of Change (MOC), 317
Michigan Department of Education, 151
Michigan Farm to Institution Network (MFIN), 156–157, 162
Michigan Food Hub Network, 163–164
Michigan Food Policy Council, 151–155
Midsized farms
  Iowa's Food-Buying Institutions, 225–227
  local food consumption by institutions, 224–227
  U.S. food systems, 222–224
Montana's beef-to-school (B2S) project
  agriculture, 197
  cattle production basics, 197–198
  Dillon
    community member involvement models, 210
    donation models, 210
  district levels, 201–202
  federal food safety standards, 201
  federal nutrition standards, 199–200
  governance of, 199
  higher cost, 212–213
  Hinsdale, vertically integrated model, 205–206
  implementing challenges, 211–215
  local beef, 199
  meal participation, 213–214
  motivations for, 211
  partnership models, 204f
  processing and yield, 198
  processor contract model

Flathead Valley region, 207
  Kalispell, 209–210
  Livingston, 206–207
  Somers, 207–208
  Whitefish, 208–209
  processors, producers participating in, 203f
  product availability, 214–215
  school size, 213–214
  state nutrition standards, 199–200
  supply chain logistics, 214–215
  technology and kitchen convenience, 213

N

National agricultural policies, 28
National Farm to School Network's (NFSN), 177–178
National food distributors, 251
National Good Food Network, 65
National School Lunch Program (NSLP), 262
  attitudes, 274–277
  cafeteria worker experience, 268–271
  findings, 267–277
  institutional foodservice work, 266–267
  procurement changes, 271–277
  purchasing changes in, 263–266
  workload, 272–273
Natural resources, 21–23, 243–244
Network Advisory Council (NAC), 179
Non-profit healthcare system, 168
Non-profit intermediaries, 317–319
Non-profit organizations, 160–163, 239–241, 245
Non-traditional food stores, 4–5
North American Free Trade Agreement (NAFTA), 94
Northeast Organic Farming Association of Vermont (NOFA-VT), 177–178
Northeast Regional Steering Committee, 178
Northern Illinois Food Bank, 299, 301
Northwest Agricultural Business Center (NABC), 254–255
Nutrition transition, 29

O

Oakland Unified School District (OUSD), nutrition services, 112
Ohio Association of Food Banks (OAFB), 298
Organic and pasture-based systems, 85
Organic food, 7–8, 37, 253–254, 319–320
Organisation of Economic Cooperation and Development (OECD), 28

P

Perishable foods, 301
Philadelphia
  compliant and noncompliant products, 133
  food systems, 129–130
  nutrition standards, 128, 142–143
  price comparisons from vendors, 133
Place-based institutional foodservice, 48–49
Plant-based diets, 308
Plant-based foods, 12–13, 308
  animal welfare, 309
  cost savings, 311
  environment, 310–311
  foodservice management companies
    American University, 316–317
    Aramark Corporation, 316
    Arizona State University, 316
    Canisius College, 315
    Compass Group North America, 314–315
  health, 309–310
  meatless product lines with private manufacturers, 319–321
  non-profit intermediaries, 317–319
  peer networks, 317–319
  self-operating foodservice facilities
    University of California, Berkeley, 312–313
    University of Massachusetts, Amherst, 314
    University of North Texas, 313–314
Plant-based proteins, 35
Processor contract model, B2S Project
  Flathead Valley region, 207
  Kalispell, 209–210
  Livingston, 206–207
  Somers, 207–208
  Whitefish, 208–209
Processor, defined, 198
Process upgrading, 56
Producer-driven chains, 55
Product availability, 214–215
Product upgrading, 56

R

Regional food distributors, 251–252
Regional food hubs, 65, 93–94, 176, 286
Regional Food Systems Working Group (RFSWG), 221, 226
Regional Steering Committee (RSC), 177–178
Requests for proposals (RFPs), 16, 117–121, 177, 188, 292
Residential Child Care Institution (RCCI), 167

Retail food buyers, 222
Retail markets, 226

S

School food authorities (SFAs), 262–263
School foodservice, 112, 199–200
School lunch programs, 214
Scratch-cooked foods, 275–276
Scratch-cooking skills, 12–13
Second Harvest Food Bank of Greater New Orleans and Acadiana, 299, 301
Self-operating foodservice facilities
  University of California, Berkeley, 312–313
  University of Massachusetts, Amherst, 314
  University of North Texas, 313–314
Shelburne Farms, 177–178
Shepherd's Pie recipe
  original recipe for, 137f
  reformulated recipe for, 138f
Small-scale producers, 26
Social equity, 32
Social justice, 243, 309
Social upgrading, defined, 57
Staphylococcus aureus, 241
State government policy, farm-to-school legislation, 158–160
State nutrition standards, 199–200
Supermarket, 27
Supplemental Nutrition Assistance Program (SNAP), 33–34
Supply chain innovation, 253
Supply chain logistics, 214–215
Supply chain partnerships, 163–164
Supply chains/value chains, 70
Supply chain transparency, 110, 117, 121–122
Sustainability, 94–95
  social movements, 33–34
  triple-bottom line approach, 39
Sustainable food system, 23, 29–30, 247–248

T

The Emergency Food Assistance Program (TEFAP), 288, 291
Traditional food stores, 4–5
Triple threat framework, 22

U

United States (U.S.)
  agriculture, 79
  agro-food system, 58
  diet, 241

Food and Drug Administration (FDA), 201
food banks, 286
food systems, 78, 222–224
   animal welfare, 88
   environmental considerations, 85–86
   health considerations, 86–87
   institutional foodservice, 79–88, 82–83*t*
   labor and equity considerations, 87–88
   market structure and rural economy considerations, 81–85
   workers, 87
seafood in, 7
United States Department of Agriculture (USDA), 31, 91, 104, 128, 150, 196, 243–244, 288, 291–292
University of California, Berkeley, 312–313
University of Massachusetts, Amherst, 314
University of North Texas (UNT), 307, 313–314
University of Washington Medical Center (UWMC)
   food purchasing, 254
   local food, 253–256

Northwest Agricultural Business Center, 254–255
   regional organic food, 256
   student farm, 255

V

Value chain analysis, 53, 55–56
   governance, 54
   lead firm, 55
Values-based institutional food purchasing, 10, 13, 261–262
Values-based institutional foodservice, 11, 64
Values-based procurement, 47–49, 61, 78–79
   limitations, 65–67
   shifting institutional foodservice to, 62–67
Values-based supply chains, 248
Vegetarian/vegan dishes, 319
Vertically integrated model, Hinsdale School District, 205–206

W

Waste-sensitive procurement strategies, 85–86

Printed in the United States
By Bookmasters